John M. Hammersley

Disorder in
Physical Systems

Disorder in Physical Systems

A volume in honour of

JOHN M. HAMMERSLEY

on the occasion of his
70th birthday

Edited by

G. R. GRIMMETT

and

D. J. A. WELSH

CLARENDON PRESS · OXFORD
1990

Oxford University Press, Walton Street, Oxford OX2 6DP
Oxford New York Toronto
Delhi Bombay Calcutta Madras Karachi
Petaling Jaya Singapore Hong Kong Tokyo
Nairobi Dar es Salaam Cape Town
Melbourne Auckland
and associated companies in
Berlin Ibadan

Published in the United States
by Oxford University Press, New York

British Library in Cataloguing in Publication Data
Disorder in physical systems.
1. Stochastic processes
I. Grimmett, Geoffrey II. Welsh, D. J. A. III. Hammersley, John M. 1920–
519.2
ISBN 0-19-853215-6

Library of Congress Cataloging in Publication Data
(Data available)

ISBN 0 19853215 6

Typeset by the editors
Printed in Great Britain by
Courier International Ltd., Tiptree, Essex

Frontispiece photograph by Geoffrey Grimmett

Preface

On 21 March 1990 John Hammersley celebrates his seventieth birthday. A number of his colleagues and friends wish to pay tribute on this occasion to a mathematician whose exceptional inventiveness has greatly enriched mathematical science.

The breadth and versatility of Hammersley's interests are remarkable, doubly so in an age of increased specialisation. In a range of highly individual papers on a variety of topics, he has theorised, and posed (and solved) problems, thereby laying the foundations for many subjects currently under study. By his evident love for mathematics and an affinity for the hard problem, he has been an inspiration to many.

If one must single out one particular area where Hammersley's contribution has proved especially vital, it would probably be the study of random processes in space. He was a pioneer in this field of recognised importance, a field abounding in apparently simple questions whose resolutions usually require new ideas and methods. This area is not just a mathematician's playground, but is of fundamental importance for the understanding of physical phenomena. The principal theme of this volume reflects various aspects of Hammersley's work in the area, including disordered media, subadditivity, numerical methods, and the like.

The authors of these papers join with those unable to contribute in wishing John Hammersley many further years of fruitful mathematical activity.

August 1989

G.R. Grimmett
D.J.A. Welsh

Contents

Contributors

N. H. BINGHAM, Department of Mathematics, Royal Holloway and Bedford New College, Egham Hill, Egham, Surrey TW20 0EX, UK.

P. CLIFFORD, Mathematical Institute, University of Oxford, 24–29 St. Giles, Oxford OX1 3LB, UK.

C. DOMB, Physics Department, Bar-Ilan University, Ramat-Gan, Israel.

P. ERDŐS, Mathematical Institute, Hungarian Academy of Sciences, Reáltanoda ul. 13–15, Budapest, Hungary.

J. W. ESSAM, Department of Mathematics, Royal Holloway and Bedford New College, Egham Hill, Egham, Surrey TW20 0EX, UK.

M. E. FISHER, Institute for Physical Science and Technology, The University of Maryland, College Park, Maryland 20742, USA.

R. J. GIBBENS, Statistical Laboratory, University of Cambridge, 16 Mill Lane, Cambridge CB2 1SB, UK.

I. J. GOOD, Department of Statistics, Virginia Polytechnic Institute and State University, Blacksburg, Virginia 24061, USA.

G. R. GRIMMETT, School of Mathematics, University of Bristol, University Walk, Bristol BS8 1TW, UK.

D. C. HANDSCOMB, Oxford University Computing Laboratory, 8–11 Keble Road, Oxford OX1 3QD, UK.

P. J. HUNT, Statistical Laboratory, University of Cambridge, 16 Mill Lane, Cambridge CB2 1SB, UK.

F. P. KELLY, Statistical Laboratory, University of Cambridge, 16 Mill Lane, Cambridge CB2 1SB, UK.

D. G. KENDALL, 37 Barrow Road, Cambridge CB2 2AR, UK.

W. S. KENDALL, Department of Statistics, University of Warwick, Coventry CV4 7AL, UK.

H. KESTEN, Department of Mathematics, Cornell University, Ithaca, New York 14853, USA.

J. F. C. KINGMAN, Senate House, University of Bristol, Tyndall Avenue, Bristol BS8 1TH, UK.

C. J. H. MCDIARMID, Department of Statistics, University of Oxford, Oxford OX1 3TG, UK.

C. M. NEWMAN, Department of Mathematics, University of Arizona, Tucson, Arizona 85721, USA.

J. S. ROWLINSON, Physical Chemistry Laboratory, University of Oxford, South Parks Road, Oxford OX1 3QZ, UK.

A. SÁRKÖZY, Mathematical Institute, Hungarian Academy of Sciences, Reáltanoda ul. 13–15, Budapest, Hungary.

R. R. P. SINGH, AT&T Bell Laboratories, Murray Hill, New Jersey 07974, USA.

C. E. SOTEROS, Department of Chemistry, University of Toronto, Toronto, Ontario M5S 1A1, Canada.

U. STADTMÜLLER, Universität Ulm, Abteilung Mathematik-III, Oberer Eselsberg, 7900 Ulm, FRG.

J. M. STEELE, Program in Statistics and Operations Research, School of Engineering and Applied Science, Princeton University, Princeton, New Jersey 08544, USA.

D. TANLAKISHANI, Department of Mathematics, Royal Holloway and Bedford New College, Egham Hill, Egham, Surrey TW20 0EX, UK.

D. J. A. WELSH, Merton College, Oxford OX1 4JD, UK.

S. G. WHITTINGTON, Department of Chemistry, University of Toronto, Toronto, Ontario M5S 1A1, Canada.

P. WHITTLE, Statistical Laboratory, University of Cambridge, 16 Mill Lane, Cambridge CB2 1SB, UK.

J. C. WIERMAN, Department of Mathematical Sciences, The Johns Hopkins University, Baltimore, Maryland 21218, USA.

D. WILLIAMS, Statistical Laboratory, University of Cambridge, 16 Mill Lane, Cambridge CB2 1SB, UK.

Speech Proposing the Toast to John Hammersley

1 October 1987

David Kendall

John Michael Hammersley, Fellow of the Royal Society, Doctor of Science of both Cambridge and Oxford, sometime Major in the Royal Regiment of Artillery, Rouse Ball Lecturer of the University of Cambridge, von Neumann Medallist of the University of Brussels, and Gold Medallist of the Institute of Mathematics and its Applications, has of course many other distinctions too numerous to list here.

My hope is that in this brief appreciation of all that I have seen him achieve during the last forty years, I can catch the spirit of his very personal contributions to mathematics and statistics on the world scene, and his equally personal contributions to the quality of mathematical and statistical life in this country. Both have been profound.

First, contributions to mathematics and statistics. I have not had time to make the bibliographical studies such a survey demands, and very probably I shall list things out of their true order, but the first startling JMH paper I remember was about some anomalies of the solutions to iterative equations of the form $x_{n+1} = f(x_n)$, which perhaps now, if we were to look at them again, might seem a partial anticipation of the current studies of chaotic deterministic systems.

Next I remember the excitement with which I first read his Royal Statistical Society paper on the estimation of integer-valued parameters, and the superefficiency that is characteristic of this situation. That piece of work was important for me in forcing me to take an interest in one of his examples: Alexander Thom's record of his careful measurements of the diameters of neolithic stone circles, leading to a claim that a unit of length had been employed in their construction. I was one of the scoffers then — and of course there were many — but eventually I came to suspend disbelief, and at last (with Simon Broadbent and Wilfrid Kendall) to take part in a statistical examination that went a long way to confirm this startling proposal. Alexander Thom is now much respected by archaeologists *because he persuaded them to think of neolithic man as a colleague rather than a savage.* One is reminded of Hardy's — or was it Littlewood's — remark, that the ancient Greek mathematicians were not scholarship candidates, but fellows of another college. Without John's intervention that revolution in archaeological thinking might never have occurred.

Another highly original contribution was his and Simon Broadbent's development of percolation theory. Gradually this has progressed from industrial concern about coal utilisation to a central problem in both probability theory and solid state physics. Closely associated with this is the work on self-avoiding random walks which again has profound implications for physics and chemistry. Each of these problems was a natural field for the application of diverse Monte Carlo techniques with which Hammersley's name will always be associated.

As John will possibly tell us himself, in the reminiscences and perhaps refutations that these random remarks will I hope spark off, 'Monte Carlo' was not exactly the phrase with which to woo the Oxford Mathematical Institute of the nineteen forties and fifties. Probability was not taught and was scarcely known in Oxford, though there were splendid exceptions like E.A. Milne who employed its techniques with great ingenuity.

One of John's special gifts was however much appreciated there. This was his skill in concocting the all but insoluble scholarship questions that were then in vogue (and which passed the test of acceptance only if they baffled one's fellow examiners).

With John's later work I am not so closely in touch, but one ought to mention a combined attack on theories about the origin of comets by Ray Lyttleton, John Hammersley and myself. John produced a computer solution to the basic integral equation, I showed that this was the minimal solution, and to this day we don't know whether it is the only solution, or not! Nor are we likely to find out, for astronomers have an irritating way of scrapping problems every year or so and moving on to some quite different topic.

One matter which brought many of us close together was the urgent need to do something about the teaching of mathematics in schools, where "A and B were still competing with C (who always lost) in various sorts of race, and honest grocers mixed their teas and made a reasonable profit". (I quote a review of about that time by a fellow Queen's man, Horace Elam, who taught mathematics with great skill and dedication at Magdalen College School.) With Jack Howlett and Harry Reuter we tried in various ways to brighten things up.

I recall going with Jack Howlett to a school in the Cotswolds to talk severally about queues and computers to an audience of children presided over by a Headmaster who concluded the formal proceedings with the remark: "Well, you won't have understood any of that, so I think we should dispense with questions and let you run off to your teas". However, as soon as the Headmaster's back was turned, there was an eager throng of boys

Jakimovski Methods and Almost-Sure Convergence

N.H. Bingham and U. Stadtmüller

1. Introduction

The classical summability methods of Borel (B) and Euler ($E(\lambda), \lambda > 0$) play an important role in many areas of mathematics. For instance, in summability theory they are perhaps the most important methods other than the Cesàro (C_α) and Abel (A) methods, and two chapters of the classic book of Hardy (1949) are devoted to them. In probability, the distinction between methods of Cesàro-Abel and Euler-Borel type may be seen from the following two laws of large numbers, the first of which extends Kolmogorov's strong law.

THEOREM I. (Lai 1974) *For $X, X_1, X_2, \ldots$ independent and identically distributed, the following are equivalent:*
(i) $E|X| < \infty$ *and* $EX = \mu$,
(ii) $X_n \to \mu$ a.s. $(n \to \infty)$ (C_α) *for some (all)* $\alpha \geq 1$,
(iii) $X_n \to \mu$ a.s. $(n \to \infty)$ (A).

THEOREM II. (Chow 1973) *For $X, X_1, X_2, \ldots$ independent and identically distributed, the following are equivalent:*
(i) $E|X|^2 < \infty$ *and* $EX = \mu$,
(ii) $X_n \to \mu$ a.s. $(n \to \infty)$ $(E(\lambda))$ *for some (all)* $\lambda > 0$,
(iii) $X_n \to \mu$ a.s. $(n \to \infty)$ (B).

Other applications in probability arise through the technique of 'Poissonization', in accordance with Kac's dictum: if you can't solve the problem exactly, then randomise (Kesten 1986, p. 1109; cf. Kac 1949, Hammersley 1950 (pp. 219–224), 1972 (§§7,8), Hammersley et al. 1975, Pollard 1984, p. 117). There are also applications along these lines to combinatorial optimisation (Steele et al. 1987, §3; Steele 1989, §3).

Often the properties of the methods are governed by the fact that their weights — the Poisson and binomial distributions — being convolutions, obey the central limit theorem. Consequently, many such properties extend to matrix methods $A = (a_{nk})$, whose weights are also given by convolutions:

$$a_{nk} = P(S_n = k), \tag{1.1}$$

for (S_n) a random walk (see e.g. Bingham 1981, 1984). There, $S_n = \sum_1^n X_k$ is a sum of independent X_k, identically distributed (and $\mathbb{Z}$-valued). Another important case is that of X_k Bernoulli $(0, 1\text{-valued})$ but not necessarily identically distributed:

$$P(X_n = 1) = p_n, \quad P(X_n = 0) = q_n := 1 - p_n.$$

Writing $p_n = 1/(1 + d_n)$, $(d_n \geq 0)$, this leads to the method $A = (a_{nk})$ defined by

$$\prod_{j=1}^{n} \left(\frac{x + d_j}{1 + d_j} \right) \equiv \sum_{k=0}^{n} a_{nk} x^k,$$

the Jakimovski method $[F, d_n]$ (Jakimovski 1959; Zeller and Beekmann 1970 (Ergänzungen, §70)). The motivating examples are:

(i) $d_n = 1/\lambda$, the Euler method $\mathrm{E}(\lambda)$ above,

(ii) $d_n = (n-1)/\lambda$, the *Karamata-Stirling method* $\mathrm{KS}(\lambda)$, (Karamata 1935). Here

$$a_{nk} = \lambda^k S_{nk}/(\lambda)_n,$$

with $(\lambda)_n := \lambda(\lambda+1)\dots(\lambda+n-1)$ and (S_{nk}) the Stirling numbers of the first kind. The Bernoulli representation (1.1) enables both local and global central limit theory to be applied; see Bender (1973) for a perspicuous treatment. In particular, unimodality of Stirling numbers and other weights follows from this; for background see e.g. Hammersley 1951, 1952, 1972 (§§18, 19), Erdős 1953, Harper 1967, Lieb 1968, Bingham 1988.

Our aim here is to extend to Jakimovski methods the law of large numbers (Theorem II), and the corresponding analogue of the law of the iterated logarithm (Lai 1974). This complements the work of Bingham (1988), which gives a similar extension to the basic Tauberian theorem ('O-K-Satz'), due in the Euler case to Knopp in 1923 and in the Borel case to Schmidt in 1925 (Hardy 1949, Theorems 156, 241, 128). For further background on almost-sure convergence behaviour and summability methods, see e.g. Stout 1974 (Chap. 4), Bingham and Goldie 1988.

2. Results

THEOREM 1. *For $X, X_0, X_1, \dots$ independent and identically distributed random variables, and (d_n) as above, the following are equivalent:*

(i) $\mathrm{var}\, X < \infty$, $EX = m$,

(ii) $X_n \to m$ a.s. $(E(\lambda)$ or $B)$,

(iii) $X_n \to m$ a.s. $(KS(\lambda))$,

(iv) $\quad X_n \to m$ a.s. $[F, d_n]$.

In what follows, we restrict the generality slightly. We assume further that $[F, d_n]$ satisfies

$$p_n \to 0 \quad (\text{or } d_n \to \infty).$$

This ensures that $\sigma_n \asymp \sqrt{\mu_n}$ can be strengthened to

$$\sigma_n \sim \sqrt{\mu_n}.$$

The Euler case ($p_n = \lambda/(1+\lambda)$, $d_n = 1/\lambda$) is thereby excluded, but can be handled separately. These two cases together (p_n constant and $p_n \to 0$) cover the cases of main interest (though the result below and its proof may be extended to cover the case $\sigma_n \sim c\sqrt{\mu_n}$, for constant c). In (i) below, 'log' in the denominator means '$\max(1, \log_+)$'.

In Theorem 2, which gives the rates of convergence in Theorem 1, the Karamata-Stirling methods diverge from those of Euler and Borel, and one obtains an iterated logarithm, as in the classical case but unlike the Euler-Borel case (Lai 1974).

THEOREM 2. *The following are equivalent:*

(i) $\quad EX = 0, \text{var } X = \sigma^2 \ (< \infty), \ E(|X|^4/\log^2|X|) < \infty,$

(ii) $$\limsup_{x\to\infty} \frac{(4\pi x)^{1/4}}{\log^{1/2} x} \left| \sum_0^\infty e^{-x} \frac{x^k}{k!} X_k \right| = \sigma \quad \text{a.s.},$$

(iii) $$\limsup_{n\to\infty} \frac{(4\pi n)^{1/4}}{\log^{1/2} n} \left| \sum_0^n \binom{n}{k} \lambda^k X_k/(1+\lambda)^n \right| = \sigma(1+\lambda)^{1/4} \quad \text{a.s.},$$

(iv) $$\limsup_{n\to\infty} \frac{(4\pi\lambda \log n)^{1/4}}{\log\log^{1/2} n} \left| \sum_0^n a_{nk} X_k \right| = \sigma \quad \text{a.s.},$$

where $A = (a_{nk})$ is the matrix of the Kamarata-Stirling method $KS(\lambda)$,

(v) $$\limsup_{n\to\infty} \frac{(4\pi\mu_n)^{1/4}}{\log^{1/2} \mu_n} \left| \sum_0^n a_{nk} X_k \right| = \sigma \quad \text{a.s.}$$

where $A = (a_{nk})$ is the matrix of $[F, d_n]$ with $d_n \to \infty$.

Here the equivalence of (i) with (ii) ('LIL for the Borel method') and (iii) ('LIL for the Euler method') is Lai's result, and is included here for comparison. The constant $(1+\lambda)^{1/4}$ in (iii) is $a^{1/4}$, where a is the mean-variance ratio of the Euler method; see Bingham (1984) for a detailed discussion of this parameter and its significance. When $d_n \to \infty$, $\sigma_n \sim \sqrt{\mu_n}$, and $a = 1$.

Our proof of Theorem 2 will involve a non-uniform local limit theorem for the sums S_n in the Bernoulli representation $a_{nk} = P(S_n = k)$. Write

$H_3(x) := x^3 - 3x$ for the third Hermite polynomial, $\kappa_{3,n} := \mu^n_{3,0}$ for the third cumulant (third central moment) of S_n:

$$\kappa_{3,n} := \sum_1^n E[(\xi_j - p_j)^3] = \sum_1^n (p_j - 3p_j^2 + 2p_j^3).$$

Thus $\kappa_{3,n} \sim \sum_1^n p_j = \mu_n$, $(n \to \infty)$, when $p_n \to 0$.

THEOREM 3. *For S_n the Bernoulli sum above, $a_{nk} = P(S_n = k)$,*

$$\sup_{k \in \mathbf{Z}} \left(1 + \left|\frac{k - \mu_n}{\sigma_n}\right|^3\right) \times$$

$$\left|\sigma_n a_{nk} - \frac{1}{\sqrt{2\pi}} \exp\left\{-\frac{1}{2}\left(\frac{k - \mu_n}{\sigma_n}\right)^2\right\}\left(1 + H_3\left(\frac{k - \mu_n}{\sigma_n}\right)\frac{\kappa_{3,n}}{3!\sigma_n^3}\right)\right|$$

$$= o(1/\sigma_n) \quad \text{as } n \to \infty.$$

This result is closely related to Petrov's non-uniform local limit theorem. The 'uniform' part (taking the '1' term) is the Bernoulli case with $k = 3$ of Theorem 12 of Petrov (1975, VII.3), except that Petrov's condition

$$(*) \qquad \liminf_{n \to \infty} \sigma_n^2 / n > 0$$

is violated when $d_n \to \infty$, as in Theorem 2 (iv), (v), since $\sigma_n^2 = \sum_1^n d_j/(1 + d_j)^2$. However, to compensate for this, we know the characteristic function of our Bernoulli sum explicitly, and this enables us to handle the error terms in the Fourier analysis of Petrov's method successfully. The 'non-uniform' part (taking the '$|(k - \mu_n)/\sigma_n|^3$' term) is similarly related to Theorem 16 of Petrov (1975, VII.3), except that he has general identical distributions and we have Bernoulli non-identical distributions.

Theorem 3 involves the first term of an expansion of Edgeworth type ($k = 3$ in Petrov's notation). Extensions to Edgeworth expansions of arbitrary length (general k) are also possible, and can be proved by Petrov's method, adapted to our Bernoulli case as in the proof of Theorem 3 below. We shall return to this in Section 4.

3. Proofs

PROOF OF THEOREM 1: We follow the argument of the proof of Theorem 1 of Bingham and Maejima (1985) — BM for short — indicating differences when these arise.

That (i) implies (ii) is Chow's result. Now if $d_n \geq \delta > 0$ for all large n, as assumed, $E(1/\delta) \subset [F, d_n]$ by a result of Meir (1963), Zeller and Beekmann (1970, Ergänzungen, §70); thus (ii) implies (iii) and (iv).

Conversely, the implication from (ii) to (i) is in BM. If (iii) or (iv) holds and $A = (a_{nk})$ denotes the relevant matrix method,

$$\sum a_{nk} X_k \to m \quad \text{a.s.}$$

Write X_k^s for the symmetrisation of X_k (difference of two independent copies of X_k):

$$\sum a_{nk} X_k^s \to 0 \quad \text{a.s.}$$

Split the sum into the sums over $k \leq \mu_n$ and $k > \mu_n$: Y_n and Z_n say. As in BM, $Y_n \to 0$ a.s. Split off the last term of Y_n: arguing as there,

$$a_{n,[\mu_n]} X^s_{[\mu_n]} \to 0 \quad \text{a.s.}$$

But (cf. Bingham 1988)

$$a_{n,[\mu_n]} \sim \frac{1}{\sigma_n \sqrt{2\pi}} \asymp \frac{1}{\sqrt{\mu_n}.\sqrt{2\pi}}$$

and hence

$$X^s_{[\mu_n]} / \sqrt{[\mu_n]} \to 0 \quad \text{a.s.} \quad (n \to \infty).$$

Write N for $[\mu_n]$:

$$X^s_N / \sqrt{N} \to 0 \quad \text{a.s.} \quad (N \to \infty).$$

From this, we obtain (i) as in BM. ■

PROOF OF THEOREM 2: The argument follows that of Theorem 2 of BM with Petrov's non-uniform local limit theorem replaced by Theorem 3.

First, note that by a Borel-Cantelli argument, our moment condition in (i) is equivalent to

$$X_n = o(n^{1/4} \log^{1/2} n) \quad \text{a.s.}$$

We have, writing $\phi(x) := e^{-x^2/2}/\sqrt{2\pi}$,

$$\sum a_{nk} X_k - \sum \phi\left(\frac{k-\mu_n}{\sigma_n}\right) X_k$$
$$= \sum \phi\left(\frac{k-\mu_n}{\sigma_n}\right) H_3\left(\frac{k-\mu_n}{\sigma_n}\right) \frac{\kappa_{3,n}}{3!\sigma_n^3} X_k + \sigma_n^{-2} \sum \frac{o(1) X_k}{\left(1 + |\frac{k-\mu_n}{\sigma_n}|^3\right)},$$

the $o(1)$ being uniform in k. Call the two terms on the right the *Edgeworth term* and the *error term*. With probability one, we may replace X_k by $o(k^{1/4} \log^{1/2} k)$ in each. We may then estimate each by the methods of

BM, obtaining $o(\mu_n^{1/4}\log^{1/2}\mu_n)$ (a.s.) in each case. This enables us to reduce (v) (which contains (iv)) to

$$\text{(v}'\text{)} \qquad \limsup_n \frac{(4\pi\mu_n)^{1/4}}{\log^{1/2}\mu_n}\left|\sum \phi\left(\frac{k-\mu_n}{\sigma_n}\right)X_k\right| = \sigma \quad \text{a.s.}$$

This is substantially contained in the paper of Lai (1974), where he uses the result ('LIL for the Valiron method') to prove his results for the Borel and Euler methods (see particularly (16) and between (26) and (27)). Two new complications arise: (a) our mean $\mu_n \to \infty$ is not integer-valued, and (b) our variance $\sigma_n^2 \to \infty$ satisfies $\sigma_n^2 \sim \mu_n$ rather than $\sigma_n^2 = \mu_n$. However, our a.s. bound $X_k = o(k^{1/4}\log^{1/2}k)$ is exactly what is required to reduce our sums to Lai's, to the required accuracy $o(\mu_n^{-1/4}\log^{1/2}\mu_n)$. It suffices to show that

$$\text{(a}'\text{)} \quad \limsup_{\lambda\to\infty}\left\{\frac{\lambda^{1/4}}{\log^{1/2}\lambda}\sum_0^\infty o(k^{1/4}\log^{1/2}k)\times \right.$$
$$\left.\left|\frac{1}{\sqrt{2\pi\lambda}}\exp\left\{-\frac{(k-\lambda)^2}{2\lambda}\right\} - \frac{1}{\sqrt{2\pi[\lambda]}}\exp\left\{-\frac{(k-[\lambda])^2}{2[\lambda]}\right\}\right|\right\} = 0,$$

$$\text{(b}'\text{)} \quad \limsup_{\lambda\to\infty}\left\{\frac{\lambda^{1/4}}{\log^{1/2}\lambda}\sum_0^\infty o(k^{1/4}\log^{1/2}k)\times \right.$$
$$\left.\left|\frac{1}{\sqrt{2\pi\lambda}}\exp\left\{-(1+o(1))\frac{(k-\lambda)^2}{2\lambda}\right\} - \frac{1}{\sqrt{2\pi\lambda}}\exp\left\{-\frac{(k-\lambda)^2}{2\lambda}\right\}\right|\right\} = 0.$$

For (a$'$), note that if

$$f(\lambda) := \frac{1}{\sqrt{2\pi\lambda}}\exp\left\{-\frac{(k-\lambda)^2}{2\lambda}\right\}$$

then

$$f'(\lambda) = \frac{f(\lambda)}{\lambda}\left\{-\frac{1}{2} + (k-\lambda) + \frac{(k-\lambda)^2}{\lambda}\right\}.$$

Replace the difference $f(\lambda) - f([\lambda])$ by $(\lambda - [\lambda])f'(\lambda_k)$, where $[\lambda] \le \lambda_k \le \lambda$, which may be estimated by

$$\lambda^{-1}f(\lambda)\left\{\frac{1}{2} + |k-\lambda| + \frac{(k-\lambda)^2}{2\lambda}\right\}.$$

The first term is negligible with respect to $f(\lambda)$. For the second, we have to show

$$\frac{\lambda^{1/4}}{\log^{1/2}\lambda}\sum_{0}^{\infty} o(k^{1/4}\log^{1/2}k)\frac{|k-\lambda|}{\lambda}\frac{1}{\sqrt{2\pi\lambda}}\exp\left\{-\frac{(k-\lambda)^2}{2\lambda}\right\} \to 0 \text{ as } \lambda\to\infty,$$

or

$$\frac{1}{\lambda^{1/4}\log^{1/2}\lambda}\int_0^{\infty} o(y^{1/4}\log^{1/2}y)\frac{|y-\lambda|}{\sqrt{\lambda}}\frac{1}{\sqrt{2\pi\lambda}}\exp\left\{-\frac{(y-\lambda)^2}{2\lambda}\right\}dy \to 0 \text{ as } \lambda\to\infty.$$

Write $(y-\lambda)/\sqrt{\lambda}=t$: thus

$$y^{1/4}=\lambda^{1/4}(1+t/\sqrt{\lambda})^{1/4},\quad \log^{1/2}y=\log^{1/2}\lambda\left(1+\frac{\log(1+t/\sqrt{\lambda})}{\log\lambda}\right)^{1/2}.$$

It remains to consider

$$\int o\left((1+t/\sqrt{\lambda})^{1/4}\left\{1+\frac{\log(1+t/\sqrt{\lambda})}{\log\lambda}\right\}^{1/2}\right)|t|e^{-t^2/2}dt,$$

which tends to 0 as $\lambda\to\infty$, as required. The remaining $((k-\lambda)^2/\lambda)$ term is handled in the same way. Finally, (b′) follows similarly. (A similar analysis is given by Hardy and Littlewood 1916, Thm. 3.4 and Proof of Lemma 2.13.)

In the converse direction, that (iv) or (v) imply (i), follows as in the implication from (ii), (iii) to (i) (Lai 1974, p. 260; BM, p. 389). ■

PROOF OF THEOREM 3: We consider separately the '1' and '$|(k-\mu_n)/\sigma_n|^3$' terms; call the two parts A and B. Write $x_{k,n}$ for $(k-\mu_n)/\sigma_n, \phi_n, \phi_{n,0}$ for the characteristic functions of S_n, S_n-ES_n, c_n for $\kappa_{3,n}/(3!\sigma_n^3)\sim 1/(3!\sigma_n)$.

A: $$a_{nk}=P(S_n=k)=\frac{1}{2\pi}\int_{-\pi}^{\pi}e^{-itk}\phi_n(t)\,dt,$$

while for constant c

$$\phi(x)\{1+cH_3(x)\}=\frac{1}{2\pi}\int_{-\infty}^{\infty}e^{-t^2/2}\{1+c(it)^3\}e^{-itx}\,dx.$$

So

$$2\pi\sigma_n a_{nk}=\int_{-\pi\sigma_n}^{\pi\sigma_n}\exp\{-itx_{k,n}\}\phi_{n,0}(t/\sigma_n)\,dt,$$

$$2\pi\sigma_n a_{nk} - \sqrt{2\pi}\exp\{-x_{k,n}^2\}\{1 + H_3(x_{k,n})c_n\}$$
$$= \int_{-\pi\sigma_n}^{\pi\sigma} \exp\{-itx_{k,n}\}\left(\phi_{n,0}(t/\sigma_n) - e^{-t^2/2}\{1 + (it)^3 c_n\}\right) dt$$
$$+ \int_{|t|\geq\pi\sigma_n} \exp\{-itx_{k,n}\}e^{-t^2/2}\{\ldots\}\, dt,$$

$$|\ldots| \leq \int_{-\pi\sigma_n}^{\pi\sigma_n} |\ldots|\, dt + \int_{|t|\geq\pi\sigma_n} |\ldots|\, dt = \mathrm{I} + \mathrm{II}, \text{ say.}$$

Expanding $\phi_{n,0}$ as far as the third cumulant, we find that for $|t| = o(\sigma_n)$ (actually $|t| = o(\sigma_n^{1/6})$ is all we need)

$$\phi_{n,0}(t/\sigma_n) = \exp\left\{-\frac{1}{2}t^2 + (it)^3 c_n + O(t^4\mu_n/\sigma_n^4)\right\}.$$

Now we choose $\epsilon_n \to 0$, and decompose I as the sum of integrals over $|t| \leq \epsilon_n\sigma_n^{1/6}$, $\epsilon_n\sigma_n^{1/6} \leq |t| \leq \sigma_n/4$ and $\sigma_n/4 \leq |t| \leq \pi\sigma_n$:

$$\mathrm{I} = \mathrm{I}_a + \mathrm{I}_b + \mathrm{I}_c, \quad \text{say.}$$

In $\mathrm{I}_a, |t| = o(\sigma_n^{1/6})$, and the integrand may be checked to be $e^{-t^2/2}o(1/\sigma_n)$. Hence $\mathrm{I}_a = o(1/\sigma_n)$. For I_b, use Lemma 12 of Petrov (1975, p. 179) on the first term. The integrand is exponentially small in σ_n, hence ('normal tails') so is the integral when $\epsilon_n \to 0$ sufficiently slowly; similarly for the second term. For I_c, the $\{\cdots\}$ term is handled as with I_b. The other term is

$$\mathrm{I}_d \leq \sigma_n \int_{1/4\leq t\leq\pi} |\phi_n(t)|dt + \sigma_n \int_{1/4\leq t\leq\pi} \exp\{-\sigma_n^2 t^2\}(1 + |t|^3\sigma_n^2)dt.$$

By direct estimation,

$$\log|\phi_n(t)| \leq -\sum_1^n p_j(1-p_j)(1-\cos t) = -\sigma_n^2(1-\cos t) \leq -c\sigma_n^2$$

in the range of integration, for some $c > 0$, so the first term is exponentially small; clearly, so is the second. Thus $\mathrm{I} = o(1/\sigma_n)$.

For II, the '1' term in ... is exponentially small as above, while the 't^3' term is $o(1/\sigma_n)$ as $c_n \sim 1/(3!\sigma_n)$.

B: $$x_{k,n}^3 2\pi\sigma_n a_{nk} = x_{k,n}^3 \int_{-\pi\sigma_n}^{\pi\sigma_n} \exp\{-itx_{k,n}\}\phi_{n,0}(t/\sigma_n)\, dt.$$

Integrating by parts three times, the right is

$$i \int_{-\pi\sigma_n}^{\pi\sigma_n} \exp\{-itx_{k,n}\} D^3 \phi_{n,0}(t/\sigma_n)\, dt.$$

Also

$$\begin{aligned}\sqrt{2\pi} x_{k,n}^3 & \exp\{-x_{k,n}^2\}(1 + H_3(x_{k,n})c_n) \\ &= x_{k,n}^3 \int_{-\infty}^{\infty} e^{-t^2/2}(1 + (it)^3 c_n) \exp\{-itx_{k,n}\}\, dt \\ &= i \int_{-\infty}^{\infty} \exp\{-itx_{k,n}\} D^3[e^{-t^2/2}(1 + (it)^3 c_n)]\, dt,\end{aligned}$$

integrating by parts three times again.

Subtract, and estimate the difference as a sum of integrals over the interval $[-\pi\sigma_n, \pi\sigma_n]$ and its complement, I and II say, as before. Write (cf. Petrov 1975, p. 209)

$$g_n(t) := \log \phi_n(t/\sigma_n) - \frac{it\mu_n}{\sigma_n} + \frac{1}{2}t^2 - (it)^3 c_n.$$

Then

$$\begin{aligned}\phi_{n,0}(t/\sigma_n) &= e^{-\frac{1}{2}t^2} \exp\{(it)^3 c_n\} \exp\{g_n(t)\} \\ &= e^{-\frac{1}{2}t^2} \left(1 + (it)^3 c_n + R_n(t)\right) \exp\{g_n(t)\}, \quad \text{say.}\end{aligned}$$

Because we know $\phi_{n,0}$ explicitly, we can calculate the first three derivatives of $g_n, \exp\{g_n\}$ (and R_n) explicitly. We can then estimate I (splitting it up as before) and II, along the lines above. All remainders are power series, so may be differentiated term-wise. The exponential estimates obtained above are at worst multiplied by polynomials. The extra detail, which is tedious, is omitted. ∎

4. Remarks

1. In BM, an alternative proof of the LIL is given, using a 'weighted l^1 version' of the local limit theorem, due to Bikyalis and Jasjunas (1967). We raise here the question of obtaining a non-identically distributed version of this result, which would provide an alternative proof of Theorem 2.
2. In the special case

$$\sum_1^{\infty} \frac{1}{(1 + d_n)^2} < \infty$$

(which covers the Karamata-Stirling methods), a quite different proof of Theorem 2 may be given, using Poisson instead of normal approximation to reduce to Lai's result for the Borel case. We use Theorem 2 of Barbour (1987) with $l = p = 1$. In (3.15), W is the Bernoulli sum S_n, so (with $h(n) := X_n = o(n^{1/4}\log^{1/2} n)$ a.s.) $Eh(W)$ is the sum $\sum a_{nk}X_k$ to be approximated. In (2.7) with $l = 1$, $\int h\,dQ_1$ is the corresponding 'discrete Borel mean'

$$\sum_0^\infty e^{-\mu_n}\frac{\mu_n^k}{k!}X_k.$$

The error term (in view of Remark 3, p. 765) is $\nu_1/\sqrt{\lambda}$, where

$$\lambda = \mu_n = \sum_1^n \frac{1}{(1+d_j)},$$

$$\nu_1 = \sum_1^n \frac{1}{(1+d_j)^2} \quad (= \mu_n - \sigma_n^2).$$

By assumption, $\nu_1 = O(1)$, so this is $O(1/\sqrt{\lambda})$. Barbour's theorem tells us that the Jakimovski and discrete-Borel means differ by an amount of order $o((\lambda^{1/4}\log^{1/2}\lambda)/\sqrt{\lambda}) = o(\lambda^{-1/4}\log^{1/2}\lambda)$ (cf. BM, p. 389), which reduces Theorem 2 to the discrete-Borel case. We then use Lai's result for the Borel case, or rather its proof (Lai 1974, p. 258), with $M := [\lambda]$ replaced by $M := [\mu_n]$.

3. Central limit theorems have been given in this context by Embrechts and Maejima (1984), complementing our results on LLN and LIL. Note that their condition (6.1) holds —

$$\sum_k a_{nk}^2 \sim \frac{1}{\sqrt{2}}\sup_k a_{nk} \quad (n \to \infty),$$

which simplifies their Theorems 2 and 3. To see the above, write $\phi_{n,k}$ for $\phi(x_{k,n})$. Then

$$\sum a_{nk}^2 - \sum \phi_{n,k}^2 \le (\sup_k a_{nk} + \sup_k \phi_{n,k})\sum |a_{nk} - \phi_{n,k}|.$$

The sum is $o(1)$ (Bingham 1988, Proposition, (iii)), while (*loc. cit.*, (ii)) each of the suprema has order $(\sigma_n\sqrt{2\pi})^{-1}$, so the right hand side is $o(1/\sigma_n)$. But

$$\sum \phi_{n,k}^2 \sim \frac{1}{2\sqrt{\pi}\sigma_n}$$

(Hardy 1949, Thm. 140).

4. Closely linked with the E(λ), B and KS(λ) methods considered here is

the Riesz mean $R(e^{\sqrt{n}}, 1)$ (or 'moving average $M(\sqrt{n})$'; see Bingham and Goldie (1988). Here a functional (Strassen) version of the LIL is available; see de Acosta and Kuelbs (1983), Chan, Csörgő and Révész (1978).

For other LIL results for weighted means, see e.g. Bingham (1986, §15).

5. The Petrov condition (*), whose failure here necessitated our Theorem 3, guarantees that normal rather than Poisson approximation is appropriate. When it fails, as for $KS(\lambda)$, we may use Poisson approximation as in Remark 2, and Lai's result. This hinges (Lai 1974, p. 258) on large-deviation results approximating Poisson to normal (Hardy 1949, p. 200). This suggests a direct use of large-deviation approximations to normality. Such results are known (Petrov 1975, p. 219, (2.5)), but again only under (*). Accordingly, we raise the question of obtaining large-deviation theorems (and non-uniform local limit theorems for general rather than Bernoulli distributions) when (*) is violated.

6. In Bingham (1984) results are obtained reducing convergence under a 'random-walk method' (a_{nk}) to Valiron convergence for sequences (s_n) of polynomial growth. Here one uses Petrov's non-uniform local limit theorem, with the number of Edgeworth terms retained depending on the degree of polynomial growth. The same method applies here, using the extension of Theorem 3 to general Edgeworth expansions mentioned in Section 2. Thus when

$$s_n = O(n^r) \quad \text{for some } r,$$

Theorem 1 there extends to give the equivalence of

$$\sum a_{nk} s_k \to s$$

and

$$\sum \phi_{n,k} s_k \to s.$$

References

Acosta, A. de and Kuelbs, J. (1983). Limit theorems for moving averages. *Zeitschrift für Wahrscheinlichkeitstheorie und Verwandte Gebiete* 64, 67–123.

Barbour, A.D. (1987). Asymptotic expansions in the Poisson limit theorem. *Annals of Probability* 15, 748–766.

Bender, E.A. (1973). Central and local limit theorems applied to asymptotic enumeration. *Journal of Combinatorial Theory A* 15, 91–111.

Bikyalis, A. and Jasjunas, G. (1967). Limit theorems in the metric of the space L_1, l_1 (in Russian). *Litovskii Matematicheskii Sbornik* 7, 195–218.

Bingham, N.H. (1981). Tauberian theorems and the central limit theorem. *Annals of Probability* 9, 221–231.

——— (1984). On Valiron and circle convergence. *Mathematische Zeitschrift* 186, 273–286.

——— (1986). Variants on the law of the iterated logarithm. *Bulletin of the London Mathematical Society* 18, 433–467.

——— (1988). Tauberian theorems for Jakimovski and Kamarata-Stirling methods. *Mathematika* 35, 216–224.

Bingham, N.H. and Goldie, C.M. (1988). Riesz means and self-neglecting functions. *Mathematische Zeitschrift* 199, 443–454.

Bingham, N.H. and Maejima, M. (1985). Summability methods and almost-sure convergence. *Zeitschrift für Wahrscheinlichkeitstheorie und Verwandte Gebiete* 68, 383–392.

Chan, A.H.C., Csörgő, M., and Révész, P. (1978). Strassen type limit points for moving averages of a Wiener process. *Canadian Journal of Statistics* 6, 57–75.

Chow, Y.-S. (1973). Delayed sums and Borel summability of independent and identically distributed random variables. *Bulletin of the Institute of Mathematics, Academia Sinica*, 1, 207–220.

Embrechts, P. and Maejima, M. (1984). The central limit theorem for summability methods of i.i.d. random variables. *Zeitschrift für Wahrscheinlichkeitstheorie und Verwandte Gebiete* 68, 191–204.

Erdős, P. (1953). On a conjecture of Hammersley. *Journal of the London Mathematical Society* 28, 232–236.

Hammersley, J.M. (1950). On estimating restricted parameters (with discussion). *Journal of the Royal Statistical Society*, Series B, 12, 192–240.

——— (1951). The sums of products of the natural numbers. *Proceedings of the London Mathematical Society* 1, 435–452.

——— (1952). Tauberian theory for the asymptotic forms of statistical frequency functions. *Proceedings of the Cambridge Philosophical Society* 48, 592–599.

——— (1972). A few seedlings of research. *Proceedings of the Sixth Berkeley Symposium for Mathematical Statistics and Probability*, Volume 1, 345–394, University of California Press.

Hammersley, J.M., Lewis, J.W.E., and Rowlinson, J.S. (1975). Relationships between the multinomial and Poisson models of stochastic processes, and between the canonical and grand canonical ensembles in statistical mechanics, with illustrations and Monte-Carlo methods for the penetrable sphere model of liquid-vapour equilibrium. *Sankhyā A* 37, 457–491.

Hardy, G.H. (1949). *Divergent Series.* Clarendon Press, Oxford.

Hardy, G.H. and Littlewood, J.E. (1916). Theorems concerning the summability of series by Borel's exponential method. *Rendiconti Circolo Matematico de Palermo* 41, 36–53. *Collected Works of G.H. Hardy*, Volume VI, 609–628, Clarendon Press, Oxford, 1974.

Harper, L.H. (1967). Stirling behaviour is asymptotically normal. *Annals of Mathematical Statistics* 38, 410–414.

Jakimovski, A. (1959). A generalisation of the Lototsky method. *Michigan Mathematical Journal* 6, 277–296.

Kac, M. (1949). On deviations between theoretical and empirical distributions. *Proceedings of the National Academy of Sciences, U.S.A.* 35, 252–257. In *Mark Kac: Probability, Number Theory and Statistical Physics, Selected Papers*, ed. K. Bacławski and M.D. Donsker, MIT Press, Cambridge, Massachusetts, 1979, 281-286.

Karamata, J. (1935). Théorèmes sur la sommabilité exponentielle et d'autres sommabilités s'y rattachant. *Mathematica (Cluj)* 9, 164–178.

Kesten, H. (1986). The influence of Mark Kac on probability theory. *Annals of Probability* 14, 1103–1128.

Lai, T. -L. (1974). Summability methods for independent, identically distributed random variables. *Proceedings of the American Mathematical Society* 45, 253–261.

Lieb, E.H. (1968). Convexity properties and a generating function for Stirling numbers. *Journal of Combinatorial Theory* 5, 203–206.

Meir, A. (1963). On the $[F, d_n]$-transformation of A. Jakimovski. *Bulletin of the Research Council of Israel* (F: Mathematics) 10, 165–187.

Petrov, V.V. (1975). *Sums of Independent Random Variables.* Springer-Verlag, Berlin.

Pollard, D. (1984). *Convergence of Stochastic Processes.* Springer-Verlag, Berlin.

Steele, J.M. (1989). Probabilistic and worst case analyses of classical problems of combinatorial optimization in Euclidean space. *Mathematics of Operations Research*, to appear.

Steele, J.M., Shepp, L.A., and Eddy, W.F. (1987). On the number of leaves of a Euclidean minimal spanning tree. *Journal of Applied Probability* 24, 809–826.

Stout, W.F. (1974). *Almost Sure Convergence.* Academic Press, New York.

Zeller, K. and Beekmann, W. (1970). *Theorie der Limitierungsverfahren.* Springer-Verlag, Berlin.

Department of Mathematics
Royal Holloway and Bedford New College
Egham Hill
Egham
Surrey TW20 0EX.

Universität Ulm
Abteilung Mathematik-III
Oberer Eselsberg
7900 Ulm.

Markov Random Fields in Statistics

Peter Clifford

1. Introduction

For nearly a century, statisticians have been intrigued by the problems of developing a satisfactory methodology for the analysis of spatial data; see Student (1914), for an early example. It is only since the early 1970's, however, that the statistical analysis of large data sets, using flexible parametric models has become a feasible proposition.

On the practical side, progress has been made possible by the availability of relatively cheap, computerised resources for the collection and analysis of data. The study of digital images and the use of satellite data for remote sensing are prominent examples in this respect. On the methodological side, substantial progress is associated with the introduction of Markov random fields (MRFs), as a class of parametric models for spatial data (Besag 1974). Shaped by these developments, spatial statistics has emerged as perhaps the most dynamic and computer intensive of all the areas of statistical endeavour; building upon models used originally in the description of physical systems and borrowing and improving upon ideas from computational physics.

Monte Carlo methods, in particular, have played a dominant role in dealing with problems of inference. The practicalities of working with high dimensional parameter sets within a Bayesian framework, have led to the invention of refreshing and novel techniques (Geman and Geman 1984), which promise to have a profound effect on the way in which Bayesian methods are used in more general contexts and which may serve to reintegrate these methods into the main body of applied statistics.

Much of physics is concerned with providing an understanding of the spatial organisation of matter and it is not suprising that many of the ideas which have become central in the theory of spatial statistics should have their origins in physical theory. The introduction of MRFs into the theory of statistics is yet another example of the continuing transfer of knowledge from the world of theoretical physics. John Hammersley whose interests include both domains of study, was ideally placed to facilitate the process of cross-fertilisation. Others who were involved in this instance include Neyman and Besag. Neyman was responsible for bringing Hammersley and a number of other visitors, including myself, to the University of California,

Berkeley in the summer of 1971. Hammersley gave an advanced course of lectures on probabilistic problems in physics, which included among other things a discussion of Spitzer's (1971) characterisation of two-state MRFs on a square lattice. This characterisation had been obtained independently by Averintsev (1970). Hammersley and I were able to generalise the results to arbitrary graphs and lattices, and to identify the central importance of the clique functions, as terms in the potential of a generalised Gibbs distribution. Hammersley returned to Oxford and sent a copy of the Berkeley paper to Besag who had already obtained partial results for rectangular lattices (Besag 1972). Besag then wrote to Hammersley with a much simpler, analytical proof of the general result, which appeared later in his very influential paper on spatial statistics (Besag 1974). Three other authors published proofs of the main theorem at about this time (Grimmett 1973; Preston 1973; Sherman 1973). A simple derivation is also possible using the factorisation theorem of Brook (1964). The basic theorem has more recently become important in non-spatial applications, most notably in the description of dependence structure for log-linear models (Ove and Strauss 1981; Darroch et al. 1980).

The Berkeley paper was never published and only a few copies were distributed. There are, however, many references to it in the literature and although the main result is stated as a named theorem in Kotz and Johnson (1983, Vol. 3, p. 570) there is, perhaps inevitably, some confusion about the exact contents. The method of proof in the unpublished paper is constructive and the operator techniques used are unusual. For these reasons it seems appropriate to take this opportunity to state the main results and to describe the methods by which they were obtained. This is done in Section 2.

The Markov property for random fields can be formulated in great generality (Preston 1974; Rozanov 1982). For statistical applications, an important step forward was the extension to point processes (Strauss 1975; Ripley and Kelly 1977). An excellent review of this topic is given by Baddeley and Møller (1989), who consider further generalisations to cover the case of marked point processes in which the neighbourhood relations for the marks are given by the graphical structure of the points.

A challenging problem is that of constructing random mosaics which are spatially Markov. A special case is the problem of subdividing two-dimensional space into regions whose boundaries are made up of line segments. In a remarkable paper, Arak (1982) showed that a time-homogeneous annihilating/birth particle system can give rise to space-time trajectories which have a two dimensional Markov property. These results were generalised by Arak and Surgailis (1989), to cover a wide class Markov polygonal fields. In Section 3 we consider how these processes might be used in the analysis of polygonal images. Some light is shed on a conjecture by Arak and Surgailis (1989) and a method of simulating the posterior distribution of a polygonal image is proposed.

2. Markov Fields on Finite Graphs

2.1. Notation

Let $G = (Z, E)$ be an undirected graph, where $Z = \{z_1, z_2, \ldots, z_n\}$ is a finite set of sites and E is a set of simple edges, i.e. a set of unordered pairs of distinct sites. Two sites which form an edge are said to be *neighbours* of each other. We use capital letters $U, V, \ldots, X, Y$ for subsets of Z and write $X + Y$ for the union of X and Y, and $X - Y$ for the set $\{x \in X : x \notin Y\}$. A lower-case letter stands for both an element of Z and also the associated singleton set. The set of all subsets of Z, including $\emptyset$ and Z itself is denoted by Ω. For any Y we define ∂Y, the *boundary* of Y by

$$\partial Y = \{x : (x, y) \in E, x \notin Y, y \in Y\}.$$

A set Y is said to be a *clique* if and only if

$$Y \subseteq y + \partial y, \quad \forall y \in Y;$$

in other words Y is a clique if and only if it is a singleton or if every member of Y is a neighbour of every other member of Y.

We associate with each site z_i, a finite set of colours $C_i, i = 1, 2, \ldots, n$. To avoid trivial cases we will assume that the cardinality of each set is greater than one. We also assume, without loss of generality, that every set contains a colour which we can agree to call *black*. Suppose that for each z_i we select a colour from $C_i, i = 1, 2, \ldots, n$. Such an assignment of colours to sites is called a *colouring* of Z. A typical colouring is denoted by χ. Let χ_Y denote the colouring obtained from χ by changing the colours on the sites in Y to black. A *partial colouring* has colours assigned on only a subset of sites. The partial colouring obtained by considering which colours have been assigned to sites in X by the colouring χ is denoted by χ^X. In particular, the colour at a site z is written as χ^z. The set of all possible colourings of Z is given by $\mathcal{C} = C_1 \times C_2 \times \cdots \times C_n$. A set Y is said to be light relative to χ if no site in Y is black under the colouring χ. We define L_χ to be the set of cliques which are light relative to χ.

Let us now consider a probability distribution on $\mathcal{C}$ with mass function P satisfying $\sum_{\chi \in \mathcal{C}} P(\chi) = 1$ and the *positivity condition* $P(\chi) > 0, \forall \chi \in \mathcal{C}$. We denote the marginal probability of the partial colouring χ^Y by $P(\chi^Y)$. This latter probability is obtained by summing P over all colourings which agree with χ on Y. We say that P is Markovian for the set X if and only if it satisfies the positivity condition and

$$P(\chi)/P(\chi^{Z-X}) = P(\chi^{X+\partial X})/P(\chi^{\partial X}), \quad \forall \chi \in \mathcal{C}.$$

We call this condition $M(X)$. If we postulate $M(z)$ for all singleton sets $z \in Z$, we say P is locally Markovian. If we postulate $M(X)$ for all $X \subseteq Z$ we say it is globally Markovian. The main theorems are as follows.

THEOREM 1. *Global and local Markov properties are equivalent.*

THEOREM 2. *P is Markovian if and only if it can be written in the form*

$$P(\chi)/P(\chi_Z) = \exp\Big(\sum_{Y \in L_\chi} Q(\chi^Y)\Big),$$

where Q is an arbitrary real-valued function of light colourings on cliques.

Furthermore, if P is Markovian then the associated function Q is given by

$$Q(\chi^Y) = \sum_{X \subseteq Y} (-1)^{|X|} \log P(\chi_{(Z-Y)+X}), \quad \forall Y \in L_\chi,$$

where $|X|$ denotes the cardinality of X.

The theorems are proved by introducing an operator algebra.

2.2. The Blackening Algebra

Let $\mathcal{R}$ be the set of all real-valued functions defined on $\mathcal{C}$. We define the *pure* blackening operator B_Y by

$$B_Y R(\chi) = R(\chi_Y), \quad R \in \mathcal{R}.$$

Since

$$B_X B_Y R(\chi) = B_X R(\chi_Y) = R(\chi_{X+Y}) = B_{X+Y} R(\chi),$$

in terms of the operators we have

$$B_X B_Y = B_Y B_X = B_{X+Y}$$

so that pure operators commute.

A *mixed* blackening operator $\alpha_1 B_{X_1} + \cdots + \alpha_m B_{X_m}$ is a finite linear combination of pure operators, where $\alpha_1, \ldots, \alpha_m$ are real-valued coefficients. For such an operator we have

$$(\alpha_1 B_{X_1} + \cdots + \alpha_m B_{X_m}) R(\chi) = \alpha_1 R(\chi_{X_1}) + \cdots + \alpha_m R(\chi_{X_m}).$$

Mixed operators multiply according to

$$\sum \alpha_i B_{X_i} \sum \beta_j B_{Y_j} = \sum \alpha_i \beta_j B_{X_i + Y_j}.$$

The identity operator is denoted by $1 = B_\emptyset$ and the zero operator by 0. With the preceding definitions, the blackening operators can be seen to form a commutative algebra.

The following lemma is a simple consequence of the definitions:

LEMMA 1. *If* $X \subseteq Y$ *then* $(1 - B_X)B_Y = 0$.

An operator which is equal to its square is called a projector. Every pure operator is a projector. In general, if B is a projector then so is $1-B$. It follows that

$$B_X + B_Y - B_{X+Y} = 1 - (1 - B_X)(1 - B_Y)$$

is also a projector. In the special case $Y = Z - (X + \partial X)$, for which $X + Y = Z - \partial X$, we denote the projector by β_X, i.e.

$$\beta_X = B_X + B_{Z-(X+\partial X)} - B_{Z-\partial X} = B_X + B_{Z-(X+\partial X)}(1 - B_X).$$

We also define $B_z^* = B_{Z-(z+\partial z)}(1 - B_z)$, so that $\beta_z = B_z + B_z^*$. Finally we define the projector $\beta = \prod_{z \in Z} \beta_z$. Writing $B_Y^* = \prod_{z \in Y} B_z^*$ and $B_\emptyset^* = 1$, we have

$$\beta = \prod_{z \in Z}(B_z + B_z^*) = \sum_{Y \in \Omega} B_{Z-Y} B_Y^*. \tag{2.1}$$

LEMMA 2. *If* $Y \neq \emptyset$ *and* Y *is not a light clique relative to* χ, *then* $B_Y^* R(\chi) = 0, \forall R \in \mathcal{R}$.

PROOF: (i) Suppose that Y is not a clique. Then Y has two distinct elements, x, y, say, such that x is not a neighbour of y, i.e. $x \in Z - (y + \partial y)$. From Lemma 1, it follows that $B_{Z-(y+\partial y)}(1 - B_x)$, and hence B_Y^*, equals 0.
(ii) Suppose that Y is not light relative to χ, then Y contains a site z which is already black, so that $(1 - B_z)R(\chi)$, and hence $B_Y^* R(\chi)$ equals 0. ■

Let us now consider the subset of $\mathcal{R}$ which is invariant under the operator β. Denoting this subset by $I(\beta)$, we have

$$I(\beta) = \{R : \beta R = R, R \in \mathcal{R}\} = \{\beta R : R \in \mathcal{R}\}.$$

If R is arbitrary then from Lemma 2 and (2.1) we have

$$\beta R(\chi) = R(\chi_Z) + \sum_{Y \in L_\chi} B_Y^* B_{Z-Y} R(\chi). \tag{2.2}$$

Furthermore, if Y is a clique then $Y \subseteq z + \partial z$ for any $z \in Y$, so that $Z - Y \supseteq Z - (z + \partial z)$ and $B_{Z-Y} = B_{Z-(z+\partial z)} B_{Z-Y}$ by Lemma 1. It follows that for arbitrary $R \in \mathcal{R}$ we have the further simplification

$$\beta R(\chi) = R(\chi_Z) + \sum_{Y \in L_\chi} \prod_{z \in Y} (1 - B_z) R(\chi_{Z-Y}). \tag{2.3}$$

Lemma 3. *The invariant subset $I(\beta)$ consists of those functions $R \in \mathcal{R}$ which have the representation*

$$R(\chi) = S(\chi_Z) + \sum_{X \in L_\chi} S(\chi_{Z-X}) \tag{2.4}$$

for some $S \in \mathcal{R}$.

Proof: (i) Let R have the representation (2.4) for some $S \in \mathcal{R}$. We will apply (2.3) to show that $\beta R = R$. Since there are no light cliques in χ_Z, we have $R(\chi_Z) = S(\chi_Z)$. Notice that if $Y \in L_\chi$, then $L_{\chi_{Z-Y}}$ is just the set of all nonempty subsets of Y. It follows that when R is given by (2.4) then

$$R(\chi_{Z-Y}) = \sum_{X \subseteq Y} S(\chi_{Z-X}), \tag{2.5}$$

Furthermore,

$$\prod_{z \in Y} (1 - B_z) R(\chi_{Z-Y}) = S(\chi_{Z-Y}), \tag{2.6}$$

since if $z \in Y$ then

$$(1 - B_z) R(\chi_{Z-Y}) = R(\chi_{Z-Y}) - R(\chi_{Z-Y+z}) = \sum_{z \subseteq X \subseteq Y} S(\chi_{Z-X}).$$

From (2.3) we therefore have $\beta R = R$.
(ii) Suppose now that $R \in I(\beta)$, i.e. $R = \beta R$. Since $\prod_{z \in Y}(1 - B_z) R(\chi_{Z-Y})$ is some function of χ_{Z-Y}, say $S(\chi_{Z-Y})$, and $R(\chi_Z)$ can be taken to be $S(\chi_Z)$, it follows immediately that R can be expressed as the right-hand side of (2.4). ■

Lemma 4. *If $X \subseteq Z$, then $I(\beta) \subseteq I(\beta_X)$.*

Proof: Let $R \in I(\beta)$, then R will have a representation as in Lemma 3. Since β_X is linear it suffices to show that

$$\beta_X S(\chi_Z) = S(\chi_Z)$$

and

$$\beta_X S(\chi_{Z-Y}) = S(\chi_{Z-Y})$$

for all cliques Y. Writing β_X as

$$\beta_X = 1 - (1 - B_X)(1 - B_{Z-(X+\partial X)})$$

the first of the equalities follows immediately. To establish the second equality it is sufficient to show that

$$(1 - B_X)(1 - B_{Z-(X+\partial X)}) B_{Z-Y} = 0. \tag{2.7}$$

But if Y is a clique it cannot be partly in X and partly in $Z-(X+\partial X)$. Suppose that $Y \subseteq X$ then

$$Z-(X+\partial X) \subseteq Z-X \subseteq Z-Y$$

and therefore (2.7) is satisfied as a consequence of Lemma 1. Alternatively, suppose that $Y \subseteq Z-(X+\partial X)$, then $Z-Y \supseteq X+\partial X \supseteq X$, so that the equation is again satisfied by Lemma 1. ■

LEMMA 5. *The invariant set* $I(\beta)$ *is given by* $\cap_{z\in Z} I(\beta_z)$.

PROOF: As a special case of Lemma 4 we have $I(\beta) \subseteq I(\beta_z), \forall z \in Z$, which implies that $I(\beta) \subseteq \cap_{z\in Z} I(\beta_z)$. On the other hand, if $R \in \cap_{z\in Z} I(\beta_z)$ then $R = \beta_z R, \forall z \in Z$, and hence $R = \prod_{z\in Z} \beta_z R$, so that $R \in I(\beta)$. ■

2.3. Proofs of Theorems 1 and 2

We show firstly, that the Markov condition $M(X)$ is equivalent to

$$P(\chi)/P(\chi_X) = P(\chi_{Z-(X+\partial X)})/P(\chi_{Z-\partial X}), \quad \forall \chi \in \mathcal{C}. \tag{2.8}$$

Under condition $M(X)$ we have

$$P(\chi) = P(\chi^{X+\partial X})P(\chi^{Z-X})/P(\chi^{\partial X}), \quad \forall \chi \in \mathcal{C}. \tag{2.9}$$

Equation (2.8) then follows by making the substitutions $\chi_X, \chi_{Z-(X+\partial X)}$ and $\chi_{Z-\partial X}$, and noting that $\chi_X^{Z-X} = \chi^{Z-X}, \chi_X^{\partial X} = \chi^{\partial X}$ etc. Conversely, if (2.8) holds, then

$$P(\chi) = P(\chi_X)P(\chi_{Z-(X+\partial X)})/P(\chi_{Z-\partial X}).$$

By summation over the appropriate subsets of $\mathcal{C}$, the marginal probabilities which appear in condition $M(X)$ can now be expressed as marginal probabilities of blackened colourings, which can be simplified as in the first part of the proof. Condition $M(X)$ is then verified by substitution.

PROOF OF THEOREM 1: From (2.8), condition $M(X)$ is equivalent to

$$R(\chi) - R(\chi_X) = R(\chi_{Z-(X+\partial X)}) - R(\chi_{Z-\partial X}), \quad \forall \chi \in \mathcal{C}$$

where $R(\chi) = \log P(\chi)$. In other words,

$$\beta_X R(\chi) = R(\chi), \quad \forall \chi \in \mathcal{C}.$$

Condition $M(X)$ is therefore equivalent to $R \in I(\beta_X)$.

Theorem 1 then follows immediately since if P is locally Markovian then $R \in \cap_{z \in Z} I(\beta_z) = I(\beta) \subseteq I(\beta_X)$ by Lemma 4 and hence P is globally Markovian. ■

PROOF OF THEOREM 2: From Lemma 3, $R \in I(\beta)$ iff

$$R(\chi) - R(\chi_Z) = \sum_{X \in L_\chi} S(\chi_{Z-X}), \quad \forall \chi \in \mathcal{C}$$

for some $S \in \mathcal{R}$. Defining $Q(\chi^X)$ to be $S(\chi_{Z-X})$, the proof of the first part of Theorem 2 is complete. For the last part, notice that $Q(\chi^Y) = S(\chi_{Z-Y})$ is given by

$$Q(\chi^Y) = \prod_{z \in Y} (1 - B_z) R(\chi_{Z-Y}),$$

as in (2.6). The result now follows since the operator $\prod_{z \in Y}(1 - B_z)$ has the expansion $\sum_{X \subseteq Y} (-1)^{|X|} B_X$. ■

3. Markov Polygonal Mosaics

The random fields described in Section 2 have proved to be useful models in the analysis of two-dimensional images (Geman and Geman 1984; Besag 1983). For image analysis, the sites of the graph, $z_1, \cdots, z_n$ correspond to pixels in a digitised picture. In the Bayesian framework χ, the unknown colouring of Z, i.e. the true scene, is treated as a realisation of a Markov random field. The observations of the pixel values $\mathcal{O} = \{O^z, z \in Z\}$ are assumed to be random corruptions of the true scene. In the simplest case, the likelihood is assumed to be proportional to

$$\exp\left(\sum_{z \in Z} h(O^z \mid \chi^z)\right). \tag{3.1}$$

Up to an additive constant, the logarithm of the posterior density of χ is therefore

$$\sum_{Y \in L_\chi} Q(\chi^Y) + \sum_{z \in Z} h(O^z \mid \chi^z),$$

which can expressed as

$$\sum_{Y \in L_\chi} Q^*(\chi^Y)$$

where the singleton clique functions $Q^*(\chi^z)$ have been modified by inclusion of terms from the likelihood. It follows that the family of MRFs is conjugate with likelihoods of the form (3.1).

Bayes estimates of the true scene can be made by a variety of techniques. Simulated annealing can be used to find maximum *a posteriori* estimates and the Gibbs sampler can be used to find estimates with minimum mean square error and estimates with minimum mis-classification error (Geman and Geman 1984).

When large artificial structures are present in the scene, it may be more natural to model true scenes as random mosaics which subdivide two-dimensional space into regions whose boundaries are made up of line segments. These random fields are defined on a continuous space rather than on the nodes of a graph.

3.1. Polygonal Colouring Measure

The simplest building block for polygonal fields is the Poisson line process (Kendall and Moran 1963). To describe the construction we introduce the following notation.

Let $T \subset \mathbb{R}^2$ be a convex bounded domain. Let $\mathcal{L}_T^n$ be the family of all sets of n distinct lines which intersect T and let $\mathcal{L}_T = \cup_{n=0}^{\infty} \mathcal{L}_T^n$, with $\mathcal{L}_T^0$ defined to be $\{\emptyset\}$, the family consisting of the empty set alone. We consider a Poisson line process defined on $\mathcal{L}_T$. To fix ideas we will assume that the process is homogeneous and isotropic with intensity λ, so that the number of lines crossing a disc of diameter d has a Poisson distribution with mean λd and the mean number of lines intersecting T is λd_T, where d_T is the mean diameter of T. We write μ_T for the Poisson line measure on $\mathcal{L}_T$, and we denote the conditional line measure on $\mathcal{L}_T^n$ by ν_T^n, so that

$$\mu_T(A) = \sum_{n=0}^{\infty} \frac{e^{-\lambda d_T}(\lambda d_T)^n}{n!} \nu_T^n(A \cap \mathcal{L}_T^n), \tag{3.2}$$

for events $A \subset \mathcal{L}_T$.

Suppose that C is a finite set of colours and χ maps T into C. The colouring is said to be polygonal if and only if the set of discontinuity points of χ is the union of intervals of a finite number of distinct lines, where each line contributes exactly one interval. We disregard intervals of zero length. Associated with each polygonal colouring χ there is the unique set of lines which contain the discontinuity points. We call this set $\langle \chi \rangle$. For an open set $S \subset T$ we define $\langle \chi^S \rangle$ to be the set of lines associated with discontinuities of χ on S. If S is not open, we define $\langle \chi^S \rangle$ as the limit for a sequence of diminishing open neighbourhoods of S. We denote the set of all polygonal colourings χ such that $\langle \chi \rangle = \ell$ by $\Omega_T^{\ell}, \ell \in \mathcal{L}_T$ and write $\Omega_T = \cup_{\ell \in \mathcal{L}_T} \Omega_T^{\ell}$.

The polygonal colouring measure is then defined to be

$$\gamma_T(A) = \int_{\mathcal{L}_T} |A \cap \Omega_T^{\ell}| \, \mu_T(d\ell), \tag{3.3}$$

where A is a measurable subset of Ω_T and $|\cdot|$ denotes cardinality. We consider distributions on Ω_T which are absolutely continuous with respect to γ_T. These can be specified by a density $f : \Omega_T \to [0, \infty)$. The associated probability measure is therefore given by

$$P_T^f(A) = \frac{\int_A f(\chi)\gamma_T(d\chi)}{\int_{\Omega_T} f(\chi)\gamma_T(d\chi)}, \tag{3.4}$$

provided that the denominator is finite.

3.2. The Uniform Density

Arak and Surgailis (1989) conjectured that it might be possible for γ_T to be finite, i.e. for $P_T^f(A)$ to be a probability measure, when f is constant. The following theorem gives a sufficient condition for this to be so.

THEOREM 3. *If* $|C| = 2$ *and* $\lambda d_T < 1$, *then* $\int_{\mathcal{L}_T} |\Omega_T^\ell| \mu_T(d\ell) < \infty$.

Before proving the theorem, we must introduce a little more notation. An extended polygonal colouring is a function $\chi^+ : \mathbb{R}^2 \to C$, whose discontinuity points are the union of intervals of lines in $\mathcal{L}_T$, but here the intervals are either semi-infinite, infinite, or of finite positive length. We write $\langle\chi^+\rangle$ for the line set associated with χ^+. The set of all extended colourings for which $\langle\chi^+\rangle = \ell$, is denoted by $\Theta_T^\ell, \ell \in \mathcal{L}_T$, and $\Theta_T = \cup_{\ell\in\mathcal{L}_T} \Theta_T^\ell$. We now restrict our attention to the case $|C| = 2$. Since $|\Theta_T^\ell|$ depends only on the cardinality of ℓ (say n) we will write it simply as $2c_n$; the factor 2 arising from the two possible colourings for a given discontinuity set.

LEMMA 6. *The sequence* $\{c_n\}$ *satifies the recurrence relation*

$$c_{n+1} = c_n + 4nc_{n-1} + \sum_{j=2}^{n} \frac{n!}{(n-j)!} \left(2j + \tfrac{5}{2}\right) c_{n-j}$$

with $c_0 = c_1 = 1$. *Furthermore, the power series*

$$g(u) = \sum_{n=0}^{\infty} \frac{c_n u^n}{n!}$$

has radius of convergence 1 *and is given by*

$$4 \log g(u) = -6u - u^2 + \frac{8u}{1-u} - 2\log(1-u).$$

For brevity the proof of this lemma is omitted. The proof of the theorem follows from the lemma by noting that $|\Omega_T^\ell| \le |\Theta_T^\ell|$, $\forall \ell \in \mathcal{L}_T$, so that from (3.2) and (3.3)

$$\gamma_T(\Omega_T) \le 2g(\lambda d_T)e^{-\lambda d_T}, \ \lambda d_T < 1.$$

Polygonal fields do not necessarily have a Markov property. However, Arak and Surgailis (1989) have established a sufficient condition for this to be so, namely that f is of the form $f(\chi) = e^{-F(\chi)}$, where $F : \Omega_T \to \mathbb{R} \cup \{\infty\}$ is an additive function (Rozanov 1982).

Examples of additive functions are: the total length of the intercolour boundary of χ, the number of times that the boundary between two colours turns so as to circle a particular colour and most importantly functions of the form

$$F(\chi) = \int_T k(\chi(t))\alpha(dt),$$

where α is a measure on T and $k : C \to \mathbb{R}$.

3.3. Statistical Applications

Additive functions arise naturally in statistical contexts. Thus, if χ is the true scene and observations of χ are limited to realisations of a spatial Poisson process whose intensity at point t is $\eta(\chi(t))$, $t \in T$, then the likelihood of the data is proportional to

$$\exp\left(-\int_T \eta(\chi(t))dt + \int_T \log \eta(\chi(t))N(dt)\right) \qquad (3.5)$$

where $N(A), A \subset T$ is the counting measure of the observed point pattern. If the prior density of the true scene is proportional to $\exp(-F_0(\chi))$, then applying Bayes Theorem, the posterior density of χ is proportional to $\exp(-F^*(\chi))$, where

$$F^*(\chi) = F_0(\chi) + \int_T \eta(\chi(t))dt - \int_T \log \eta(\chi(t))N(dt). \qquad (3.6)$$

It follows that if F_0 is additive then so is F^*. In other words, polygonal Markov fields are conjugate with Poisson sampling. It is therefore important to be able to simulate Markov polygonal fields, in particular the posterior distributions within this family.

3.4. Conditional Distributions

Polygonal Markov fields are Markov in the following sense. Let $S \subset T$ be an open set with a smooth boundary $\partial S = \bar{S} - S$ and let $\xi = (\chi^{\partial S}, \langle \chi^{\partial S} \rangle)$ then the distribution of χ^S given ξ is the same as that of χ^S given χ^{T-S}. Convex sets are of particular interest, and we will assume that S is convex from now on. Note that the information in ξ consists of the colouring on the boundary $\chi^{\partial S}$ and also the identification of those lines, intersecting ∂S, which separate different boundary colours. The conditional polygonal measure on S is

$$\gamma_S(A \mid \xi) = \int_{\mathcal{L}_S} \left|A \cap \Omega_S^\ell(\xi)\right| \mu_S(d\ell),$$

where A is a measurable subset of Ω_S and $\Omega_S^\ell(\xi)$ is the set of polygonal colourings on S which have discontinuity lines $\ell \cup \langle\chi^{\partial S}\rangle$ and which are consistent with the boundary conditions ξ. For fields which are specified by an additive function F, Arak and Surgailis (1989) have shown that the conditional distribution of χ^S is absolutely continuous with respect to $\gamma_S(\cdot \mid \xi)$, with density proportional to $\exp(-F(\chi^S))$.

The principal advantage of having an explicit form for the conditional density, is that Monte Carlo methods, such as the Gibbs sampler can be applied to simulate the process.

3.5. Monte Carlo Simulation of Markov Polygonal Fields

The idea is to run a Markov process on the state space Ω_T, whose equilibrium will be the desired field. The set T can be taken to be a rectangle. The procedure is as follows.

Let χ be the current state of the Markov process.

(a) Select a rectangle S, at random in T.

(b) Put down a realisation of a Poisson line process with intensity ρ in S. Suppose that the lines of the process are ℓ and that they are n in number.

(c) Calculate $K(\xi, \ell)$ given by

$$[K(\xi,\ell)]^{-1} = \sum_{\omega \in \Omega_S^\ell(\xi)} e^{-F(\omega)},$$

and select a new colouring for S, from the distribution with probability mass function $K(\xi,\ell)\exp(-F(\omega))$, $\omega \in \Omega_S^\ell(\xi)$.

(d) Let

$$q = \left(\frac{\rho}{\lambda}\right)^{n_0 - n} \frac{K(\xi,\ell_0)}{K(\xi,\ell)},$$

where $\ell_0 = \langle\chi^S\rangle - \langle\chi^{\partial S}\rangle$ and n_0 is the cardinality of ℓ_0. Change the colouring on S to χ_*^S if q is greater than 1. If q is less than 1, then with probability q change the colouring to χ_*^S and with probability $1-q$ leave the colouring unchanged.

(e) Go to (a).

The algorithm is a special case of the general class of algorithms discussed by Hastings (1970). If we let

$$P(A \mid \ell) = K(\xi,\ell) \sum_{\omega \in \Omega_S^\ell(\xi)} 1_A(\omega) e^{-F(\omega)},$$

then the probability distribution for the candidate colouring χ_*^S is

$$\sum_{n=0}^{\infty} \frac{e^{-\rho d_S}(\rho d_S)^n}{n!} \int_{\mathcal{L}_S^n} P(A \mid \ell)\, \nu_S^n(d\ell),$$

which has density

$$e^{-(\rho-\lambda)d_S}(\rho/\lambda)^n e^{-F(x_*^S)}K(\xi,\ell),$$

with respect to $\gamma_S(\cdot \mid \xi)$. The expression in (d) is therefore the appropriate ratio of required and sampled densities. The parameter ρ can be adjusted to maximise the acceptance probability.

REFERENCES

Arak, T. (1982). On Markovian random fields with a finite number of values. *4th USSR-Japan Symposium on Probability Theory and Mathematical Statistics: abstracts of communications*, Tbilisi.

Arak, T. and Surgailis, D. (1989). Markov fields with polygonal realisations. *Probability Theory and Related Fields* 80, 543–579.

Averintsev, M.B. (1970). On a method of describing complete parameter fields. *Problemy Peredaci Informatsii* (in Russian) 6, 100–109.

Baddeley, A. and Møller, J. (1989). Nearest-neighbour Markov point processes and random sets. *International Statistical Review*, to appear.

Besag, J.E. (1972). Nearest-neighbour systems and the auto-logistic model for binary data. *Journal of the Royal Statistical Society B* 34, 75–83.

——— (1974). Spatial interaction and the statistical analysis of lattice systems. *Journal of the Royal Statistical Society B* 36, 192–236.

——— (1983). Discussion of paper by P. Switzer. *Bulletin of the International Statistical Institute* 47, 77–92.

Brook, D. (1964). On the distinction between the conditional probability and joint probability approaches in the specification of nearest-neighbour systems. *Biometrika* 51, 481–483.

Darroch, J.N., Lauritzen, S.L., and Speed, T.P. (1980). Markov fields and log-linear interaction for contingency tables. *Annals of Statistics* 8, 522–539.

Geman, S. and Geman, D. (1984). Stochastic relaxation, Gibbs distributions and the Bayesian restoration of images. *I.E.E.E. Transactions of Pattern Analysis and Machine Intelligence* 6, 721–741.

Grimmett, G.R. (1973). A theorem about random fields. *Bulletin of the London Mathematical Society* 5, 81–84.

Hastings, W.K. (1970). Monte Carlo sampling methods using Markov chains and their applications. *Biometrika* 57, 97–109.

Kendall, M.G. and Moran, P.A.P. (1963). *Geometrical Probability.* Charles Griffin, London.

Kotz, S. and Johnson, N.L. (1983). *Encyclopædia of Statistical Science.* Wiley, New York.

Ove, F. and Strauss, D.J. (1981). Markov graphs. *Journal of the American Statistical Association* 86, 832–842.

Preston, C.J. (1973). Generalised Gibbs states and Markov random fields. *Advances in Applied Probability* 5, 242–261.

——— (1974). *Gibbs States on Countable Sets.* Cambridge University Press, Cambridge.

Ripley, B.D. and Kelly, F.P. (1977). Markov point processes. *Journal of the London Mathematical Society* 15, 188–192.

Rozanov, Yu.A. (1982). *Markov Random Fields.* Springer-Verlag, New York.

Sherman, S. (1973). Markov random fields and Gibbs random fields. *Israeli Journal of Mathematics* 14, 92–103.

Spitzer, F. (1971). Markov random fields and Gibbs ensembles. *American Mathematical Monthly* 78, 142–154.

Strauss, D.J. (1975). A model for clustering. *Biometrika* 62, 467–475.

Student (1914). The elimination of spurious correlation due to position in time or space. *Biometrika* 10, 179–180.

Mathematical Institute
24–29 St. Giles
Oxford OX1 3LB.

On Hammersley's Method for One-Dimensional Covering Problems

Cyril Domb

---

1. Personal History

I first met John Hammersley in Oxford during the years 1949–1952 when I held an ICI Fellowship at the Clarendon Laboratory. David Kendall used to run (with the aid of Pat Moran) a regular probability seminar in which he encouraged research workers in widely differing disciplines to participate. Hammersley then held an appointment in a department with the intriguing title *Lectureship in the Design and Analysis of Scientific Experiment.* David Finney was the lecturer from 1948–1955, and Michael Sampford and John Hammersley were his assistants. Their job was to provide mathematical, statistical, and computational advice to any of the science departments in Oxford that requested it. Hammersley had thus already begun his fruitful practice of "keeping open shop to all customers" and whenever he delivered a talk at the seminar, one could be sure of encountering a variety of stimulating new problems and ideas.

My wartime experience in radar research had introduced me to problems in geometrical probability, and whilst a graduate student at Cambridge I had published papers on the covering of a line by random intervals, and on the statistics of particle counters. It was useful for me to meet others who shared my interest, and a number of the problems discussed at the seminar were subsequently described in the monograph on *Geometrical Probability* by Maurice Kendall and Pat Moran (1963).

But my major research interest had moved to problems of lattice statistics, the Ising model, and order-disorder transitions in alloys. Moran (1947) had considered the statistical problem of the distribution of black-white joins in a lattice whose points could be black or white independently with probabilities p, $(1-p)$, and had proved that the distribution is normal. I felt it important to draw attention to the difference between the requirements of statistics and those of statistical mechanics. For the latter, the normality of the distribution gives *little* information of physical importance; the physicist, surprisingly, needs to determine all the higher moments and cumulants, and it is on the asymptotic form of these that the interesting critical behaviour depends.

For me personally one of the great benefits which I derived from the Oxford seminar was the introduction to the bright group of young statisticians who were active in organizing the research section of the Royal Statistical Society. Their public discussions and symposia were lively and challenging, and they cast their net widely. Two of the papers which Hammersley read to this section were *On Estimating Restricted Parameters* (1950) which dealt with problems for example in which the parameter sought was known to be an integer; and *Poor Man's Monte Carlo* (with K.W. Morton 1954) which discussed a Monte Carlo technique which did not require the use of large machines.

Much of the latter paper was devoted to a lattice model of a polymer molecule which took the excluded volume into account in a realistic way. In this paper Hammersley attributed the model of a random walk on a lattice which is not allowed to visit any site more than once to Meyer. He gave no reference and I was unable to trace to which Meyer he referred. In subsequent correspondence he suggested that it might have been J.E. Mayer the architect of the well known cluster integral theory of a condensing gas. I myself had been introduced to the model by G.S. Rushbrooke who presumably heard of it from his supervisor R.H. Fowler; in a letter to me Hammersley agreed that this may also have been his source.

Hammersley coined the term *self-avoiding walk* for the model, and this was adopted universally. Previously such walks had been described by a variety of names — *non-intersecting*, *non self-intersecting* and even *simple* (but of course they are far from simple). Curiously enough the terminology was challenged nearly thirty years later by Amit and his collaborators (1983). The original model envisaged selecting from the total ensemble of random walks these with no double or multiple points and giving equal weight to each of them. Amit et al. (1983) generated and analysed walks, which do not visit any site more than once, but whose probability of taking a step at any point is inversely proportional to the number of *unoccupied* neighbouring lattice sites; these they called *true self-avoiding walks.* The early Monte Carlo workers were careful *not* to general walks of this type (see e.g. Rosenbluth and Rosenbluth 1955) by weighting appropriately at the vertices. The argument is clearly one of semantics.

In a discussion remark following the above paper of Hammersley and Morton, Broadbent drew attention to a novel problem in which the randomness is associated with the *medium* rather than with the fluid. Subsequently he collaborated with Hammersley (1957) in providing a comprehensive formulation of this new class of problems which Hammersley described as *percolation processes*, a term which also gained universal acceptance. Some fifteen years later, when the important applications to solid state physics became apparent, the literature on percolation processes grew at an incredible rate.

My own approach to problems of lattice statistics had been to generate exact series expansions of substantial length for the logarithm of the partition function (the analogue of the cumulant generating function) and to use them to assess the asymptotic behaviour of the coefficients. This method had proved quite successful for the Ising model, for which a number of exact results were available by which the method could be checked. The same approach could be used to explore the behaviour of self-avoiding walks and percolation processes, and my research group at King's College established striking analogies between various features of these systems and thermodynamic properties of magnetic models.

I presented a paper to the Royal Statistical Society on these topics entitled *Some Statistical Problems Connected with Crystal Lattices* (1964) and was grateful for Hammersley's support in the discussion. Any talk of drawing conclusions from extrapolation arouses suspicion in the mind of the statistician. It was important to emphasize that our method was not just conjectured extrapolation; we made use of physical knowledge and insight to *postulate* an asymptotic form, and this postulate was tested and its parameters fitted by statistical data in a fairly standard manner.

By a simple argument involving sub-additive functions Hammersley had proved that the total number of self-avoiding walks of n steps on an infinite lattice was asymptotically of order μ^n and he called μ the *connective constant* (another term which gained wide acceptance). We were able to provide convincing statistical evidence that the total number of self-avoiding polygons of n steps was also asymptotically of order μ^n, and Hammersley subsequently established this result rigorously (1961).

For me one of the most amazing results of later research was the formulation in exact terms of the analogy we had discovered between self-avoiding walk models and percolation models and magnetic systems. In the n-vector model of ferromagnetism each site is occupied by an elementary magnetic spin which is free to rotate isotropically in n dimensions. $n = 1$ corresponds to the Ising model, $n = 2$ is called the x–y model, $n = 3$ the classical Heisenberg model. In 1972 de Gennes showed that $n = 0$ corresponds to the self-avoiding walk model.

One of my outstanding graduate students during my stay at Oxford was an Australian Rhodes Scholar named R.B. Potts. I had drawn his attention to a magnetic model with three orientations in a plane which had some properties analogous to those of the two orientation $\updownarrow$ Ising model. I thought the model might generalize to q-orientations in a plane. Potts demonstrated to me that the generalization which I sought was not as I had thought, q vectors in a plane, but in space, and the vectors must be such that the angle between any pair of them is the same. Potts published his results in a paper in the Proceedings of the Cambridge Philosophical Society (1952) since we considered the investigation to be an abstract mathematical exercise with little chance of physical application.

For nearly twenty years the Potts model was ignored. Then interest began to focus on magnetic models with different types of symmetry, and the number of papers on the Potts model grew with amazing rapidity. I can echo Hammersley's remarks (1983) "When children become adults, they embark on ideas and activities of which their parents are only dimly aware". Most surprising of all Kasteleyn and Fortuin (1969) demonstrated that the Potts model with $q = 1$ corresponds precisely to the percolation model.

In the present article I shall discuss problems arising from random intervals on a line in which Hammersley and I were interested in the late 1940's and early 1950's. I will relate these problems to one dimensional continuum percolation, a subject which has attracted interest and attention recently.

2. Statistics of Counters

The following problem arises when the finite resolving time of a recording apparatus is taken into account. Events are divided into two classes, recorded and unrecorded. Any recorded event is followed by a dead interval of length τ, during which any other event which occurs will be unrecorded. A typical example is an α-particle counter; a recorded particle causes the chamber to ionize, and no other particle can be recorded until the chamber has de-ionized. I dealt with this problem (Domb 1948) in the following manner.

Assume that the events are defined by a Poisson process, the probability of an event occurring in the interval $[y, y + dy]$ being λdy. Let $z_n(y)$ be the probability that n recorded events occur in $[0, y]$. It can be divided into mutually exclusive groups: (i) Those in which the recorder is live at point y, probability $z_{n1}(y)$. This means that no recorded event occurs in $[y - \tau, y]$, and hence n recorded events occur in $[0, y - \tau]$. (ii) Those in which the recorder is dead at point y, probability $z_{n2}(y)$. In this case a recorded event occurs in $[y - \tau, y]$.

It is now easy to construct equations for $z_n(y + dy)$ in terms of $z_{n1}(y)$ and $z_{n2}(y)$ leading to the following differential equation:

$$z'_n(y) = \lambda\big[z_{n-1}(y) - z_{n1}(y)\big]. \tag{2.1}$$

Thus, the function $z_{n1}(y)$ plays a key role in the structure of the equation.

When we look at $z_{n1}(y + dy)$ we see that all possibilities are covered by two cases:

(a) The recorder is live at y and remains live at $y + dy$; n recorded events occur in $[0, y]$, no event occurs in $[y, y + dy]$.

(b) The recorder is dead at y but becomes live at $y + dy$; $(n-1)$ recorded events occur in $[0, y-\tau]$, one event occurs in $[y-\tau, y-\tau+dy]$, and no event occurs in $[y, y+dy]$.

This gives rise to the differential equation:

$$z'_{n1}(y) = -\lambda z_{n1}(y) + \lambda z_{(n-1)1}(y-\tau). \tag{2.2}$$

I then showed that equations (2.1) and (2.2) are amenable to treatment by Laplace transforms, and that an explicit solution can readily be derived for $Z_n(p)$ the Laplace transform of $z_n(y)$. Moreover, the treatment can be generalized to a stochastic distribution of intervals $u(\tau)d\tau$. The only change required for this is the replacement of the second term on the right of (2.2) by the integral

$$\lambda \int_0^y u(\tau) z_{(n-1)1}(y-\tau)\, d\tau$$

and such a faltung can equally easily be handled by Laplace transforms.

A second type of instrument was used for recording events of a different kind which remains dead as long as events follow one another at intervals less than τ. This is closely related to the problem of covering a line by random intervals, which I had discussed previously (Domb 1947), again by means of Laplace transforms. I used my previous analysis to derive the distribution of recorded events for this second type of counter (Domb 1950), but noted that it was no longer a simple matter to generalize to a stochastic distribution. "The possibility of one interval completely covering another which follows it causes considerable mathematical complications."

A few years later Hammersley (1953) became interested in this second type of counter in connection with a device for counting blood cells electronically which had been developed in the Clinical Pathology Department of the Radcliffe Infirmary at Oxford. "A large number of blood cells, contained in a shallow chamber, are scanned by a photoelectric cell. The depth of the chamber and the concentration of blood cells in solution therein allow blood cells (supposed distributed at random through the chamber) to overlap when viewed from above the scanner. The field of view of the scanner at any instant is somewhat larger than the size of a blood cell, but is, nevertheless, of much the same order of magnitude. With passage of time the chamber moves underneath the photocell so that the field of view traces out a long narrow path not crossing or overlapping itself and only embracing a portion of the whole chamber. The blood cells have no motion relative to the chamber. As each blood cell comes under the photocell it produces an electrical impulse, whose duration depends upon the size and shape and orientation of the blood cell. These impulses go to a counter, which counts them except that it will not count any impulse which is overlapped by a previous impulse. The problem is to determine the number of blood cells in the chamber from a knowledge of the recorded count and the distribution of the lengths of individual impulses."

Hammersley came to discuss the problem with me, and I pointed out to him that there was no difficulty in calculating the mean, mean-square or any other moment; but I could not see how to provide a closed form solution. Hammersley worked on the problem and did in fact produce a complete solution. He made a handsome acknowledgement to me "I am very much indebted to Domb, who showed me how to surmount these difficulties by a brilliant application of the elementary theorem that the expectation of the sum of several (possibly independent) quantities is the sum of their expectations". In fact, he had achieved far more than I had ever thought possible.

A few years later Walter L. Smith (1957) re-derived Hammersley's results more neatly and concisely using the powerful methods of renewal theory. The Cambridge mathematician A.S. Besicovitch used to say "A mathematician's reputation rests on his bad proofs" (Burkill 1971). He wished to convey the idea that the originator of a result in mathematics usually establishes it by long and complicated proofs. This paves the way for the shorter and simpler proofs of later workers.

I wish to focus attention on one particular aspect of the solution, the probability that the portion $[0, y]$ of the line is completely covered. Hammersley incidentally provides a formal solution to this problem, but the expression he gives is complicated, and it seems to me that a direct attack on the problem itself, using his approach, yields a solution more readily.

3. Covering of a Line or Circle by Random Intervals[1]

When I returned to Cambridge in 1946 after radar-research for the Admiralty in World War 2, I brought with me the above covering problem with equal intervals. I needed to know whether anyone had tackled the problem previously, and Herman Bondi (who had been one of my colleagues at the Admiralty) referred me to Harold Jeffreys, whom he described as a mine of information on miscellaneous mathematical problems. Jeffreys immediately thought of the 'bicycle wheel problem' which he himself had formulated a few years previously as follows: A man is cycling along a road and passes through a region strewn with tacks; he wishes to know whether one has entered his tyre. Because of the traffic, he can only snatch glances at random times. At each glance he covers a fraction x of the wheel. What

[1] The remaining sections are adapted from a recent paper by the author in the Journal of Statistical Physics (Domb 1989). Detailed reference to other work on one-dimensional continuum percolation will be found in this paper. The author is indebted to Plenum Publishing Corporation for permission to reproduce some of this work.

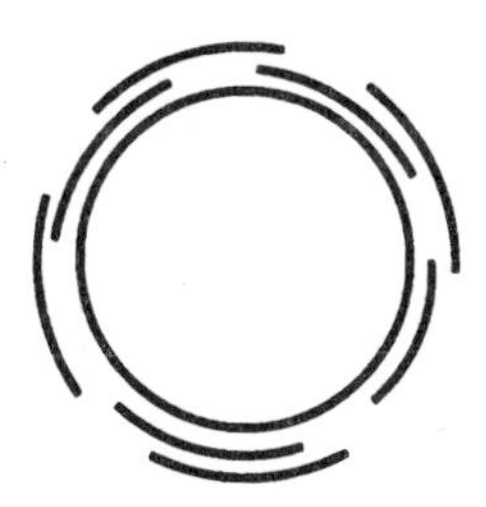

FIG. 1. The bicycle wheel problem.

is the probability that after n glances he has covered the whole wheel? In mathematical terminology: n intervals are placed randomly on a circle, each covering a fraction x of the circle. What is the probability that the circle is completely covered (Figure 1)?

Jeffrey's drew my attention to a paper published by W.L. Stevens in 1939 in the Annals of Eugenics, entitled *Solution to a Geometrical Problem in Probability*, in which his problem was solved. Using a neat combinatorial argument, Stevens found for the probability $F(0)$ of complete coverage

$$F(0) = 1 - \binom{n}{1}(1-x)^{n-1} + \binom{n}{2}(1-2x)^{n-1} - \binom{n}{3}(1-3x)^{n-1} + \cdots \quad (3.1)$$

the series terminating at the kth term, k being the integral part of $1/x$. Stevens also derived a formula for $F(i)$, the probability that there are i gaps on the circle.

In 1929, R.A. Fisher published an article entitled *Tests of Significance in Harmonic Analysis*, in which he calculated the probability that the largest interval in the random division of a circle is less than x (Figure 2). When Stevens's solution for $F(0)$ appeared, Fisher noted that it was identical with his, and a moment's reflection is enough to convince one that the two problems are identical. Fisher pointed this out in a note published in 1940.

But surprisingly, R.A. Fisher, one of the founders of the modern theory of statistics, was unaware that the distribution of length of the largest interval in the random division of a line had been correctly solved by Whitworth many years before, and was reproduced in his classic book, *Choice and Chance* (solutions to problems 666 and 667 published in 1897).

Problem 666: A line of length c is divided into n segments by $n-1$ random points. Find the chance that no segment is less than a given length a, where $c > na$ (say, $c - na = ma$).

Problem 667: In the last question find the chance that r of the segments shall be less than a and $n - r$ greater than a.

FIG. 2. Random division of a circle.

More precise dating of the solutions will be discussed in the next section.

4. Whitworth's Choice and Chance

We will preface this section with a few biographical details relating to Whitworth, taken from the Dictionary of National Biography (1901–1911, p. 655) and will continue with some comments on the different editions of his famous publication *Choice and Chance.*

William Allen Whitworth was born in 1840, and entered St. John's College as a undergraduate in October 1858. His performance in the Mathematics Tripos was not distinguished — he was 16th Wrangler in 1862 — but this does not seem to have represented his true ability. While still an undergraduate he was principal editor of the *Oxford, Cambridge and Dublin Messenger of Mathematics*, started at Cambridge in November 1861. The publication was continued as *The Messenger of Mathematics*; Whitworth remained one of the editors till 1880, and was a frequent contributor.

After leaving Cambridge in 1862 he was successively chief mathematics master at Portarlington School and Rossal School, and professor of mathematics at Queen's College, Liverpool (1862, 1864); he was a fellow of St. John's College from 1867 to 1882. At the same time Whitworth followed a second career of distinction in the Church, being ordained deacon in 1865 and priest in 1866. He held appointments as a curate at three churches in Liverpool from 1865 to 1875, and as vicar of two churches in London from 1875 until his death in 1905.

The first edition of *Choice and Chance* was published in 1867 while he was in Liverpool, and was a reproduction of lectures given to ladies in Queen's College Liverpool in the Michaelmas term of 1866. The book was subtitled *Two Chapters in Arithmetic*, and its aims, as described in the Preface, were modest enough:

I had already discovered that the usual method of treating ques-

> tions of selection and arrangement was capable of modification and so great simplification, that the subject might be placed on a purely arithmetical basis; and I deemed that nothing would better serve to furnish the exercise which I desired for my classes, and to elicit and encourage a habit of exact reasoning, than to set before them, and establish as an application of arithmetic, the principles on which such questions of "choice and chance" might be solved.

He expressed the hope that his publication might be of service "in conducing to a more thoughtful study of arithmetic than is common at present; extending the perception and recognition of the important truth, that arithmetic, or the art of counting, demands no more science than good and exact common sense".

Chapter 1 was devoted to "Choice", and was followed by 24 questions; Chapter 2 to "Chance", followed by 20 questions. The questions were all arithmetical in character. An appendix was devoted to *Permutations and Combinations Treated Algebraically*: "In my experience as a teacher I have found the proofs here set forth more intelligible to younger students than those given in the text books in common use". Whitworth here derived a number of standard elementary combinatorial formulae, and ended with a new combinatorial proof of the binomial theorem.

The second edition, published only three years later (1870) from St. John's College, Cambridge, added three appendices containing more sophisticated material. Appendix II was devoted to *Distributions* (into different groups or parcels), Appendix III to *Derangements*: "a series of propositions are given which are not usually found in text books of algebra. But I can see no reason why examples of such simple propositions ... should be excluded from elementary treatises in which more complex but essentially less important theorems find place". Appendix IV was concerned with the celebrated St. Petersburg problem and its background. More than 100 miscellaneous new examples were added.

In the third edition, published in 1878, the material in the appendices was revised and enlarged, and incorporated into the main text. There were now four chapters on *Choice* and four chapters on *Chance*, the final, brief eighth chapter carrying the title, *The Geometrical Representation of Chances*, the number of examples was increased to 300, and they were divided into different classes. The Preface contained the proclamation, "Questions requiring the application of the Integral Calculus are not included in the book, which only fulfills its title to be an Elementary Treatise".

In the fourth edition, published in 1886, the number of examples grew to 640, and a new chapter in the *Choice* section was added dealing with problems where the order in which gains and losses occur is relevant, e.g., if there is a condition that losses must never exceed gains. A short additional chapter in the *Chance* section entitled, *The Rule of Succession*, was devoted to a precise treatment of situations in which the probability of an event is supposed completely unknown, but the results of a number of trials are available. What can now be predicted about future trials?

The fifth and final edition was not published until 1901. But in 1897 there appeared a volume entitled, *DCC Exercises in Choice and Chance*. which provided fairly detailed solutions to the 640 examples of the fourth edition, and to 60 new examples, several of which were concerned with the random division of a line by a number of points. Questions 667 and 668, which were quoted in the previous section, are included among the latter. The preface to the fifth edition, which now contained 1000 examples, described the new category as follows: "A new feature will be recognized in a class of problems which found scarcely any place in former editions; the class which includes investigations into the mean value of the largest part, (or the smallest, or any other in order of magnitude) or of functions of such a part, when a magnitude is divided at random".

It is clear that Whitworth was actively working on this type of problem at the time. Quoting again from the same preface, "the most important addition in the body of the work is the very far-reaching theorem ... which enables us to write down at sight the mean value of such functions as α^3, $\alpha^3\beta^4$, $\alpha\beta\gamma$ etc. when $\alpha, \beta, \gamma, \ldots$ are the parts into which a given magnitude is divided at random. I first published this theorem in a pamphlet in the year 1898". The calculations of quantities of this type given in the *DCC Exercises* volume did not make use of the theorem, and were much longer.

From the above discussion it is clear that the problem with which we are concerned was tackled by Whitworth at some date between 1886 and 1897, most probably close to the latter date.

5. Whitworth's Solution

Whitworth divided the line into a number of discrete segments, which would eventually be allowed to become very large. He then used standard combinatorial formulae which he had developed in the text to enumerate various cases outlined in examples 666 and 667 (see Section 3).

We shall retain Whitworth's notation for historical reasons, but shall find it convenient to use generating functions to reproduce his combinatorial formulae. Whitworth assumed that the line of length c was divided into ωc equal elements. The given length a would then contain ωa elements. Take a dummy variable x_1 to enumerate the possible configurations of the first segment, x_2 the second segment, ..., x_n the nth segment. Then the generating function which enumerates all configurations in any division of the line by $n-1$ points is

$$F(t, x_1, x_2, \ldots, x_n) = (tx_1 + t^2x_1^2 + \cdots)(tx_2 + t^2x_2^2 + \cdots)\cdots(tx_n + t^2x_n^2 + \cdots) \tag{5.1}$$

assuming no two points are identical. The total number of segments is ωc and therefore all possible configurations are enumerated by the coefficient of $t^{\omega c}$ in $F(t; x_1, \ldots, x_n)$. If we need the total number of configurations, we put $x_1 = x_2 = \cdots = x_n = 1$ and find the coefficient $t^{\omega c - n}$ in $(1-t)^{-n}$, which is

$$\binom{\omega c - n + n - 1}{n-1} = \binom{\omega c - 1}{n-1} = \frac{(\omega c - 1)(\omega c - 2)\cdots(\omega c - n + 1)}{(n-1)!}. \tag{5.2}$$

For problem 666 one needs to enumerate all configurations with each of the segments containing ωa or more elements, and Whitworth realized that this was identical with finding all possible configurations which divide a line of length $c - n\omega a$ into n parts. This is clear from the generating function approach, since the appropriate enumerator is now

$$t^{\omega a}x_1^{\omega a}(1 + tx_1 + t^2x_1^2 + \cdots)t^{\omega a}x_2^{\omega a}(1 + tx_2 + t^2x_2^2 + \cdots)\cdots \times t^{\omega a}x_n^{\omega a}(1 + tx_n + t^2x_n^2 + \cdots). \tag{5.3}$$

We therefore require the coefficient of $t^{\omega(c-na)}$, i.e., of $t^{\omega ma}$ $(ma = c - na)$ in $(1-t)^{-n}$, which is

$$\binom{\omega ma + n - 1}{n-1} = \frac{(\omega ma + n - 1)(\omega ma + n - 2)\cdots(\omega ma + 1)}{(n-1)!}. \tag{5.4}$$

Hence the probability that no segment is less than a is found by taking the quotient of (5.4) by (5.2) and is equal to

$$\frac{(\omega ma + n - 1)(\omega ma + n - 2)\cdots(\omega ma + 1)}{(\omega c - 1)(\omega c - 2)\cdots(\omega c - n + 1)}. \tag{5.5}$$

When ω increases indefinitely, this reduces to

$$[ma/c]^{n-1} \tag{5.5$'$}$$

For example 667, Whitworth pointed out that all orders of choice of the r segments less than a, and the $n-r$ segments greater than a, give rise to the same number of configurations, and we can therefore deal with the case in which the r segments are at the beginning and the $n-r$ at the end, and multiply by $\binom{n}{r}$. The enumerating generating function is then

$$\begin{aligned}&(tx_1 + t^2x_1^2 + \cdots + t^{\omega a-1}x_1^{\omega a-1})(tx_2 + t^2x_2^2 + \cdots + t^{\omega a-1}x_2^{\omega a-1})\cdots \\ &\times (tx_r + t^2x_r^2 + \cdots + t^{\omega a-1}x_r^{\omega a-1})t^{\omega a}x_{r+1}^{\omega a}(1 + tx_{r+1} + t^2x_{r+1}^2 + \cdots) \\ &\times t^{\omega a}x_{r+2}^{\omega a}(1 + tx_{r+2} + t^2x_{r+2}^2 + \cdots)\cdots t^{\omega a}x_n^{\omega a}(1 + tx_n + t^2x_n^2 + \cdots).\end{aligned} \tag{5.6}$$

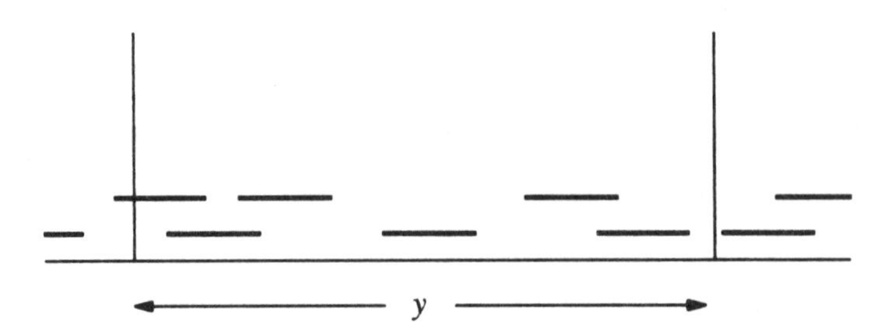

FIG. 3. Random intervals on a line.

The total number of configurations is the coefficient of $t^{\omega ma+\omega ra-r}$ in

$$\frac{(1-t^{\omega a-1})^r}{(1-t)^r}(1-t)^{-(n-r)} = (1-t^{\omega a-1})^r(1-t)^{-n}. \tag{5.7}$$

Expanding the first factor by the binomial theorem, we derive the series

$$\begin{aligned}\binom{n+\omega(m+r)a-r-1}{n-1} - \binom{r}{1}\binom{n+\omega(m+r-1)a-r}{n-1}\\ +\binom{r}{2}\binom{n+\omega(m+r-2)a-r+1}{n-1} - \cdots\\ +(-1)^s\binom{r}{s}\binom{n+\omega(m+r-s)a-r+s-1}{n-1} + \cdots .\end{aligned} \tag{5.8}$$

In the limit of very large ω this simplifies very considerably; dividing by (5.2) and taking the limit, we obtain

$$\begin{aligned}\left(\frac{m+r}{m+n}\right)^{n-1} - \binom{r}{1}\left(\frac{m+r-1}{m+n}\right)^{n-1} + \binom{r}{2}\left(\frac{m+r-2}{m+n}\right)^{n-1} + \cdots\\ +(-1)^s\binom{r}{s}\left(\frac{m+r-s}{m+n}\right)^{n-1} + \cdots + (-1)^r\left(\frac{m}{m+n}\right)^{n-1}.\end{aligned} \tag{5.9}$$

Expression (5.9) must be multiplied by $\binom{n}{r}$ to obtain the complete solution.

Although (5.8) looks complicated, the generating function (5.7) from which it is derived is quite simple, and the calculation of averages and higher moments can be undertaken by standard routine.

The probability of complete coverage, with which we have been concerned, corresponds to $r = n$, and is given by

$$\begin{aligned}1 - \binom{n}{1}\left(\frac{c-a}{c}\right)^{n-1} + \binom{n}{2}\left(\frac{c-2a}{c}\right)^{n-1} + \cdots\\ +(-1)^s\binom{n}{s}\left(\frac{c-sa}{c}\right)^{n-1} + \cdots\end{aligned} \tag{5.10}$$

the series terminating at the last term before $c - sa$ becomes negative.

The solutions given above are the same as those derived later by Fisher (1929) and Stevens (1939), with the slight adaptation needed for a problem on a circle rather than on a line.

6. Use of a Poisson Process: Equal Intervals

The problem to be considered is the following (Figure 3).

Events occur at random on a line in a Poisson distribution, the probability of an occurrence in $[y, y+dy]$ being λdy. Each event is the left-hand end of an interval of length τ. Choose any section $[0, y]$ of the line. Calculate the probability $z(y)$ that the section is completely covered.

We divide $z(y)$ into mutually exclusive classes $z(y, \xi)$ in which the last event occurred between $y - \xi$ and $y - \xi - d\xi$. Then if $y > \tau$, ξ cannot be greater than τ or the section $[0, y]$ would not be covered. Also, $z(y, \xi)$ can be decomposed into three independent contributions: (i) No event occurs in $[y-\xi, y]$; (ii) an event occurs in $[y-\xi-d\xi, y-\xi]$; (iii) the section $[0, y-\xi]$ is covered. Hence, we deduce that

$$z(y) = \int_0^\tau z(y, \xi)\, d\xi = \int_0^\tau \lambda e^{-\lambda\xi} z(y-\xi)\, d\xi \quad (y > \tau). \tag{6.1}$$

If $y \le \tau$, we must take into account the additional possibility that an event occurs in $[y-\tau, 0]$, and no event occurs in $[0, y]$; we easily find that

$$z(y) = \int_0^y \lambda e^{-\lambda\xi} z(y-\xi)\, d\xi + e^{-\lambda y} - e^{-\lambda\tau} \quad (y \le \tau). \tag{6.2}$$

Taking Laplace transforms in y in (6.1) and (6.2), we derive for the Laplace transform $Z(p)$ of $z(y)$,

$$Z(p) = \frac{p(1-e^{-\lambda\tau}) - \lambda e^{-\lambda\tau}(1-e^{-p\tau})}{p + \lambda e^{-(p+\lambda)\tau}}. \tag{6.3}$$

If the denominator is expanded as $[1 + (\lambda/p)e^{-(p+\lambda)\tau}]^{-1}$ and the terms are interpreted individually, the combinatorial solution is obtained. If further the solution is broken down into mutually exclusive classes in which exactly n events occur in $[0, y]$, the identity

$$z(y) = \sum_{n=0}^{\infty} \frac{\lambda^n}{n!} e^{-\lambda} f_n(y) \tag{6.4}$$

can be deduced, where $f_n(y)$ is the probability for n events. In this way the solution of Whitworth, Fisher, and Stevens can be simply derived.

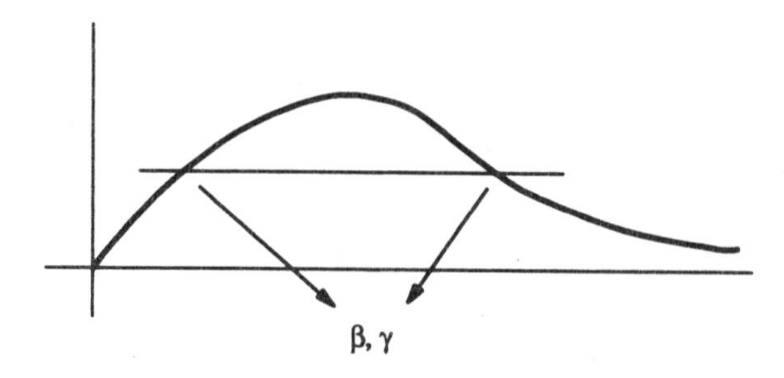

FIG. 4. Solution of $xe^{-x} = \beta e^{-\beta}$ giving asymptotic decay.

But if we are interested in large y/τ, the asymptotic behaviour of $z(y)$ is determined by the zeros of the denominator of (6.3), i.e. by solutions $-\gamma$ of

$$q + \beta e^{-(\beta+q)} = 0 \quad (q = p\tau, \beta = \lambda\tau). \tag{6.5}$$

There is only one real root, $-\gamma$, which dominates the asymptotic behaviour, the complex roots providing transients which rapidly decay. γ is the solution other than β of the equation

$$xe^{-x} = \beta e^{-\beta} \tag{6.6}$$

(see Figure 4). We then find the asymptotic solution

$$z(y) \sim \frac{e^{-\beta}(\beta - \gamma)}{\gamma(1-\gamma)} e^{-\gamma\nu} \quad (y = \nu\tau). \tag{6.7}$$

When β is large (high density of events), γ is small, and when β is small, γ is large. The probability of an infinite cluster of overlapping intervals in a one-dimensional percolating system is zero; equation (6.7) describes the approach to zero as a finite system grows large.

The calculation for $z_k(y)$, the probability that the line contains k clusters, follows similar lines. The integral equation is now

$$z_k(y) = \begin{cases} \int_0^y \lambda e^{-\lambda\xi} z_k(y-\xi)\, d\xi & (\xi \le \tau) \\ \int_0^y \lambda e^{-\lambda\xi} z_{k-1}(y-\xi)\, d\xi & (\xi > \tau) \end{cases} \tag{6.8}$$

with special treatment for $k = 1$. Taking Laplace transforms, we find

$$Z_k(p) = \frac{\lambda e^{-\tau(p+\lambda)}}{p + \lambda e^{-\tau(p+\lambda)}} Z_{k-1}(p) = \left(\frac{\lambda e^{-\tau(p+\lambda)}}{p + \lambda e^{-\tau(p+\lambda)}}\right)^{k-1} Z_1(p). \tag{6.9}$$

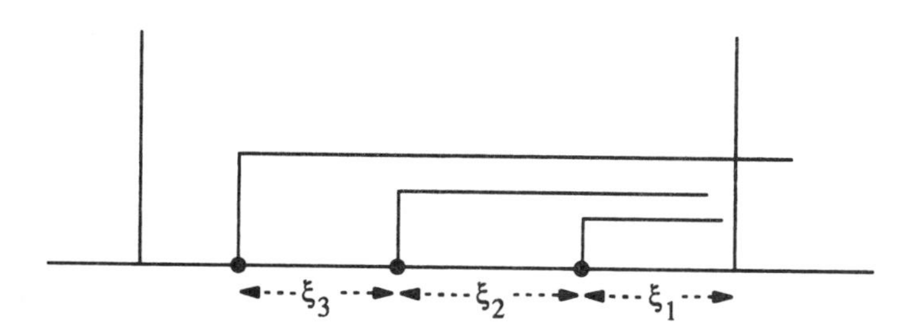

FIG. 5. Stochastic distribution: covering of intervals.

From this it can be deduced that the asymptotic distribution of clusters, in the limit of large ν ($= y/\tau$), is normal with mean $\nu\beta e^{-\beta}$ and variance $\nu[\beta e^{-\beta} - 2\beta^2 e^{-2\beta}]$.

The calculation of $W(x, y)dx$, the probability that the covered portion of the line is between x and $x+dx$, is more complicated, and the distribution contains δ-function terms corresponding to various discrete probabilities. The moments of the distribution can be calculated in a straightforward manner. For example,

$$\begin{aligned}\langle x\rangle &= y(1-e^{-\beta})\\ \langle x^2\rangle &= y^2(1-e^{-\beta}) - e^{-\beta}\left(y^2 - \frac{2y}{\lambda} + \frac{2}{\lambda^2}\right)\\ &\quad + e^{-2\beta}\left[(y-\tau)^2 - \frac{2(y-\tau)}{\lambda} + \frac{2}{\lambda^2}\right].\end{aligned} \tag{6.10}$$

7. Stochastic Distribution of Intervals

When the intervals are not all equal the previous method breaks down because an early event can overlap a later one (Figure 5). The behaviour at the point y is no longer dependent only on the latest event at $y - \xi_1$, but all previous events at $y - \xi_1, y - \xi_1 - \xi_2, \ldots,$ must be considered. The way in which to deal with this new situation was demonstrated by Hammersley, as we mentioned above in Section 2.

Assume a probability distribution of intervals $u(\tau)d\tau$, and divide $z(y)$ into mutually exclusive classes as follows:

$$z(y) = z(y;\xi_1) + z(y;\xi_1,\xi_2) + z(y;\xi_1,\xi_2,\xi_3) + \cdots + z_0(y) \tag{7.1}$$

where $z(y;\xi_1)$ represents the class in which the point y is covered by the last event at $y - \xi_1$, $z(y;\xi_1,\xi_2)$ represents the class in which the point is *not covered* by the *last event* at $y - \xi_1$, but is covered by the *last but one* at $y - \xi_1 - \xi_2$; $z(y; \xi_1, \xi_2, \xi_3)$ represents the class in which the point y is *not covered* by the *last two events*, but is covered by the *last but two* at $y - \xi_1 - \xi_2 - \xi_3$; $z_0(y)$ represents the class in which *no covering event* occurs in $[0, y]$ but the point y is covered by an event occurring *before*. Write

$$U(\tau) = \int_0^\tau u(t)\, dt \tag{7.2}$$

which represents the probability of an interval of length not exceeding τ; $1 - U(\tau)$ then represents the probability of an interval greater than τ. It is easy to derive the following relations (Figure 5):

$$\begin{aligned} z(y; \xi_1) &= \int_0^y \lambda e^{-\lambda\xi}[1 - U(\xi_1)] z(y - \xi_1)\, d\xi_1 \quad (0 < \xi_1 < y) \\ z(y; \xi_1, \xi_2) &= \iint \lambda e^{-\lambda\xi_1} U(\xi_1)\, d\xi_1\, \lambda e^{-\lambda\xi_2}[1 - U(\xi_1 + \xi_2)] \\ &\quad \times d\xi_2\, z(y - \xi_1 - \xi_2) \quad (0 < \xi_1, \xi_2 < y, \xi_1 + \xi_2 < y) \\ z(y; \xi_1, \xi_2, \xi_3) &= \iiint \lambda e^{-\lambda\xi_1} U(\xi_1)\, d\xi_1\, \lambda e^{-\lambda\xi_2} U(\xi_2)\, d\xi_2\, \lambda e^{-\lambda\xi_3} \\ &\quad \times [1 - U(\xi_1 + \xi_2 + \xi_3)]\, d\xi_3\, z(y - \xi_1 - \xi_2 - \xi_3) \\ &\qquad (0 < \xi_1, \xi_2, \xi_3 < y, \xi_1 + \xi_2 + \xi_3 < y). \end{aligned} \tag{7.3}$$

To see the structure of these relations, it is convenient to transform to new variables,

$$\eta_1 = \xi_1, \ \eta_2 = \xi_1 + \xi_2, \ \eta_3 = \xi_1 + \xi_2 + \xi_3, \ldots \tag{7.4}$$

so that the limits of integration in the new variables are

$$0 < \eta_1 < \eta_2 < \eta_3 < \cdots < y. \tag{7.5}$$

We then find

$$\begin{aligned} z(y; \eta_1) &= \int_0^y \lambda e^{-\lambda\eta_1}[1 - U(\eta_1)] z(y - \eta_1)\, d\eta_1 \\ z(y; \eta_1, \eta_2) &= \iint \lambda^2 U(\eta_1)\, d\eta_1\, e^{-\lambda\eta_2}[1 - U(\eta_2)] z(y - \eta_2)\, d\eta_2 \\ z(y; \eta_1, \eta_2, \eta_3) &= \iiint \lambda^3 U(\eta_1) d\eta_1 U(\eta_2) d\eta_2\, e^{-\lambda\eta_2}[1 - U(\eta_3)] z(y - \eta_3) d\eta_3. \end{aligned} \tag{7.6}$$

The integration in η_1 in $z(y; \eta_1, \eta_2)$ yields a function of η_2. Similarly, the integrations of η_1, η_2 in $z(y; \eta_1, \eta_2, \eta_3)$ yield a function of η_3. The structure of equation (7.1) is therefore

$$z(y) = \int_0^y v(\eta) z(y - \eta)\, d\eta + z_0(y) \tag{7.7}$$

which is still of the form amenable to Laplace transforms. The function $v(\eta)$ can be calculated by summing the successive contributions in (7.6).

However, we shall use a shortcut to evaluating $v(\eta)$ by considering a related problem, the probability $\zeta(y)$ that the point y is *covered* by an event occurring in $[0, y]$. We can decompose $\zeta(y)$ in a similar manner to (7.1)–(7.6) and we obtain the same integrals without the $z(y-\eta)$ factors, i.e.,

$$\zeta(y) = \int_0^y v(\eta)\, d\eta. \tag{7.8}$$

But the probability $1 - \zeta(y)$ that the point y is *not covered* by an event occurring in $[0, y]$ was calculated in an elementary manner by Hammersley (1953) to be

$$\exp\left[-\lambda y + \lambda \int_0^y U(t)\, dt\right]. \tag{7.9}$$

The derivation is straightforward. Let us call an event which occurs in $[0, y]$ and covers the point y a *covering event.* The probability that a covering event does not occur in the interval $[y - \xi - d\xi, y - \xi]$ is

$$\exp\{-\lambda[1 - U(\xi)]\, d\xi\}. \tag{7.10}$$

But all such intervals from $\xi = 0$ to $\xi = y$ are independent. Hence the probability that *no* covering event occurs in $[0, y]$ is the product of factors of type (7.10) from $\xi = 0$ to $\xi = y$, and this leads directly to (7.9). Hence we can derive $v(y)$ by differentiating (7.8),

$$v(y) = \lambda e^{-\lambda y}[1 - U(y)] \exp\left[\lambda \int_0^y U(t)\, dt\right]. \tag{7.11}$$

On examining (7.11) and comparing with (7.5) and (7.6), it is not difficult to see how the formula could be derived directly, the successive terms in (7.6) corresponding to successive terms in the expansion of $\exp[\lambda \int_0^y U(t)dt]$.

It is convenient to introduce a function $\overline{U}(y)$ which is the complement of $U(y)$,

$$U(y) + \overline{U}(y) = 1. \tag{7.12}$$

Relations (7.9) and (7.11) assume a simplified form in terms of $\overline{U}(y)$ as follows:

$$1 - \zeta(y) = \exp\left[-\lambda \int_0^y \overline{U}(t)\, dt\right], \tag{7.13}$$

$$v(y) = \lambda \overline{U}(y) \exp\left[\lambda \int_0^y \overline{U}(t)\, dt\right]. \tag{7.14}$$

For a distribution $u(\tau)d\tau$ which is zero for $\tau \geq \tau_0$, $\overline{U}(y)$ is also zero for $\tau \geq \tau_0$; for a long-range distribution, $\overline{U}(y)$ provides a direct representation of the tail.

The solution of (7.7) by Laplace transforms is very simple in principle, and gives for the Laplace transform $Z(p)$ of $z(y)$

$$Z(p) = \frac{Z_0(p)}{1 - V(p)} \tag{7.15}$$

where $V(p)$ is the Laplace transform of $v(y)$. As in Section 6 the asymptotic behaviour of $z(y)$ is determined by the roots of the denominator of (7.15), and we shall find close parallels to the behaviour for equal intervals.

8. Distributions with a Finite Mean Value

It is important to discuss the general behaviour of the function $V(p)$ as p decreases from $+\infty$ through zero to $-\infty$. First note that $v(y)$ is positive for all y. Hence

$$V(p) = \int_0^\infty v(y)e^{-py}\, dy \tag{8.1}$$

increases monotonically as p decreases. Thus, there can be only one real root of the equation $V(p) = 1$.

We illustrate this behaviour by reconsidering the case of equal intervals, for which

$$u(t) = \delta(t - \tau) \tag{8.2}$$

$$v(y) = \begin{cases} \lambda e^{-\lambda y} & y \leq \tau \\ 0 & y > \tau \end{cases} \tag{8.3}$$

$$V(p) = \frac{\lambda}{p + \lambda}\left[1 - e^{-\tau(p+\lambda)}\right]. \tag{8.4}$$

For large positive p, $V(p)$ is small; as p decreases to zero, $V(p)$ rises to $(1 - e^{-\lambda\tau})$; and at $p = 0$, it is therefore less than 1; for negative p, it continues its steady increase, becoming 1 at a unique negative value $-\gamma/\tau$; it then increases exponentially for large p.

Let us now consider a general distribution with a finite cutoff τ_0. From (7.14) we see that $v(y)$ is zero for $y > \tau_0$. The general pattern of behaviour is similar to that for equal intervals, the value for $p = 0$ being given, from (8.1), by

$$V(0) = \int_0^\infty v(y)\, dy. \tag{8.5}$$

Using (7.8) and (7.13), we find that

$$V(0) = 1 - \exp\left[-\lambda \int_0^\infty \overline{U}(t)\, dt\right]. \tag{8.6}$$

But

$$\int_0^\infty \overline{U}(t)\,dt = \left[t\overline{U}(t)\right]_0^\infty + \int_0^\infty tu(t)\,dt = \overline{\tau} \tag{8.7}$$

which is the average length of interval. Therefore

$$V(0) = 1 - e^{-\lambda\overline{\tau}} \tag{8.8}$$

which is again less than 1. Hence $V(p)$ reaches the value 1 for a negative value of $p = -\overline{\gamma}/\overline{\tau}$, and by analogy with (6.7) the asymptotic behaviour of $z(y)$ is an asymptotic decay, $\exp(-\overline{\gamma}y/\overline{\tau})$. The probability of the line $[0, y]$ being covered tends to zero for large y, i.e., there is no percolating cluster.

Now consider a distribution with a long tail of the form

$$u(\tau) \sim \frac{A}{\tau^s}. \tag{8.9}$$

Then

$$\overline{U}(y) = \int_y^\infty u(\tau)\,d\tau \sim \frac{A}{(s-1)y}. \tag{8.10}$$

Reverting to equation (8.7), the integral on the left-hand side exists if $s > 2$, $\overline{\tau}$ is defined, and the equation remains valid. Hence the argument of the previous paragraph can be repeated, and there is no percolating cluster.

The argument can be extended to a distribution of the form

$$u(\tau) \sim \frac{A}{\tau^2(\ln\tau)^s} \quad (s > 1) \tag{8.11}$$

for which the integral of $tu(t)$ converges to give a finite mean value $\overline{\tau}$. We now have

$$\overline{U}(y) \sim \frac{A}{(s-1)y(\ln y)^{s-1}} \tag{8.12}$$

and equation (8.7) is still valid. Again there is no percolating cluster for large y. The argument applies equally for

$$u(\tau) \sim \frac{A}{\tau^2(\ln\tau)(\ln\ln\tau)^s}, \frac{A}{\tau^2\ln\tau(\ln\ln\tau)(\ln\ln\ln\tau)^s}, \dots \quad (s > 1) \tag{8.13}$$

the general conclusion being that as long as the mean interval of the distribution is finite, $z(y)$ decays exponentially for large y.

9. Distributions with an Infinite Mean Value

For a distribution $u(\tau)$ for which the integral of $tu(t)$ does not converge, i.e., for which $\overline{\tau}$ becomes infinite, the argument of the previous section would indicate that $V(0)$, which is equal to $1 - e^{-\lambda\overline{\tau}}$, becomes equal to 1. Hence, from (7.15) the dominating term in the asymptotic behaviour of $z(y)$ will no longer be an exponential decay, but a constant. Therefore the system should now have a percolating cluster.

We can use the argument of Section 7 to specify in more detail what happens. Consider the probability that the point y is not covered by an event which has occurred in $[-y_0, 0]$. Using equation (7.10), we see that this probability is given by

$$\exp\left[-\lambda \int_y^{y+y_0} \overline{U}(\xi)\, d\xi\right]. \tag{9.1}$$

But for any of the distributions of the previous section for which $\overline{\tau}$ is infinite [(8.9) with $s \leq 2$; (8.11) and (8.13) with $s \leq 1$] the integral of $\overline{U}(\xi)$ diverges, and by choosing y_0 sufficiently large, (9.1) can be made as small as we please. Hence there is probability 1 that the point y is covered by an event occurring before 0, i.e., that the interval $[0, y]$ is completely covered by such an event. This corresponds to a percolating cluster.

We therefore find that with such distributions percolation occurs however small the value of λ, so that the system becomes critical however small the percolation probability.

References

Amit, D.J., Parisi, G., and Peliti, L. (1983). Asymptotic behaviour of the "true" self-avoiding walk. *Physical Review B* 27, 1635–1645.

Broadbent, S.R., and Hammersley, J.M. (1957). Percolation processes I. Crystals and mazes. *Proceedings of the Cambridge Philosophical Society* 53, 629–641.

Burkill, J.C. (1971). Abram Samoilovitch Besicovitch. *Biographical Memoirs of the Fellows of the Royal Society* 17, 1–16.

Domb, C. (1947). The problem of random intervals on a line. *Proceedings of the Cambridge Philosophical Society* 43, 329–341.

——— (1948). Some probability distributions connected with recording apparatus. *Proceedings of the Cambridge Philosophical Society* 44, 335–341.

——— (1950). Some probability distributions connected with recording apparatus II. *Proceedings of the Cambridge Philosophical Society* 46, 429–435.

——— (1964). Some statistical problems connected with crystal lattices. *Journal of the Royal Statistical Society* 26, 367–397.

——— (1989). Covering by random intervals and one-dimensional continuum percolation. *Journal of Statistical Physics* 55, 441–460.

Fisher, R.A. (1929). Tests of significance in harmonic analysis. *Proceedings of the Royal Society A* 125, 54–59.

——— (1940). Similarity of the distributions found for tests of significance in harmonic analysis, and in Stevens' problem in geometrical probability. *Annals of Eugenics* 10, 14–17.

Gennes, P.G. de (1972). Exponents for the excluded volume problem as derived by the Wilson method. *Physics Letters* 38A, 339–340.

Hammersley, J.M. (1950). On estimating restricted parameters. *Journal of the Royal Statistical Society* 12, 192–240.

——— (1953). On counters with random dead time I. *Proceedings of the Cambridge Philosophical Society* 49, 623–637.

——— (1961). The number of polygons on a lattice. *Proceedings of the Cambridge Philosophical Society* 57, 516–523.

——— (1983). Origins of percolation theory. In *Percolation Structures and Processes*, Annals of the Israeli Physical Society 5, 47–57, Adam Hilger, Bristol.

Hammersley, J.M. and Morton, K.W. (1954). Poor man's Monte Carlo. *Journal of the Royal Statistical Society B* 16, 23–38.

Kasteleyn, P.W. and Fortuin, C.M. (1969). Phase transitions in lattice systems with random local properties. *Journal of the Physical Society, Japan Supplement* 26, 11–14.

Kendall, M.G. and Moran, P.A.P. (1963). *Geometrical Probability.* Charles Griffin, London.

Moran, P.A.P. (1947). Random associations on a lattice. *Proceedings of the Cambridge Philosophical Society* 43, 321–328.

Potts, R.B. (1952). Some generalized order-disorder transformations. *Proceedings of the Cambridge Philosophical Society* 48, 106–109.

Rosenbluth, M.N. and Rosenbluth, A.W. (1955). Monte Carlo calculation of the average extension of molecular chains. *Journal of Chemical Physics* 23, 356–359.

Smith, W.L. (1957). On renewal theory, counter problems, and quasi-Poisson processes. *Proceedings of the Cambridge Philosophical Society* 53, 175–193.

Stevens, W.L. (1939). Solution to a geometrical problem in probability. *Annals of Eugenics* 9, 315–320.

Whitworth, W.A. (1897). *D.C.C. Exercises*, including hints for the solution of all the questions in *Choice and Chance*. Cambridge University Press, Cambridge, reprinted Hafner (1965).

Physics Department
Bar-Ilan University
Ramat-Gan
Israel.

On a Problem of Straus

P. Erdős and A. Sárközy[1]

---

1. Introduction and Results

Throughout this paper we use the following notation: $\{x\}$ denotes the fractional part of x. The distance from x to the nearest integer is denoted by $\|x\|$: $\|x\| = \min(\{x\}, 1 - \{x\})$. The cardinality of the finite set $\mathcal{S}$ is denoted by $|\mathcal{S}|$. The counting functions of the finite sets $\mathcal{A}, \mathcal{B}, \ldots$ of non-negative integers are denoted by $A(x), B(x), \ldots$ so that, e.g., $A(x) = |\mathcal{A} \cap \{1, 2, \ldots, [x]\}|$. If $\mathcal{A}$ is a finite or infinite set of integers, then let $\mathcal{P}(\mathcal{A})$ denote the set of distinct integers n that can be represented in the form $n = \sum_{a \in \mathcal{A}} \epsilon_a a$ where $\epsilon_a = 0$ or 1 for all a and $0 < \sum_{a \in \mathcal{A}} \epsilon_a < \infty$. If $\mathcal{A}$ is a set of integers such that no a_i is the average of any subset of $\mathcal{A}$ consisting of two or more elements, then $\mathcal{A}$ is said to be non-averaging.

Erdős and Straus raised the question of deciding the maximum cardinality $f(N)$ of a non-averaging subset of $\{0, 1, 2, \ldots, N\}$? This problem has been studied by Abbott (1976, 1980, 1986), Bosznay (1989), Straus (1968), and Erdős and Straus (1970), and the best estimates are due to Bosznay (1989) and Erdős and Straus (1970) who proved that

$$f(N) \gg N^{1/4} \tag{1.1}$$

and

$$f(N) \ll N^{2/3}, \tag{1.2}$$

respectively. Furthermore, Straus (1968) reduced the upper estimate of $f(N)$ to the following problem: what is the maximum number $k = F(N)$ such that there exist two subsets $\mathcal{A} = \{a_1, a_2, \ldots, a_k\}, \mathcal{B} = \{b_1, b_2, \ldots, b_k\}$ of $\{0, 1, 2, \ldots, N\}$ so that the sums of non-empty subsets of $\mathcal{A}$ are different from the sums of non-empty subsets of $\mathcal{B}$, i.e., $\mathcal{P}(\mathcal{A}) \cap \mathcal{P}(\mathcal{B}) = \emptyset$? He conjectured that the maximum number $F(N)$ is attained when $\mathcal{A}, \mathcal{B}$ are of the form $\mathcal{A} = \{0, 1, \ldots, k-1\}$ and $\mathcal{B} = \{N-k+1, N-k+2, \ldots, N\}$ for an optimal k. This construction leads to

$$F(N) \geq [(2N)^{1/2}] - 1. \tag{1.3}$$

---

[1]Research partially supported by Hungarian National Foundation for Scientific Research grant no. 1811.

Furthermore, he proved that

$$f(N) \le 2F(N) + 1. \tag{1.4}$$

In this paper, our goal is to give an upper bound for $F(N)$ and thus, by (1.4), also for $f(N)$. In fact, we will prove

THEOREM 1. *For $N > N_0$ we have*

$$F(N) < 201(N \log N)^{1/2}. \tag{1.5}$$

This is only by a factor $(\log N)^{1/2}$ worse than the conjectured $F(N) \ll N^{1/2}$ and probably also this $(\log N)^{1/2}$ factor could be eliminated by improving on a lemma (Lemma 1) in our proof; we will return to this problem.

Combining Theorem 1 with (1.4) we obtain:

COROLLARY 1. *For $N > N_0$ we have*

$$f(N) < 403(N \log N)^{1/2}.$$

(Compare with (1.2).)

In the second half of this paper our goal is to study the infinite analogue of this problem: if $\mathcal{A}, \mathcal{B}$ are infinite sets of positive integers such that

$$\mathcal{P}(\mathcal{A}) \cap \mathcal{P}(\mathcal{B}) = \emptyset, \tag{1.6}$$

then how large can $\min(A(x), B(x))$ be? Of course, Theorem 1 implies that $\min(A(x), B(x)) \ll x^{1/2}$. We conjecture that (1.6) implies

$$\liminf_{x \to \infty} \frac{\min(A(x), B(x))}{x^{1/2}} = 0 \tag{1.7}$$

and, perhaps, even

$$\liminf_{x \to \infty} \frac{A(x)B(x)}{x} = 0$$

holds; unfortunately, we have not been able to prove this. On the other hand, we will prove that $x^{1/2}$ in the denominator in (1.7) cannot be replaced by $x^{1/2}(\log x)^{-1/2-\epsilon}$:

THEOREM 2. *Let $\beta_1, \beta_2, \ldots$ be an infinite sequence of positive real numbers with*

$$\sum_{n=1}^{\infty} \beta_n < \frac{1}{2}. \tag{1.8}$$

Then there exist two infinite sets $\mathcal{A} = \{a_1, a_2, \dots\}, \mathcal{B} = \{b_1, b_2, \dots\}$ *of distinct positive integers such that*

$$\max(a_n, b_n) \le 8n\beta_n^{-1} \quad \text{for } n = 1, 2, \dots \tag{1.9}$$

and

$$\mathcal{P}(\mathcal{A}) \cap \mathcal{P}(\mathcal{B}) = \emptyset. \tag{1.10}$$

Thus, e.g., choosing $\beta_n = c\big(n \log n (\log\log n)^{1+\epsilon}\big)^{-1}$ here (where $\epsilon > 0$ and c is a positive constant small enough in terms of ϵ) we obtain that there exist infinite sets $\mathcal{A}, \mathcal{B}$ of positive integers such that

$$\liminf_{x\to\infty} \frac{\min(A(x), B(x))}{x^{1/2}(\log x)^{-1/2}(\log\log x)^{-1/2-\epsilon}} > 0$$

and (1.10) holds.

On the other hand, it is easy to see that there exist infinite sets $\mathcal{A}, \mathcal{B}$ of positive integers such that (1.10) holds and both $\mathcal{A}$ and $\mathcal{B}$ have positive upper density. In fact, to see this let $x_n = 2^{2^n}$, and let

$$\mathcal{A} = \bigcup_{k=1}^{\infty} \{x_{2k}, x_{2k} + 1, \dots, [3x_{2k}/2]\}$$

and

$$\mathcal{B} = \bigcup_{k=1}^{\infty} \{x_{2k+1}, x_{2k+1} + 1, \dots, [3x_{2k+1}/2]\}.$$

It can be shown easily that these sets $\mathcal{A}, \mathcal{B}$ have the desired properties.

Also, it would be interesting to decide how fast $A(x)B(x)$ can grow for infinite sets $\mathcal{A}, B$ satisfying (1.10). In the construction above we have

$$A(x)B(x) \gg x^{3/2}$$

for infinitely many x. Perhaps, this inequality is nearly best possible.

Let $\mathcal{A} = \{a_1, a_2, \dots\}, \mathcal{B} = \{b_1, b_2, \dots\}$ be infinite increasing sequences of positive real numbers with the property that

$$\left| \sum_{i=1}^{\infty} \epsilon_i a_i - \sum_{i=1}^{\infty} \epsilon_i' b_i \right| \ge 1$$

whenever $\epsilon_i = 0$ or 1 for all i, $\epsilon_i' = 0$ or 1 for all i, $0 < \sum_{i=1}^{\infty} \epsilon_i < \infty$ and $0 < \sum_{i=1}^{\infty} \epsilon_i' < \infty$. Perhaps, these assumptions imply that

$$\liminf_{x\to\infty} \frac{\max(A(x), B(x))}{x} = 0$$

and

$$\lim_{x\to\infty} \frac{\min(A(x), B(x))}{x} = 0$$

(where $A(x) = |\{i : a_i \le x\}|$, $B(x) = |\{i : b_i \le x\}|$); we have not been able to prove this.

2. Two Lemmas

The proof of Theorem 1 will be based on the following result of Sárközy (1989):

LEMMA 1. *Let N be a positive integer with $N > 2500$, let $\mathcal{A} \subset \{1, 2, \dots, N\}$ and*

$$|\mathcal{A}| > 100(N \log N)^{1/2}.$$

Then there are integers d, y, z such that

$$1 \le d < 10^4 N |\mathcal{A}|^{-1},$$

$$z > 7^{-1} 10^{-4} |\mathcal{A}|^2,$$

$$y < 7 \cdot 10^4 N z |\mathcal{A}|^{-2}$$

and

$$\{yd, (y+1)d, \dots, zd\} \subset \mathcal{P}(\mathcal{A}).$$

We need one more lemma:

LEMMA 2. *Let M, N, t, d be positive integers with $M \le N$,*

$$d \le t \le N, \tag{2.1}$$

and let

$$\mathcal{A} \subset \{M, M+1, \dots, N\}, \tag{2.2}$$

$$|\mathcal{A}| = t. \tag{2.3}$$

Then for every integer u with

$$0 \le u \le M(t-d), \tag{2.4}$$

there is an integer s such that

$$u \le s < u + Nd, \tag{2.5}$$

$$d \mid s \tag{2.6}$$

and s can be written in the form

$$s = \sum_{a \in \mathcal{A}} \epsilon_a a \quad \text{where} \quad \epsilon_a = 0 \;\; \text{or} \;\; 1 \;\; \text{for all} \;\; a \tag{2.7}$$

(so that either $s = 0$ or $s \in \mathcal{P}(\mathcal{A})$).

PROOF OF LEMMA 2: It suffices to show that there are integers $x_0, x_1, \ldots, x_r$ such that $x_0 = 0$,

$$x_{i-1} < x_i \leq x_{i-1} + Nd \quad \text{for} \quad i = 1, 2, \ldots, r, \tag{2.8}$$

$$x_r > M(t-d),$$

$$d \mid x_i \quad \text{for} \quad i = 0, 1, \ldots, r \tag{2.9}$$

and

$$x_i \in \mathcal{P}(\mathcal{A}) \quad \text{for} \quad i = 1, 2, \ldots, r. \tag{2.10}$$

In fact, if $x_0, x_1, \ldots, x_r$ are defined in this way and u satisfies (2.4), then there is an x_i with $u \leq x_i < u + Nd$ so that (2.5), (2.6), and (2.7) hold with x_i in place of s.

These numbers $x_0, x_1, \ldots, x_r$ can be defined recursively. Let $x_0 = 0$. Assume that $x_0, x_1, \ldots, x_i$ $(i \geq 0)$ have been defined with the desired properties and

$$x_i \leq M(t-d). \tag{2.11}$$

Then by (2.10) (and $x_0 = 0$) there is a subset $\mathcal{A}_1 \subset \mathcal{A}$ $(\mathcal{A}_1 = \emptyset$ for $i = 0)$ such that

$$\sum_{a \in \mathcal{A}_1} a = x_i. \tag{2.12}$$

By (2.1), this implies

$$x_i \geq \sum_{a \in \mathcal{A}_1} M = |\mathcal{A}_1| M. \tag{2.13}$$

It follows from (2.11) and (2.13) that

$$|\mathcal{A}_1| \leq t - d. \tag{2.14}$$

Let us write $\mathcal{A}_2 = \mathcal{A} \setminus \mathcal{A}_1$ so that, by (2.3) and (2.14),

$$|\mathcal{A}_2| = |\mathcal{A}| - |\mathcal{A}_1| = t - |\mathcal{A}_1| \geq d.$$

Let $\mathcal{A}_3$ be a subset of $\mathcal{A}_2$ with $|\mathcal{A}_3| = d$. Then there is a non-empty subset $\mathcal{A}_4$ of $\mathcal{A}_3$ with

$$d \mid \sum_{a \in \mathcal{A}_4} a. \tag{2.15}$$

(In fact, if $\mathcal{A}_3 = \{a_1, a_2, \dots, a_d\}$, then either there is a k with $a_1 + a_2 + \dots + a_k \equiv 0 \pmod{d}$ so that we may choose $\mathcal{A}_4 = \{a_1, a_2, \dots, a_k\}$, or there are k, l with $k < l$, $a_1 + a_2 + \dots + a_l \equiv a_1 + a_2 + \dots + a_k \pmod{d}$ so that $\mathcal{A}_4 = \{a_{k+1}, a_{k+2}, \dots, a_l\}$ can be chosen.) Let

$$x_{i+1} = x_i + \sum_{a \in \mathcal{A}_4} a. \tag{2.16}$$

Then by (2.9) and (2.15) we have $d \mid x_{i+1}$. Furthermore, $x_{i+1} \in \mathcal{P}(\mathcal{A})$ follows from (2.10) and (2.16). Finally, by (2.2) and (2.16) we have

$$\begin{aligned} x_i < x_{i+1} &\le x_i + \sum_{a \in \mathcal{A}_4} N \\ &= x_i + N|\mathcal{A}_4| \le x_i + N|\mathcal{A}_3| \\ &= x_i + Nd \end{aligned}$$

and this completes the proof of Lemma 2. ∎

3. Completion of the Proof of Theorem 1

We have to show that $N > N_0$,

$$\begin{aligned} \mathcal{A} &= \{a_1, a_2, \dots, a_k\} \subset \{1, 2, \dots, N\}, \\ \mathcal{B} &= \{b_1, b_2, \dots, b_k\} \subset \{1, 2, \dots, N\} \end{aligned} \tag{3.1}$$

(where $a_1 < a_2 < \dots < a_k$, $b_1 < b_2 < \dots < b_k$) and

$$k \ge 201(N \log N)^{1/2} \tag{3.2}$$

imply that

$$\mathcal{P}(\mathcal{A}) \cap \mathcal{P}(\mathcal{B}) \ne \emptyset. \tag{3.3}$$

We may assume that $a_{[k/2]+1} \le b_{[k/2]+1}$. Let us write

$$\begin{aligned} M &= a_{[k/2]+1}, \\ \mathcal{A}' &= \{a_2, a_3, \dots, a_{[k/2]+1}\}, \\ \mathcal{B}' &= \{b_{[k/2]+1}, b_{[k/2]+2}, \dots, b_k\} \end{aligned}$$

so that, in view of (3.1) and (3.2) for $N > N_0$ we have

$$N > M = a_{[k/2]+1} \ge k/2 > 100(N \log N)^{1/2}, \tag{3.4}$$

$$\mathcal{A}' \subset \{1, 2, \dots, M\}, \tag{3.5}$$

$$|\mathcal{A}'| = [k/2] > 100(N \log N)^{1/2} \ge 100(M \log M)^{1/2}, \tag{3.6}$$

$$\mathcal{B}' \subset \{M, M+1, \dots, N\} \tag{3.7}$$

and

$$|\mathcal{B}'| = k - [k/2] \geq \frac{k}{2}. \tag{3.8}$$

By (3.4), (3.5) and (3.6), for large N we may apply Lemma 1 with M and $\mathcal{A}'$ in place of N and $\mathcal{A}$, respectively. We obtain that there exist integers d, y, z such that

$$1 \leq d < 10^4 M |\mathcal{A}'|^{-1}, \tag{3.9}$$

$$z > 7^{-1} 10^{-4} |\mathcal{A}'|^2, \tag{3.10}$$

$$y < 7 \cdot 10^4 M z |\mathcal{A}'|^{-2} \tag{3.11}$$

and

$$\{yd, (y+1)d, \ldots, zd\} \subset \mathcal{P}(\mathcal{A}'). \tag{3.12}$$

To prove (3.3), it suffices to show that

$$\mathcal{P}(\mathcal{A}') \cap \mathcal{P}(\mathcal{B}') \neq \emptyset. \tag{3.13}$$

If there is a positive integer s such that

$$yd \leq s \leq zd, \tag{3.14}$$

$$d \mid s \tag{3.15}$$

and

$$s \in \mathcal{P}(B'), \tag{3.16}$$

then by (3.12), also $s \in \mathcal{P}(\mathcal{A}')$ holds so that $s \in \mathcal{P}(\mathcal{A}') \cap \mathcal{P}(\mathcal{B}')$ whence (3.13) follows. Thus it suffices to show that there is a positive integer s satisfying (3.14), (3.15) and (3.16). To prove this, we are going to apply Lemma 2 with $\mathcal{B}'$, $|\mathcal{B}'| = k - [k/2]$ and yd in place of $\mathcal{A}$, t and u, respectively. Then (2.2) holds by (3.7). Furthermore, by (3.1), (3.2), (3.6), (3.8) and (3.9) we have

$$t = |\mathcal{B}'| \geq \frac{k}{2} \tag{3.17}$$

and

$$\begin{aligned} d &< 10^4 M |\mathcal{A}'|^{-1} \leq 10^4 N \big(100 (N \log N)^{1/2}\big)^{-1} \\ &= 100 N^{1/2} (\log N)^{-1/2} = o(k) \end{aligned} \tag{3.18}$$

so that also (2.1) holds. Finally, it follows from (3.5), (3.6), (3.11) and (3.12) that

$$\begin{aligned} u = yd &< 7 \cdot 10^4 M z |\mathcal{A}'|^{-2} d \\ &= 7 \cdot 10^4 M |\mathcal{A}'|^{-2} (zd) < 7 \cdot 10^4 M |\mathcal{A}'|^{-2} |\mathcal{A}'| M \\ &= 7 \cdot 10^4 M^2 |\mathcal{A}'|^{-1} < 7 \cdot 10^4 M^2 \big(100 (M \log M)^{1/2}\big)^{-1} \\ &= 700 M^{3/2} (\log M)^{-1/2} \end{aligned} \tag{3.19}$$

and by (3.2), (3.17) and (3.18), for large N we have

$$\begin{aligned} t-d &\geq \frac{k}{2}-o(k)>\frac{k}{3} \\ &> 60(N\log N)^{1/2} \\ &\geq 60(M\log M)^{1/2}. \end{aligned} \tag{3.20}$$

(2.4) follows from (3.19) and (3.20). Thus, in fact, all the assumptions in Lemma 2 hold so that the lemma can be applied. We obtain that there is an integer s such that

$$u=yd\leq s<yd+Nd, \tag{3.21}$$

$$d \mid s \tag{3.22}$$

and

$$s\in\mathcal{P}(\mathcal{B}'). \tag{3.23}$$

(Note that $s\neq 0$ by $s\geq yd>0$.)

It follows from (3.6), (3.10), (3.11) and (3.21) that

$$\begin{aligned} s &< (y+N)d < \left(7\cdot 10^4 Mz|\mathcal{A}'|^{-2}+N\right)d \\ &< \left\{(7\cdot 10^4 Nz\left(100(N\log N)^{1/2}\right)^{-2}+10^{-4}(\log N)^{-1}|\mathcal{A}'|^2\right\}d \\ &= \left(o(z)+o(z)\right)d = o(zd). \end{aligned} \tag{3.24}$$

(3.15) and (3.16) hold by (3.22) and (3.23) while (3.14) follows from (3.21) and (3.24), and this completes the proof of Theorem 1. ■

4. Proof of Theorem 2

We are going to define the sequences $a_1, a_2, \dots$ and $b_1, b_2, \dots$ recursively.

Let $\alpha=(\sqrt{5}+1)/2$. Let a_1 and b_1 be the least positive integers a and b such that $0<\{a\alpha\}<\beta_1$ and $1-\beta_1<\{b\alpha\}<1$, respectively. If $a_1, a_2 \dots, a_{n-1}$ and $b_1, b_2, \dots, b_{n-1}$ have been defined, then let a_n and b_n be the least positive integers a and b such that

$$0<\{a\alpha\}<\beta_n, \quad a\notin\{a_1,a_2,\dots,a_{n-1}\}$$

and

$$1-\beta_n<\{b\alpha\}<1, \quad b\notin\{b_1,b_2,\dots,b_{n-1}\},$$

respectively.

First we are going to prove (1.10). If $\mathcal{A}'$ is a finite (non-empty) subset of $\mathcal{A}$, then in view of (1.8) we have

$$0 < \sum_{a_i \in \mathcal{A}'} \{a_i \alpha\} < \sum_{a_i \in \mathcal{A}'} \beta_i < \sum_{i=1}^{\infty} \beta_i < \frac{1}{2}$$

whence

$$0 < \sum_{a_i \in \mathcal{A}'} \{a_i \alpha\} = \left\{ \Big(\sum_{a_i \in \mathcal{A}'} a_i \Big) \alpha \right\} < \frac{1}{2}. \tag{4.1}$$

Furthermore, it follows from (1.8) and the definition of the set $\mathcal{B}$ that if $\mathcal{B}'$ is a finite (non-empty) subset of $\mathcal{B}$, then we have

$$\begin{aligned} |\mathcal{B}'| = \sum_{b_i \in \mathcal{B}'} 1 > \sum_{b_i \in \mathcal{B}'} \{b_i \alpha\} > \sum_{b_i \in \mathcal{B}'} (1 - \beta_i) \\ = |\mathcal{B}'| - \sum_{b_i \in \mathcal{B}'} \beta_i > |\mathcal{B}'| - \sum_{i=1}^{\infty} \beta_i > |\mathcal{B}'| - \frac{1}{2} \end{aligned}$$

whence

$$\frac{1}{2} < \left\{ \sum_{b_i \in \mathcal{B}'} \{b_i \alpha\} \right\} = \left\{ \Big(\sum_{b_i \in \mathcal{B}'} b_i \Big) \alpha \right\} < 1. \tag{4.2}$$

It follows from (4.1) and (4.2) that

$$\sum_{a_i \in \mathcal{A}'} a_i \neq \sum_{b_i \in \mathcal{B}'} b_i$$

which proves (1.10).

To prove (1.9), we need the following lemma:

LEMMA 3. *Let $\alpha = (\sqrt{5}+1)/2$. If δ is a real number with $0 < \delta < 1$ and x, y are arbitrary real numbers, then there is an integer m such that*

$$x < m \le x + 4\delta^{-1} \tag{4.3}$$

and

$$\|m\alpha - y\| < \delta. \tag{4.4}$$

PROOF: This can be proved by using standard tools of the theory of continued fractions (see, e.g., Hardy and Wright 1960); for the sake of completeness we give the proof. Let $q_o = 1, q_1 = 1, \dots, q_n = q_{n-1} + q_{n-2}, \dots$ denote the Fibonacci numbers. These numbers are the denominators of the convergents of the continued fraction expansion of α so that for all n there is an integer p_n such that

$$\left|\alpha - \frac{p_n}{q_n}\right| < q_n^{-2} \quad (\text{for } n = 0, 1, 2, \dots). \tag{4.5}$$

Clearly, $q_n = q_{n-1} + q_{n-2} \le 2q_{n-1}$ for $n \ge 2$. Thus there is an integer k with

$$\frac{2}{\delta} < q_k \le \frac{4}{\delta}. \tag{4.6}$$

Then we have

$$x < [x] + i \le x + 4\delta^{-1} \quad \text{for} \quad i = 1, 2, \dots, q_k. \tag{4.7}$$

Write $j = \big[q_k(y - [x]\alpha)\big]$ so that

$$\left|\frac{j}{q_k} - (y - [x]\alpha)\right| < \frac{1}{q_k}. \tag{4.8}$$

Define the integer i_j by

$$i_j p_k \equiv j \pmod{q_k}, \ 1 \le i_j \le q_k \tag{4.9}$$

and write $m = [x] + i_j$. Then (4.3) holds by (4.7), and it follows from (4.5), (4.6), (4.8) and (4.9) that

$$\begin{aligned}
\|m\alpha - y\| &= \big\|([x] + i_j)\alpha - y\big\| \\
&= \left\|\frac{i_j p_k}{q_k} + i_j\left(\alpha - \frac{p_k}{q_k}\right) + ([x]\alpha - y)\right\| \\
&= \left\|\frac{j}{q_k} - (y - [x]\alpha) + i_j\left(\alpha - \frac{p_k}{q_k}\right)\right\| \\
&\le \left|\frac{j}{q_k} - (y - [x]\alpha)\right| + |i_j|\left|\alpha - \frac{p_k}{q_k}\right| \\
&< \frac{1}{q_k} + q_k \cdot \frac{1}{q_k^2} \\
&= \frac{2}{q_k} < \delta
\end{aligned}$$

so that also (4.4) holds and this completes the proof of the lemma. ∎

Now we are going to prove (1.9). By the construction of the sets $\mathcal{A}, \mathcal{B}$, it suffices to show that there are at least n integers a and at least n integers b such that

$$\big|\{a : 0 < a \le 8n\beta_n^{-1}, \ 0 < \{a\alpha\} < \beta_n\}\big| \ge n \tag{4.10}$$

and

$$\left|\{b : 0 < b \leq 8n\beta_n^{-1},\ 1 - \beta_n < \{b\alpha\} < 1\}\right| \geq n, \tag{4.11}$$

respectively.

To prove (4.10), it suffices to show that for $i = 0, 1, \ldots, n-1$, there is an integer a such that

$$8i\beta_n^{-1} < a \leq 8(i+1)\beta_n^{-1},\ 0 < \{a\alpha\} < \beta_n. \tag{4.12}$$

In fact, applying Lemma 3 with $\beta_n/2, 8i\beta_n^{-1}$ and $\beta_n/2$ in place of δ, x and y, respectively, we obtain that there is an integer m such that

$$x = 8i\beta_n^{-1} < m \leq x + 4\delta^{-1} = 8i\beta_n^{-1} + 4(\beta_n/2)^{-1} = 8(i+1)\beta_n^{-1}$$

and

$$\left\| m\alpha - \frac{\beta_n}{2} \right\| < \frac{\beta_n}{2}$$

whence

$$\{m\alpha\} < \beta_n$$

(and $\{m\alpha\} > 0$ since α is irrational) which proves (4.12). Similarly, applying Lemma 3 with $\beta_n/2, 8i\beta_n^{-1}$ and $1 - \beta_n/2$ in place of δ, x and y, respectively, we obtain that there is an integer b with

$$8i\beta_n^{-1} < b \leq 8(i+1)\beta_n^{-1},\ 1 - \beta_n < \{b\alpha\} < 1$$

which implies (4.11) and this completes the proof of the theorem. ■

Acknowledgement

We would like to thank Professor Abbott for his valuable remarks.

REFERENCES

Abbott, H.L. (1976). On a conjecture of Erdős and Straus on non-averaging sets of integers. *Proceedings of the 5th British Combinatorial Conference*, Aberdeen, 1975, 1–4.

——— (1980). Extremal problems on non-averaging and non-dividing sets. *Pacific Journal of Mathematics* 91, 1–12.

——— (1986). On the Erdős-Straus non-averaging set problem. *Acta Mathematica Hungarica* 47, 117–119.

Bosznay, Á.P. (1989). On the lower estimation of non-averaging sets. *Acta Mathematica Hungarica*, to appear.

Erdős, P. and Straus, E.G. (1970). Nonaveraging sets, II. *Colloquia Mathematica Societatis János Bolyai* 4, 405–411.

Hardy, G.H. and Wright, E.M. (1960). *The Theory of Numbers*. Clarendon Press, Oxford.

Sárközy, A. (1989). Finite addition theorems, II. *Journal of Number Theory*, to appear.

Straus, E.G. (1971). Nonaveraging sets, in *Combinatorics* (Proceedings of the Symposia of Pure Mathematics, University of California, Los Angeles, vol. XIX, 1968), American Mathematical Society, Providence, R.I., 215–222.

Mathematical Institute
Hungarian Academy of Sciences
Reáltanoda ul. 13–15
Budapest.

Directed Compact Percolation II: Nodal Points, Mass Distribution, and Scaling

J.W. Essam and D. Tanlakishani

Abstract

Directed compact percolation is a limiting case of a cellular automaton model which also includes directed site and bond percolation. Existing results for the latter are reviewed and previous calculations for compact percolation are extended so that comparison of several critical exponents may be made. New results are obtained for the probability distribution of the number of nodal points and for the centre of mass and moments of inertia of compact percolation clusters. Also for this model scaling is verified for the cluster size distribution and for the longitudinal moments of the pair connectedness.

1. Introduction

A percolation process, as introduced by Broadbent and Hammersley (1957) in the first published work on percolation theory, is the passage of *fluid* through a *random medium*. The terms fluid and medium are capable of a very broad interpretation and as foreseen by Frisch and Hammersley (1963) such processes have been found to occur widely in all branches of science. The early history of the subject is described by Hammersley in an article in the book edited by Deutscher et al. (1983) where a collection of recent applications of percolation theory may also be found.

The medium was originally modelled by a random maze derived from a *crystal* of *atoms* between which local connections were made by *bonds*. The bonds could either be directed or undirected and each had independently probability p of being open to the passage of fluid. This model has become known as bond percolation. It was shown by Broadbent and Hammersley (1957) that under certain conditions there is a critical value of $p \in (0,1)$, called the *critical probability* p_c, below which only finitely many atoms are wet from any source with probability 1. Above p_c there is positive probability $P(p)$, the percolation probability, that the set of atoms which are wet is unbounded and percolation is said to take place.

Percolation processes now form an important branch of critical phenomena theory. This connection was first noted by Domb (1959) who observed that if the crystal atoms are of two species, A and B, and species A occurs with probability p, then the expected size $S(p)$ of A-clusters will become infinite at a value of p which he identified with Hammersley's critical probability, except that the randomness was now in the atoms rather than the bonds. This type of percolation has become known as site percolation since the sites of a crystal lattice are randomly occupied by species of two types. Clusters can also be defined for bond percolation as maximal sets of atoms connected by open bonds. This yields an alternative definition of critical probability for bond percolation as the point at which the cluster size becomes infinite. (These definitions have recently been shown to coincide in cases of physical interest, see Kesten 1982 for references.) Subsequently Hammersley (1961) showed that the critical probability for bond percolation cannot exceed that for site percolation on the same crystal lattice.

Here we consider percolation on a square lattice with nearest neighbour bonds and all bonds which are parallel are oriented in the same direction. For analytical work on this oriented percolation model see Durrett (1984) where references to earlier work may also be found. As anticipated by Frisch and Hammersley (1963) it turns out to be useful to allow both bonds and sites to be random elements and we denote by p_b and p_s the probabilities that a given bond or site is open respectively. A site can only become wet if it can be reached by a path from the source of fluid, all elements of which are open (including the site itself). For bond percolation on this lattice it was shown by Hammersley (1959) that $\frac{1}{2} \leq p_c \leq 0.85$. These bounds have subsequently been improved but the exact value of p_c still remains unknown. For the general model the critical probability becomes a critical curve in the p_b–p_s plane.

The lattice sites will be labelled by cartesian co-ordinates t and y chosen in such a way that the two types of directed bond make angles $\pm 45°$ with the positive t-axis. A line source perpendicular to the t-axis will be considered which has m sites and the origin of co-ordinates is chosen as the centre of this line. For m even, the lattice sites are (t, y) where t and y are any pair of integers with odd sum and the source sites have co-ordinates $(0, y)$ with $y = \pm 1, \pm 3, \ldots, \pm(m-1)$. For m odd, the sum of t and y is even and the source sites have $y = 0, \pm 2, \pm 4, \ldots, \pm(m-1)$. Fluid arriving at site (t, y) may come from either of the sites $(t-1, y \pm 1)$. The probability p_1, that (t, y) is wet given that exactly one of these sites is wet, is $p_b p_s$ and the probability p_2, that (t, y) is wet given that both of them are wet, is $(2p_b - p_b^2)p_s$. We denote the cluster which is wet from this source by c_m.

As p_b and p_s range over $[0, 1]$ it can be seen that given p_1, p_2 must lie in the interval $[p_1, 2p_1 - p_1^2]$. Durrett and Schonmann (1987) have considered a generalised percolation process in which the probabilities of the bonds leading to (t, y) being open are not necessarily independent. In this model p_2 lies in the extended range $[p_1, 2p_1]$. Domany and Kinzel (1984) considered an even more general cluster growth model in which p_1 and p_2 both range over $[0, 1]$. They considered this to be a one-dimensional stochastic cellular automaton model in which each cell has two states (wet and dry) and t is the time variable and also noted that this automaton could be mapped (Verhagen 1976) onto a triangular lattice Ising model with three-spin interactions in every other triangle (in addition to the usual pair interactions). All of these models are examples of the stochastic growth model of Durrett and Schonmann (1987) who have obtained very general analytical results on the shape of the infinite cluster above p_c. Their work has recently been extended to $p \geq p_c$ and all space dimensions by Bezuidenhout and Grimmett (1989).

The special case $p_2 = 1$, $p_1 = p$ has the simplifying feature that a given site will always be wet whenever its two predecessors are wet irrespective of the value of p. The cluster c_m therefore cannot branch and hence is free from holes; for this reason we shall call it a compact percolation cluster. Finite compact clusters also have a unique terminal site. In this case a complete description of c_m can be obtained by specifying, for each t, the number of wet sites n_t in the tth column and the centre of mass of these sites. Domany and Kinzel (1984) noted that n_t could be thought of as the position of a one-dimensional random walker after t steps with transition probabilities p^2, $(1-p)^2$ and $2p(1-p)$ to positions $n_t + 1$, $n_t - 1$ and n_t respectively. Using this they deduced that $p_c = \frac{1}{2}$ and also determined the probability distribution of the cluster length by calculating the probability that the walker reaches the origin for the first time after t steps. In terms of the walk model the percolation probability is the probability that the walker never reaches the origin. The clusters under consideration are known by Delest and Viennot (1984) as parallelogram polyominoes and Domany and Kinzel's result can also be deduced from an enumeration formula given by these authors (see Essam 1989). Another source of combinatorial information on compact percolation clusters is the work of Huse et al. (1983) on the enumeration of domain walls in a chiral clock model.

In Section 2.1 we will define critical exponents and we shall see that the exponents for bond and site percolation are numerically equal but are quite different from those for compact percolation. There is a critical curve in the p_1–p_2 plane on which the critical points of all three models lie. Critical phenomena theory suggests that critical exponents will normally remain constant along such curves but that there may be special crossover points at which they may change discontinuously. We believe that compact percolation corresponds to such a point and here we investigate this model in detail as a preliminary to investigation of the crossover phenomenon. The special nature of this point is suggested by the fact that it is the point at which $p_b \to 0$ and $p_s \to \infty$ such that $p_b p_s \to \frac{1}{2}$. Also it is the point at which the clusters become qualitatively different in that branching and holes will occur for any value of $p_2 < 1$.

In a previous paper (Essam 1989) the work of Domany and Kinzel was extended to asymmetric compact percolation by which we mean the following. Denote the sites $(t-1, y+1)$ and $(t-1, y-1)$ by A and B respectively. If both A and B are wet then as before (t, y) is wet with probability 1. However if A is dry and B is wet then (t, y) is wet with probability p_u in which event the upper edge of the cluster moves upwards. Finally if A is wet and B is dry then (t, y) is wet with probability p_d the subscript denoting downward motion of the lower edge. Of course if A and B are both dry then (t, y) is certainly dry. Domany and Kinzel's results were for $p_u = p_d = p_1$ and their walk model may still be used by defining the transition probabilities from n_t to $n_t + 1$, $n_t - 1$ and n_t as $c = p_u p_d$, $d = (1-p_u)(1-p_d)$ and $1-c-d$ respectively. Percolation now occurs with positive probability when $c > d$ and the critical curve in the p_u–p_d plane is $p_u + p_d = 1$. A duality relation between $c < d$ and $c > d$ was shown to exist. The moment generating function for the cluster length distribution and the first two moments of the cluster size distribution were also obtained by solving recurrence relations.

In this paper we investigate the probability distribution of the number of nodal sites and bonds. We also rederive the previous results for the cluster size distribution by a method which enables the general moment to be considered and also the expected values of the centre of mass and moments of inertia to be calculated.

2. Definitions and Previous Results

2.1. Definitions

In numerical work on directed percolation using series expansion techniques (see Essam et al. 1988 for references) the mean size and moments of the mass distribution of the cluster c_m have been investigated in the case $m = 1$. These functions are conveniently defined in terms of the *pair connectedness* $C_m(t, y)$ which is the probability that the atom (t, y) is wet from a line source of m sites given that c_m is finite, thus the moments of the mass distribution of finite clusters about the y-axis are given by

$$\mu_k^{(t)}(p) = \sum_{t=0}^{\infty} \sum_{y=-y_{\max}(t)}^{y_{\max}(t)} t^k C_m(t, y) \tag{2.1}$$

with a similar definition of $\mu_k^{(y)}(p)$, the moments about the t-axis. The *mean cluster size* $S_m(p)$ is the expected number of sites which are wet and is given by the same expression with $k = 0$. In the case of compact percolation, p represents the pair of variables p_u, p_d. The position of the centre of mass of finite clusters is estimated by $(\langle t \rangle, \langle y \rangle)$ and the radii of gyration about the y and t axes by $\langle t^2 \rangle^{1/2}$ and $\langle y^2 \rangle^{1/2}$ respectively where

$$\langle t^k \rangle = \mu_k^{(t)}(p)/S_m(p) \quad \text{and} \quad \langle y^k \rangle = \mu_k^{(y)}(p)/S_m(p). \tag{2.2}$$

We shall picture the formation of the cluster, c_m, which is wet from a source of width m, as taking place in a number of growth stages in which at the tth stage column t is wet from column $t-1$. The probability that c_m has t growth stages (i.e. $t+1$ is the first dry column) will be denoted by $r_t(m)$. In terms of the random walk problem described in the introduction this is also the probability that a walker starting at position $m > 0$ reaches the origin for the first time on the $(t+1)$th step. In terms of $r_t(m)$ we may obtain $Q_m(p)$, the probability that c_m is finite, and hence the *percolation probability* $P_m(p) = 1 - Q_m(p)$:

$$Q_m(p) = \sum_{t=0}^{\infty} r_t(m). \tag{2.3}$$

The *length* L_m of c_m will be defined as the number of atoms in a path from the source to a terminal point and the *probability distribution of the length of finite clusters* is determined by $r_t(m)$ and has moments which are given by:

$$E(L_m^k) = \sum_{t=0}^{\infty} (t+1)^k r_t(m)/Q_m(p). \tag{2.4}$$

The probability $p_s(m)$, that c_m has s atoms (sites) will be known as the *cluster size distribution* and has the same normalisation factor as $r_t(m)$, i.e. $Q_m(p)$. The normalised kth moment of this distribution is

$$m_k(p) = \sum_{s=1}^{\infty} s^k p_s(m)/Q_m(p) \tag{2.5}$$

and in particular $m_1(p) = S_m(p)$ is the mean size.

For compact clusters the moments $m_k(p)$ may be related (see Section 4) to the transition probability $r_{mn}(t)$ that a cluster with source of width m has width n after t growth stages. In terms of walks, $r_{mn}(t)$ is the probability that the walker moves from position $m > 0$ to position n in t steps without visiting 0. Notice that

$$r_{m1}(t) = r_t(m)/d \tag{2.6}$$

and so the transition probability for $m = 1$ is determined by previous results (Essam 1989). An explicit formula for the following moment generating function of $r_{mn}(t)$ is given in Section 4:

$$R_{mn}(z) = \sum_{t=0}^{\infty} e^{-zt} r_{mn}(t). \tag{2.7}$$

A *nodal (articulation) point* of c_m in the compact case is an intermediate site of c_m through which all open paths from the source to the terminal point must pass. At a nodal point c_m has width one. In Section 3 we obtain the probability $g_a(p, m)$ that c_m has exactly a nodal points given that it is finite. We shall see that this is determined in terms of the transition probability $w_{mn}(t)$ which is defined in the same way as $r_{mn}(t)$ except that there must be no intermediate values of t at which the cluster width is one. In the walk analogy intermediate visits to the point distant 1 from the origin must also be avoided. It is convenient to take $w_{mn}(0) = 0$. We shall see that $w_{mn}(t)$ is simply related to $r_{mn}(t)$ by a duality argument. The moment generating function, $W_{mn}(z)$, of $w_{mn}(t)$ is defined in the same way as $R_{mn}(z)$.

For $p_2 < 1$ there is no unique terminal point and nodal points are defined relative to each site of the cluster. A *nodal point relative to the site* (t, y) of c_m is an intermediate site of c_m through which all open paths from the source to (t, y) must pass. An estimator of the number of nodal points on any open path from the source to a given lattice site averaged over all such sites is

$$\langle a \rangle = \sum_{t=0}^{\infty} \sum_{y=-y_{\max}(t)}^{y_{\max}(t)} a_m(t, y)/S_m(p) \tag{2.8}$$

where $a_m(t, y)$ is the expected number of nodal points on an open path to (t, y) (if there is no path the number of nodal points is zero).

Nodal bonds are defined similarly to nodal sites in terms of bonds through which all open paths must pass. The average of the expected number of nodal bonds over all sites will be denoted by $\langle b \rangle$.

Critical exponents are widely used in critical phenomena theory to characterise the divergence of functions as the critical point (critical probability in the present context) is approached. Thus for a typical function $F(p)$ of this kind we write

$$F(p) \cong A|1 - p/p_c|^{-\epsilon} \tag{2.9}$$

to denote that the ratio of $F(p)$ to the right-hand side approaches unity as $p \to p_c$. Here ϵ is the *critical exponent* and A is the *amplitude*. In the case of asymmetric compact percolation we shall find that u, defined by (4.13), is an appropriate variable rather than p and $u_c = 1$. In the symmetric case $p_u = p_d = p$, $1 - u$ may be replaced by $4(1 - p/p_c)$, where $p_c = \frac{1}{2}$, as p_c is approached so that only the amplitude changes in going from u to p. Notice also that in the symmetric case $d - c = 1 - 2p = 1 - p/p_c$. In a case where the exponent but not the amplitude has been determined we replace $\cong$ by $\sim$.

Again from critical phenomena theory we expect that the various distribution functions above will have *scaling forms* which imply that the ratio of moment k to moment $k-1$ has a critical exponent which is independent of k. We shall verify that this is the case for compact percolation and write

$$\begin{gathered}\langle s^k\rangle/\langle s^{k-1}\rangle \sim \sigma(p), \quad \langle t^k\rangle/\langle t^{k-1}\rangle \sim \xi_{\Uparrow}(p), \\ \langle y^{2k}\rangle/\langle y^{2k-2}\rangle \sim \xi_{\perp}(p)^2\end{gathered} \tag{2.10}$$

where in the latter case we have taken even moments since we shall consider only the symmetric case, for which the odd moments are zero. These functions are known respectively as the scaling size and the parallel and perpendicular connectedness lengths. The notation used for their critical exponents and those of other functions is given in the table below. The percolation probability exponent β describes its vanishing at p_c rather than divergence. We shall see that the ratio $E(L_m^k)/E(L_m^{k-1})$ has the same critical exponent as $\xi_{\Uparrow}(p)$.

2.2. Previous Results for Bond and Site Percolation

Recent numerical work on bond and site percolation (Essam et al. 1988) has given extremely accurate estimates of the critical probabilities for both directed square and triangular lattices. For example on the square lattice $p_c(\text{bond}) = 0.644701 \pm 0.000001$ and $p_c(\text{site}) = 0.705489 \pm 0.000004$. The critical exponent γ of $S_1(p)$ for all four percolation processes was estimated to be 2.278 ± 0.002. Assuming that γ is a 'simple' rational, the value $41/18$ was chosen and then biased estimates of the exponents $\nu_{\Uparrow}$ and $\nu_{\perp}$ of the connectedness lengths were obtained from the moments $\mu_1^{(t)}$, $\mu_2^{(t)}$, $\mu_2^{(y)}$. The existence of the scaling length $\xi_{\Uparrow}(p)$ was strongly supported by the results. A search for simple rational values of all exponents which were consistent with the data and also with scaling relations failed. A value of $\beta = 199/720$ predicted by the relation $\beta = \frac{1}{2}(\nu_{\Uparrow} + \nu_{\perp} - \gamma)$ was considered not to be simple but is supported by a direct numerical estimate (Baxter and Guttmann 1988). The rational values quoted for bond and site percolation in the table below reproduce the estimated exponents to three decimal places.

It has been shown by Coniglio (1982) that for bond percolation $\langle b\rangle = p\,d(\ln S_m(p))/dp$ so that this function diverges with exponent 1. A similar argument shows that for site percolation $\langle a\rangle = p\,d(\ln S_m(p))/dp - 1$ which therefore also has a simple pole at the critical point. On the other hand we shall show that for compact clusters $\langle a\rangle$ is finite on the critical line. This marked difference may be attributed to the ramified nature of non-compact clusters.

function	exponent		
	usual symbol	compact	bond and site
$P_m(p)$	β	1	199/720
$\langle L_m \rangle$		1	
$S_m(p)$	γ	2	41/18
$\xi_{\Uparrow}(p)$	$\nu_{\Uparrow}$	2	26/15
$\xi_{\perp}(p)$	$\nu_{\perp}$	1	79/72
$\sigma(p)$	Δ	3	
$\langle a \rangle$	ζ_a	$\langle a \rangle$ finite	1 (site)
$\langle b \rangle$	ζ_b	$\langle b \rangle$ finite	1 (bond)

2.3. Previous Results for Compact Clusters

The probability $r_m(t)$ that c_m has t growth stages satisfies (Essam 1989) the recurrence relation

$$r_t(m) = cr_{t-1}(m+1) + (1-c-d)r_{t-1}(m) + dr_{t-1}(m-1) \qquad (2.11)$$

with boundary conditions $r_0(1) = d$, together with $r_t(m) = 0$ for $m \geq t+2$ and $m = 0$. This leads to an explicit formula for the moment generating function of $r_t(m)$, namely

$$R_m(z) = e^z[\lambda(z)]^m. \qquad (2.12)$$

where $\lambda(z)$ is the root of

$$c\lambda^2 + (1-c-d-e^z)\lambda + d = 0 \qquad (2.13)$$

which tends to zero as $z \to \infty$, i.e.

$$\lambda(z) = \frac{1}{2c}\left\{c + d + e^z - 1 - \sqrt{(1-c-d-e^z)^2 - 4cd}\right\}. \qquad (2.14)$$

Using (2.6) and (2.7) the generating function $R_{m1}(z)$ is given by

$$R_{m1}(z) = R_m(z)/d = e^z\lambda(z)^m/d. \qquad (2.15)$$

Similar analysis in the case $c + d = 1$ may be found in Feller (1968).

The percolation probability

From (2.3) and (2.7) the probability that c_m is finite is equal to $R_m(0)$ and hence

$$Q_m(p) = \begin{cases} 1 & \text{for } c < d \\ (d/c)^m & \text{for } c \geq d, \end{cases} \tag{2.16}$$

and the percolation probability $P_m(p) = 1 - Q_m(p)$, just above the critical curve $c = d$, has asymptotic form

$$P_m(p) \cong \frac{m(c-d)}{c} = \frac{m(p_u + p_d - 1)}{p_u p_d} \tag{2.17}$$

and hence the critical exponent $\beta = 1$ for all m and all points on the curve as expected.

Moments of the cluster length distribution

The generating function for the cumulants of the distribution of L_m is $\ln[e^{-z} R_m(z)] = m \ln \lambda(z)$. Thus

$$E(L_m) = mG(0) \tag{2.18}$$

where

$$\begin{aligned} G(z) &= -(d/dz) \ln \lambda(z) \\ &= e^z[(1 - c - d - e^z)^2 - 4cd]^{-1/2}, \end{aligned} \tag{2.19}$$

which is a symmetric function of c and d which we will use later. Setting $z = 0$,

$$G(0) = \frac{1}{|d - c|} \tag{2.20}$$

which diverges on the critical line with exponent 1 which is therefore the critical exponent of the mean length. Further

$$\text{Var}(L_m) = -mG'(0) = m[(c+d)G(0)^3 - G(0)] \tag{2.21}$$

which therefore has critical exponent 3. The kth order cumulant average of L_m is equal to $(-1)^{k-1} G^{(k-1)}(0)$ and it follows from (2.19) that near the critical line

$$G^{(k)}(0) \cong (-c)^k (2k)_k |d - c|^{-2k-1}. \tag{2.22}$$

The critical exponent of the kth moment of L_m is therefore $2k+1$, for all m and all points on the critical curve in agreement with Essam (1989) for the symmetric case. This implies the existence of a scaling length with critical exponent 2 which we shall see is the same as that for the radius of gyration about the y-axis, i.e. $\nu_{\Uparrow} = 2$.

Moments of the cluster size distribution

The first two moments of the cluster size distribution of compact clusters have also been determined previously (Essam 1989). In particular the mean cluster size is

$$S_m(p) = \tfrac{1}{2}m\left[\frac{c+d}{(d-c)^2} + \frac{m}{|d-c|}\right] \tag{2.23}$$

a result we will need in the next section. Thus the critical exponent $\gamma = 2$. We rederive these moments in Section 4 by a method which enables the critical exponent of the general moment to be obtained.

3. Nodal Points and Bonds in Compact Clusters

In this section we break new ground in the theory of compact percolation and determine the probability distributions of nodal points and bonds.

3.1. The Probability that c_m has no Nodal Points

The probability that no nodal points (bonds) occur in c_m is the probability of finding at least two paths from the source to the terminal point of c_m which have no intermediate site (bond) in common. We first consider clusters with t growth stages ($t \geq 1$) and note that

$$\text{pr}(c_m \text{ has } t \text{ growth stages and no nodal vertices}) = w_{m1}(t)d \tag{3.1}$$

and we will calculate this quantity first by relating it to $r_{m1}(t)$. For $m = 1$ and $t \geq 2$

$$w_{11}(t) = cdr_{11}(t-2) \tag{3.2}$$

since any cluster contributing to $w_{11}(t)$ may be associated with the maximal cluster on the dual lattice which it contains (see Figure 1) and all clusters which contribute to $r_{11}(t-2)$ occur as such dual clusters. For $m > 1$ and $t \geq 1$ a similar bijection exists between clusters contributing to $w_{m1}(t)$ and $r_{m-1,1}(t-1)$ which gives

$$w_{m1}(t) = dr_{m-1,1}(t-1). \tag{3.3}$$

These results also have simple interpretations in terms of translating walks of $r_{m1}(t)$ one step to the right.

Summing (3.1) over all $t \geq 1$, remembering that $w_{mn}(0)$ is defined as zero, we obtain

$$\text{pr}(c_m \text{ is finite, has at least one growth stage and no nodal points}) = W_{m1}(0)d \tag{3.4}$$

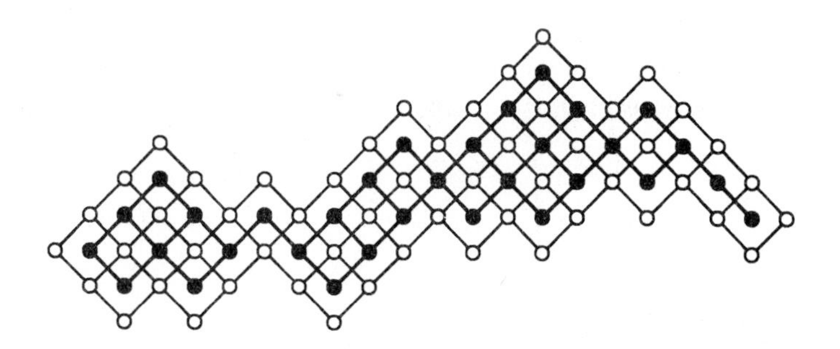

FIG. 1. The open circles are the vertices in a compact cluster contributing to $w_{11}(21)$, and the filled circles are the vertices in the maximal cluster on the dual lattice which it contains. The latter has length 20, size 32, source width $m = 1$, maximum width 3, 7 nodal points and 4 nodal bonds.

and

$$
\begin{aligned}
g_0(p) &= \mathrm{pr}(c_m \text{ has no nodal points} \mid c_m \text{ finite}) \\
&= \frac{W_{m1}(0)d + d\delta_{m,1}}{Q_m(p)},
\end{aligned}
\tag{3.5}
$$

where the term $d\delta_{m,1}$ is from the cluster having just one site which can occur only when $m = 1$ and is not included in $W_{11}(0)$. The above results give the following generating function relations

$$
\begin{aligned}
W_{11}(z) &= w_{11}(1)e^{-z} + cdR_{11}(z)e^{-2z} \\
&= [(1-c-d) + c\lambda(z)]e^{-z},
\end{aligned}
\tag{3.6}
$$

$$
W_{m1}(z) = \lambda(z)^{m-1} \quad \text{for } m > 1, \tag{3.7}
$$

which for $c < d$ leads to

$$
g_0(p) = \begin{cases} (2-d)d & \text{for } m = 1 \\ d & \text{for } m > 1, \end{cases} \tag{3.8}
$$

and for $c > d$, d is replaced by c.

3.2. The Number of Nodal Points and Bonds in c_m

For $a > 0$ the probability distribution of the number of nodal points in c_m is

$$
\begin{aligned}
g_a(p, m) &\equiv \mathrm{pr}(c_m \text{ has exactly } a \text{ nodal vertices given that it is finite}) \\
&= \frac{W_{m1}(0)W_{11}(0)^a d}{Q_m(p)}
\end{aligned}
\tag{3.9}
$$

since a cluster with a nodal points may be constructed as the series combination of $a+1$ non-nodal subclusters which have width 1 at their initial and final stages, except for the first subcluster which is initially of width m. The W factors arise from the repeated convolution of $w_{m1}(t)$ with $w_{11}(t)$ and the final factor d is the probability that the cluster terminates after the last subcluster is complete. Substituting from (3.6) and (3.7) gives, for $c < d$

$$g_a(p,m) = \begin{cases} (1-d)^{a+1}d & \text{for } m=1, a>0 \\ (2-d)d & \text{for } m=1, a=0 \\ (1-d)^a d & \text{for } m>1, \end{cases} \tag{3.10}$$

and computing the moments gives

$$E(a) = \begin{cases} (1-d)^2/d & \text{for } m=1 \\ (1-d)/d & \text{for } m>1 \end{cases} \tag{3.11}$$

and

$$\text{Var}(a) = \begin{cases} (1-d)^2/d^2 & \text{for } m=1 \\ (1-d)/d^2 & \text{for } m>1. \end{cases} \tag{3.12}$$

For $c > d$, d should be replaced by c. Notice that these quantities have only a cusp on the critical line $c = d$ whereas we saw in Section 2 that for site percolation $\langle a \rangle$ is infinite at p_c. For compact clusters we can show that the expected number of growth stages, T, between nodal points is infinite when $c = d$, thus

$$\begin{aligned} T &= -(d/dz)\ln[W_{11}(z)]|_{z=0} \\ &= \frac{d(1-d+c)}{(1-d)(d-c)} \quad \text{for } c<d \end{aligned} \tag{3.13}$$

and again c and d are interchanged for $c > d$. This is consistent with the divergence of the expected cluster length and in fact

$$E(L_m) = 1 + [E(a+1) - d]T. \tag{3.14}$$

The probability distribution for the number of nodal bonds b for clusters with a point source is for $b > 0$, $m = 1$ and $c < d$:

$$h_b(p,1) = (1-c-d)^b(1+c)^{b+1}d \tag{3.15}$$

with c and d interchanged for $c > d$.

3.3. Spatial Average of the Number of Nodal Vertices for $m = 1$

Let s_i, $i = 0, \ldots, a$, be the number of sites in the ith non-nodal subcluster in the series combination constituting c_1, not counting the initial site. The total number of sites in c_1 is therefore

$$s = 1 + \sum_{i=0}^{a} s_i. \tag{3.16}$$

For $i > 0$ let S_{11} be the expected value of s_i; this is independent of i by translational symmetry. The expected value of s for c_1 given that it has a articulation points is therefore $1 + (a+1)S_{11}$ except when $a = 0$ and $s = 1$. The unconditional expectation of s which we have previously calculated and denoted by $S_1(p)$ is therefore:

$$S_1(p) = 1 + [E(a+1) - d]S_{11} \tag{3.17}$$

which determines S_{11} in terms of $S_1(p)$ which is given by (2.23) and using (3.11), for $c < d$:

$$S_{11} = \frac{d}{1-d}\left[\frac{d}{(d-c)^2} - 1\right]. \tag{3.18}$$

The un-normalised spatial average of the number of nodal points defined in Section 2.1 may be found by (i) counting, for each site of c_m, the number of nodal sites on a path from the origin to the chosen site, (ii) summing over all sites of c_m, (iii) averaging over all clusters. For a cluster with a articulation points the result of (ii) may be written in the form $s_1 + 2s_2 + 3s_3 + \cdots + as_a$ which has conditional expectation value $\frac{1}{2}a(a+1)S_{11}$ and hence

$$\begin{aligned} \langle a \rangle &= E[a(a+1)]\frac{S_{11}}{2S_1(p)} \\ &= \frac{1-d}{d}\left[1 - \frac{1}{S_1(p)}\right]. \end{aligned} \tag{3.19}$$

Again c replaces d for $c > d$ and $\langle a \rangle$ is finite and continuous on the critical curve but using (2.23) the approach is quadratic in contrast to $E(a)$ for which the approach to the critical curve was linear with a discontinuous first derivative.

4. The Width Distribution, Cluster Size, Centre of Mass, and Moments of Inertia

In this section we calculate the quantities which were defined in Section 2.1 and verify the existence of the scaling size $\sigma(p)$, with exponent $\Delta = 3$, and the scaling length $\xi_{\Uparrow}(p)$, with exponent $\nu_{\Uparrow} = 2$.

4.1. The Cluster-Width Distribution and Moments of the Cluster-Size Distribution

We now turn to the moments of the cluster-size distribution. The size s of the cluster c_m is the number of vertices it contains and if n_t is the number of vertices which are added at stage t of the growth then

$$s = \sum_{t=0}^{\infty} n_t. \tag{4.1}$$

We first obtain formal expressions for the moments in terms of the following cumulative distribution which gives the cluster width after t growth stages:

$$\begin{aligned} \rho_{mn}(t) &= \mathrm{pr}(c_m \text{ grows for at least } t \text{ stages and has width } n \text{ after stage } t \\ &\qquad\qquad \text{given that it is finite}) \\ &= r_{mn}(t) Q_n / Q_m \end{aligned} \tag{4.2}$$

where $r_{mn}(t)$ and $Q_m(p)$ are defined in Section 2.1 and we have suppressed the argument of the latter. The following moments of $\rho_{mn}(t)$ will be important in all subsequent calculations.

$$M_k = \sum_{t=0}^{\infty} \sum_{n=1}^{\infty} n^k \rho_{mn}(t). \tag{4.3}$$

It follows, using (2.5) and (4.1), that

$$E(L_m) = M_0 \quad \text{and} \quad S_m(p) = M_1. \tag{4.4}$$

The second moment $m_2(p)$ of the size distribution is the expected value of

$$s^2 = \sum_{t=0}^{\infty} \sum_{t'=0}^{\infty} n_t n_{t'}. \tag{4.5}$$

Separating the diagonal and off-diagonal terms

$$m_2(p) = 2U_m^{(2)} - M_2 \tag{4.6}$$

where

$$U_m^{(k+1)} = \sum_{t=0}^{\infty} \sum_{n=1}^{\infty} n \rho_{mn}(t) U_n^{(k)} \tag{4.7}$$

and $U_m^{(1)} = S_m(p)$. We shall see that $U_m^{(2)}$ is expressible in terms of the M_k as far as M_3.

Carrying out the t-summation in (4.3), using (2.7), we see that the M_k are the x-derivatives at $x = 0$ of the generating function:

$$M(x) = \sum_{n=1}^{\infty} e^{-nx} R_{mn}(0) Q_n / Q_m. \tag{4.8}$$

It may be shown using the reflection principle (Feller 1968) that the moment generating function $R_{mn}(z)$ is given by

$$R_{mn}(z) = G(z) H_{mn}(\lambda(z)) \tag{4.9}$$

where $G(z)$ is given by (2.19) and

$$H_{mn}(\lambda) = \begin{cases} \lambda^{m-n} - \lambda^m \lambda^{\star n} & \text{for } n \leq m \\ \lambda^{\star n-m} - \lambda^m \lambda^{\star n} & \text{for } n \geq m \end{cases} \tag{4.10}$$

where $\lambda^{\star}(z)$ is the reciprocal of the second root of (2.13),

$$\lambda^{\star}(z) = (c/d)\lambda(z). \tag{4.11}$$

A lengthy but straightforward calculation using (4.8) and (4.9) yields

$$M(x) = \frac{G(0)(1-u)(1-e^{-mx})}{(e^x - u)(1 - e^{-x})} \tag{4.12}$$

where

$$u = \lambda(0)\lambda^{\star}(0) = \begin{cases} c/d & \text{for } c < d \\ d/c & \text{for } c > d \end{cases} \tag{4.13}$$

and has the value 1 on the critical curve. The critical exponents associated with the M_k may be deduced by expanding $M(x)$ in the form

$$M(x) = G(0) \sum_{j=0}^{\infty} \sum_{k=0}^{\infty} (-1)^{j+k} v_k (e^x - 1)^j (1-u)^{-j} \frac{x^k}{k!} \tag{4.14}$$

where $v_0 = m$ and for $k > 0$

$$v_k = \sum_{n=1}^{m-1} n^k \tag{4.15}$$

which is a polynomial in m of degree $k+1$. Collecting together powers of x in (4.14) we find

$$M_0 = mG(0) = E(L_m), \tag{4.16}$$

where $G(0)$ is given explicitly by (2.18) and so M_0 has critical exponent 1.

$$S_m(p) = M_1 = M_0 \left[\frac{1}{1-u} + \tfrac{1}{2}(m-1) \right], \tag{4.17}$$

which thus has critical exponent $\gamma = 2$ and rederives (2.23).

$$M_2 = M_0 \left[\frac{1+u}{(1-u)^2} + \frac{m-1}{1-u} + \tfrac{1}{6}(m-1)(2m-1) \right], \tag{4.18}$$

which has critical exponent 3 and

$$M_3 = M_0 \left[\frac{1+4u+u^2}{(1-u)^3} + \frac{3(m-1)(1+u)}{2(1-u)^2} + \frac{(m-1)(2m-1)}{2(1-u)} + \tfrac{1}{4}m(m-1)^2 \right] \tag{4.19}$$

which has critical exponent 4. In general, from (4.14), M_k/M_0 is the product of a polynomial in u and a factor $(1-u)^{-k}$ and hence M_k has critical exponent $k+1$.

Using (4.6), (4.7), (4.17) and (4.3) we see that the second moment of the cluster-size distribution is given by

$$m_2(p) = G(0) \left[M_3 + M_2 \frac{1+u}{1-u} \right] - M_2, \tag{4.20}$$

and hence $m_2(p)$ has critical exponent 5 which comes from the first term since the second term, which came from the diagonal contributions in (4.5), has the smaller exponent 3. We have checked that for $m = 1$ this result correctly reproduces equation (4.19) of Essam (1989).

In general we can show the critical exponent of $m_k(p)$ is that of $U_m^{(k)}$. Also we can prove by induction that the terms in $U_m^{(k)}$ with dominant critical exponent are of the form

$$\frac{G(0)^{k-1} M_i}{(1-u)^{2k-i-1}} \quad \text{with } i = 2, 3, \ldots, 2k-1 \tag{4.21}$$

and since M_i has exponent $i+1$ it follows that $U_m^{(k)}$ has exponent $3k-1$. The existence of $\sigma(p)$, the scaling size with critical exponent $\Delta = 3$ is thus established. The induction uses the fact that M_i is a combination of terms with singularities of the form

$$\frac{G(0)\nu_{ij}}{(1-u)^{i-j}} \quad \text{with } j = 0, \ldots, i, \tag{4.22}$$

where, from (4.14), ν_{ij} is a linear combination of the ν_k in (4.15), with $k \leq j$. It also uses the fact that, since ν_{ij} is a polynomial in m of degree $j+1$, when $U_m^{(k)}$ is substituted in the right hand side of (4.7) a term involving ν_{ij} produces an M_{j+2}.

4.2. The Centre of Mass and Moments of Inertia

The first moment of the mass distribution of c_m about the y-axis may be written

$$\mu_{1t}(p) = \sum_{t=0}^{\infty} \sum_{n=1}^{\infty} nt\rho_{mn}(t). \tag{4.23}$$

In calculating the moments about the t-axis we shall assume, for simplicity, the symmetric case $p_u = p_d = p$ so that the first moment is zero. It may be shown that the second moment is then given by

$$\mu_{2y}(p) = 2p(1-p) \sum_{t=0}^{\infty} \sum_{n=1}^{\infty} n(t+1)\rho_{mn}(t) + \tfrac{1}{3}[M_3 - M_1] \tag{4.24}$$

where the first term arises from the second moment of the centre of mass of column t about the t-axis and the second is the second moment of column t about its centre of mass. It is therefore possible to express this moment in terms of the M_k and $\mu_{1t}(p)$ which we now calculate:

$$\mu_{2y}(p) = 2p(1-p)[\mu_{1t}(p) + M_1] + \tfrac{1}{3}[M_3 - M_1]. \tag{4.25}$$

Now using (4.9)

$$\begin{aligned} \mu_{1t}(p) &= \sum_{n=1}^{\infty} n \left[-\frac{dR_{mn}(z)}{dz} \right]_{z=0} \frac{Q_n}{Q_m} \\ &= \left[-\frac{d}{dz} \ln G(z) \right]_{z=0} S_m(p) + \left[-\frac{d}{dz} \ln \lambda(z) \right]_{z=0} G(0) A_m \end{aligned} \tag{4.26}$$

where

$$A_m = \sum_{n=1}^{\infty} n\lambda \frac{dH_{mn}(\lambda)}{d\lambda} \frac{Q_n}{Q_m}. \tag{4.28}$$

After some lengthy calculation we find

$$A_m = \tfrac{1}{6} m(m^2 - 1) \tag{4.28}$$

and

$$\mu_{1t}(p) = [(c+d)G(0)^2 - 1] S_m(p) + \tfrac{1}{6} m(m^2-1) G(0)^2 \tag{4.29}$$

which has critical exponent 4 and it follows that the mean length $\langle t \rangle = \mu_{1t}(p)/S_m(p)$ has critical exponent 2. Assuming the existence of the scaling length $\xi_{\Uparrow}(p)$ this gives $\nu_{\Uparrow} = 2$. This assumption will be verified later in the case $m = 1$.

Combining this result with (4.25) gives

$$\mu_{2y}(p) = 2p(1-p)(c+d)G(0)^2 S_m(p) + \tfrac{1}{6}m(m^2-1)G(0)^2 - \tfrac{1}{3}[M_3 - M_1] \tag{4.30}$$

which has critical exponent 4 and therefore the mean square displacement $\langle y^2 \rangle = \mu_{2y}(p)/S_m(p)$ of the cluster mass from the t-axis has critical exponent 2. Assuming the existence of the scaling length $\xi_{\perp}(p)$ gives $\nu_{\perp} = 1$. This agrees with the value given by Domany and Kinzel (1984) which they deduced from a result of Verhagen (1976) for a triangular lattice Ising model with three spin interactions in alternate triangles. Our derivation is direct and uses the same definition as is used in series expansion calculations (2.2).

In the case $m = 1$ we have the explicit formula

$$R_{1n}(z) = \frac{1}{c} e^z \lambda^\star(z)^n \tag{4.31}$$

and using this we may show that the kth moment of $t+1$ may be written

$$\begin{aligned} \langle (t+1)^k \rangle &= \sum_{n=1}^{\infty} n \left[\frac{d^k(e^{-z}R_{1n}(z))}{d(-z)^k} \right]_{z=0} \frac{Q_n}{Q_1 S_1} \\ &= \sum_{n=1}^{\infty} n\alpha_k(n) R_{1n}(0) \frac{Q_n}{Q_1 S_1} \end{aligned} \tag{4.32}$$

where, with $C_k = n(-1)^{k-1}G^{(k-1)}(0)$,

$$\begin{aligned} \alpha_1(n) = C_1, \quad \alpha_2(n) &= C_2 + C_1^2, \\ \alpha_3(n) &= C_3 + 3C_1C_2 + C_1^3, \\ \alpha_4(n) &= C_4 + 3C_1C_3 + 3C_2^2 + 6C_1^2C_2 + C_1^4, \end{aligned} \tag{4.33}$$

and in general the subscripts are the possible partitions of k and the coefficients are the number of partitions of each type. Carrying out the n-summation using (4.8) we find

$$\langle (t+1) \rangle = G(0)M_2/S_1 \tag{4.34}$$

$$\langle (t+1)^2 \rangle = [-G'(0)M_2 + G(0)^2 M_3]/S_1 \tag{4.35}$$

$$\langle (t+1)^3 \rangle = [G''(0)M_2 - 3G'(0)G(0)M_3 + G(0)^3 M_4]/S_1 \tag{4.36}$$

and so on. Equation (4.34) may be seen to agree with (4.29) when $m = 1$ by using (4.18) and $\langle t+1 \rangle = \langle t \rangle + 1$. Using our previous results that M_k has exponent $k+1$ and that $G^{(k)}(0)$ has exponent $2k+1$ it can be seen that each of the terms has the same critical exponent and hence that $\langle t^k \rangle$ has critical exponent $2k$ which verifies the existence of the scaling length $\xi_{\Uparrow}(p)$ having critical exponent $\nu_{\Uparrow} = 2$ the same as found from the cluster length distribution in Section 2.3.

References

Baxter, R. and Guttmann, A.J. (1988). Series expansion of the percolation probability for the directed square lattice. *Journal of Physics A: Mathematical and General* 21, 3193–3204

Bezuidenhout, C.E. and Grimmett, G.R. (1989). The critical contact process dies out. Preprint.

Broadbent, S.R. and Hammersley, J.M. (1957). Percolation processes I. Crystals and mazes. *Proceedings of the Cambridge Philosophical Society* 53, 629–641.

Coniglio, A. (1982). Cluster structure near the percolation threshold. *Journal of Physics A: Mathematical and General* 15, 3829–3844.

Delest, M.-P. and Viennot, G. (1984). Algebraic languages and polyominoes enumeration. *Theoretical Computer Science* 34, 169–206

Deutscher, G., Zallen, R., and Adler, J. (1983). *Percolation Structures and Processes*, editors. Adam Hilger Ltd. and *Annals of the Israel Physical Society* 5.

Domany, E. and Kinzel, W. (1984). Equivalence of cellular automata to Ising models and directed percolation. *Physical Review Letters* 53, 311–314.

Domb, C. (1959). Physical Society conference on fluctuation phenomena and stochastic processes. *Nature* 184, 509–512.

Durrett, R. (1984). Oriented percolation in two dimensions. *Annals of Probability* 12, 999–1040.

Durrett, R. and Schonmann, R.H. (1987). Stochastic growth models. In *Percolation Theory and Ergodic Theory of Infinite Particle Systems*, ed. H. Kesten, Springer-Verlag, New York, 85–119.

Essam, J.W. (1989). Directed compact percolation I: Cluster size and hyperscaling. Submitted to *Journal of Physics A: Mathematical and General.*

Essam, J.W., Guttmann, A.J., and De'Bell, K. (1988). On two-dimensional directed percolation. *Journal of Physics A: Mathematical and General* 21, 3815–3832.

Feller, W. (1968). *Introduction to Probability Theory and Its Applications*, Volume 1. 3rd edition, Chapter 14, John Wiley & Sons, New York.

Frisch, H.L. and Hammersley, J.M. (1963). Percolation processes and related topics. *Journal of the Society for Industrial Applications of Mathematics* 11, 894–918.

Hammersley, J.M. (1959). Bornes supérieures de la probabilité critique dans un processus de filtration. *Proceedings of the 87th International Colloquium*, Centre National de la Recherche Scientifique, Paris, 17–37.

——— (1961). Comparison of atom and bond percolation processes. *Journal of Mathematical Physics* 2, 728–733.

Huse, D.A., Szpilka, A.M., and Fisher, M.E. (1983). Melting and wetting transitions in the three-state chiral clock model. *Physica* 121A, 363–398.

Kesten, H. (1982). *Percolation Theory for Mathematicians.* Birkhäuser, Boston.

Mauldon, J.G. (1961). Asymmetric oriented percolation on a square lattice. *Proceedings of the 4th Berkeley Symposium on Mathematical Statistics and Probability* 2, 337–346.

Verhagen, A.M.W. (1976). An exactly soluble case of the triangular Ising model in a magnetic field. *Journal of Statistical Physics* 15, 219–231.

Department of Mathematics
Royal Holloway and Bedford New College
Egham Hill
Egham
Surrey TW20 0EX.

Critical Points, Large-Dimensionality Expansions, and the Ising Spin Glass

Michael E. Fisher and Rajiv R.P. Singh

Abstract

Critical points for the percolation process, for statistical-mechanical models of ferromagnets, and for self-avoiding and self-interacting walks are briefly discussed. The construction of expansions for such critical points in powers of $1/d$, where d is the dimensionality of the underlying hypercubic lattices, is reviewed. Corresponding expansions for the transition points, $T_c(d)$, of Ising model spin glasses with arbitrary symmetric distributions of couplings are derived to order $1/d^3$; for the $\pm J$ model results correct to fifth order are obtained. Numerical results are presented for $d = 3, 4, \ldots, 8$; the lower critical dimensionality appears to be about $d_< = 2.5$.

1. Introduction: Walks and Ferromagnets

Hammersley's pioneering work in formulating and analysing the *bond percolation problem* on a lattice (Broadbent and Hammersley 1957; Hammersley 1957a,b) laid a foundation for the systematic study of the statistics and statistical physics of random media. Our main aim here is to report some recent results concerning phase transitions in random media, specifically, for spin glasses; but, to set the topic in context, we first review the background in a little detail.

In the simplest percolation problem the nearest-neighbour bonds of a uniform space lattice, $\mathcal{L}$, are occupied (or present) with probability p and vacant (or absent) with probability $1-p$. We will mainly focus on the d-dimensional hypercubic lattices $\mathcal{L}_d$ ($\equiv \mathbb{Z}^d$) of coordination number

$$q \equiv \sigma + 1 = 2d. \tag{1.1}$$

Occupied bonds which are connected via common sites form clusters; above a *percolation threshold*, $p_c(\mathcal{L})$, an infinite cluster of connected bonds stretches across the lattice with probability one.

A relation between percolation and other statistical problems on lattices was established by Hammersley already in his earliest studies, via the general lower bound

$$p_c(\mathcal{L}) \geq 1/\mu(\mathcal{L}) \tag{1.2}$$

in which $\mu(\mathcal{L})$, the self-avoiding walk limit, was also first defined precisely by Hammersley. Specifically, if $c_m(\mathcal{L})$ is the number of distinct *self-avoiding* (i.e. nonself-intersecting) random walks of m nearest-neighbour steps starting at the origin of $\mathcal{L}$, then

$$\mu(\mathcal{L}) = \lim_{m\to\infty} |c_m(\mathcal{L})|^{1/m}, \tag{1.3}$$

where the existence of the limit follows by a subadditive argument (Hammersley and Morton 1954; Hammersley 1957a). Self-avoiding walks on lattices form natural, somewhat crude but nonetheless informative models of polymer molecules. In particular, the self-avoidance requirement represents the crucial *excluded volume* constraint which is the main theoretical obstacle to be faced in studying the statistical mechanics of polymeric systems.

Subsequently, Fisher and Sykes (1959) pointed out that there was a close parallel between the behavior of self-avoiding walks on a lattice $\mathcal{L}$ and the statistical mechanics of an Ising model of a ferromagnet on the same lattice[1]. An Ising ferromagnet is specified by its Hamiltonian

$$\mathcal{H} = -J \sum_{(i,j)} s_i s_j, \tag{1.4}$$

in which $J > 0$ represents the strength of the coupling between the spin variables, $s_i, s_j, \ldots$ at lattice sites i and j while the sum runs over all nearest-neighbour pairs of spins, i.e., over the lattice bonds. In the standard Ising model each spin takes just two values, $s_i = \pm 1$ (all i). The basic control parameter is the temperature, T, which enters only in the dimensionless combination

$$K = J/k_B T, \tag{1.5}$$

where k_B is Boltzmann's constant. The spin-spin correlation functions are defined, as usual, by the thermodynamic expectation values

$$\langle s_i s_k \rangle = \mathrm{Tr}_s\{s_i s_k \exp(-\mathcal{H}/k_B T)\}/\mathrm{Tr}_s\{\exp(-\mathcal{H}/k_B T)\} \tag{1.6}$$

[1]See also Fisher 1966.

in which the trace operation here corresponds to summing over all the values $\{s_i = \pm 1\}$. In terms of the correlation functions the (reduced) susceptibility is given by

$$\chi(T) = \sum_k \langle s_0 s_k \rangle, \tag{1.7}$$

the sum running over all sites k in $\mathcal{L}$.

An Ising ferromagnet in $d \geq 2$ dimensions displays a phase transition at a *critical point*, $T_c(\mathcal{L})$. Above T_c the correlation functions decay to zero exponentially fast with r_{ik}, the distance between sites, and the sum in (1.7) is absolutely convergent. As $T \to T_c+$, however, the susceptibility, $\chi(T)$, diverges strongly to ∞. Below T_c the system displays spontaneous magnetization and long-range order — the correlations do not decay.

An analogy with self-avoiding walks is obtained by considering the generating function

$$C(z;\mathcal{L}) = \sum_{m=0}^{\infty} c_m(\mathcal{L}) z^m. \tag{1.8}$$

One finds that $C(z;\mathcal{L})$ is similar in many ways to χ; in particular, $C(z)$ diverges strongly as $z \to z_c-$ where the critical point is simply

$$z_c(\mathcal{L}) = 1/\mu(\mathcal{L}). \tag{1.9}$$

More concretely one can establish the bound

$$\chi(T;\mathcal{L}) \leq C(\tanh K;\mathcal{L}); \tag{1.10}$$

see Fisher (1967). From this one immediately obtains a bound for the critical point analogous to (1.2), namely,

$$\tanh(J/k_{\mathrm{B}}T_c) \geq 1/\mu(\mathcal{L}) \tag{1.11}$$

(Fisher and Sykes 1959; Fisher 1967).

Ising models have been generalized in various ways important for the study of critical phenomena. In the first instance one has *rigid* or *fixed-length n-vector models* in which the simple Ising spins, s_i, are replaced by n-component vectors, $\vec{s_i}$, of magnitude which is most conveniently taken as $|\vec{s_i}| = \sqrt{n}$ (Stanley 1968, 1969). The coupling term in (1.4) is replaced by $\vec{s_i} \cdot \vec{s_j}$ and the trace operation in (1.6) becomes a product of integrations over the orientations of each $\vec{s_i}$.

A further generalization in this direction, crucial for renormalization group ϵ-expansion theory (Wilson and Fisher 1972), is to regard the spins, $\vec{s_i}$, as continuously variable in magnitude. In this *continuous* or *soft spin n-vector model* the trace operation becomes a product of integrals of the form

$$\int_{-\infty}^{\infty} ds_i^{(1)} \cdots \int_{-\infty}^{\infty} ds_i^{(n)} e^{-w(s_i^2)}, \tag{1.12}$$

in which the spin weighting function, $e^{-w(s^2)}$, decays rapidly as $s^2 \to \infty$; the form $w(s^2) = \frac{1}{2}s^2 + us^4$ with $u > 0$ is often considered.

Now it transpires, as first proposed by de Gennes (1972), that the connection between self-avoiding walks and magnetic models is much closer than originally suspected. Indeed, if one formally takes the zero-component limit, $n \to 0$, the susceptibility $\chi_n(t; \mathcal{L})$ for the rigid spin n-vector model becomes identical to the self-avoiding walk generating function, $C(z; \mathcal{L})$, with $z \propto K$. (See also Bowers and McKerrell 1973; Jasnow and Fisher 1976.)

If one studies the limit $n \to 0$ for the *continuous spin* models with weighting factor $e^{-w(s^2)}$, one obtains *self-interacting random walks*: self-intersections are now allowed but each site of the lattice which is visited r (> 0) times by the walk carries a Boltzmann factor or statistical weight given by

$$f_r = \frac{I(2r)}{(r-1)!\,I(2)} \left(\frac{e_0}{2I(2)} \right)^{r-1}, \tag{1.13}$$

in which

$$I(l) = \int_0^{\infty} e^{-w(s^2)} s^{l-1} ds, \tag{1.14}$$

while $e_0 = \exp[-w(0)]$. (Gerber and Fisher 1975; Jasnow and Fisher 1976.) Note that $f_1 \equiv 1$ always holds. When $e_0 = 0$, which is the case for rigid spins, one has $f_r = 0$ for all $r \geq 2$ so that the standard self-avoiding walk is recaptured.

Finally, we remark that the limits $n \to \infty$ and $n \to -2$ also have a special significance in that they correspond to exactly soluble models: this point is expanded in the next section.

Now all the model ferromagnets discussed above pertain, like the self-avoiding walks, to the spatially homogeneous, uniform, nonrandom medium. Considerable interest centers, however, on the study of phase transitions *in random media*. Following the example of the bond percolation problem, the simplest models to consider are *random-bond Ising models* in which, in place of (1.4), the Hamiltonian is

$$\mathcal{H} = -\sum_{(i,j)} J_{ij} s_i s_j, \tag{1.15}$$

where the interactions, J_{ij}, are independent, identically distributed random variables drawn from a specified distribution with a well defined mean and variance,

$$\bar{J} \equiv [J_{ij}]_J \quad \text{and} \quad \Delta J^2 \equiv [(J_{ij} - \bar{J})^2]_J. \tag{1.16}$$

Here and below, $[\cdot]_J$ denotes an expectation over the distribution of coupling constants. The simplest distribution is that of the so called $\pm J$ Ising model in which each bond takes the value $-J$ with probability p and $+J$ (> 0) with probability $1 - p$. A Gaussian distribution of couplings is also frequently considered.

Such random Ising models have a long history. If the mean, $\bar{J}$, is positive and the width, ΔJ, of the distribution is relatively small, one obtains a disorded or impure ferromagnet. The critical temperature, T_c, depends on the distribution of the couplings but the susceptibility still diverges strongly as $T \to T_c+$ and spontaneous magnetization arises below T_c, as it does for the uniform system. (However, the values of the critical exponents, γ, β, etc., describing the nature of the singularities at the critical point will, in general, change.)

On the other hand, if $\bar{J}$ is small enough relative to ΔJ, ferromagnetism is completely suppressed. The resulting, highly disordered system represents a *spin glass.* Real systems of this sort, made, for example, by alloying ferromagnetic metals with non-magnetic metals, show freezing phenomena into disordered states reminiscent of the behavior of ordinary glasses. A central question is whether or not this freezing behavior in a spin glass reflects the presence of a true, equilibrium phase transition of some sort: see the reviews by Binder and Young (1986) and Fisher, Grinstein and Khurana (1988). For this purpose it probably suffices to focus, as we will, on a *symmetric spin glass* for which $\bar{J}$ (along with all other odd moments of the coupling distribution) vanishes identically; the only parameter is then the width, ΔJ, or the reduced width, $\Delta J/k_{\mathrm{B}}T$.

If there is a transition to a low-temperature spin-glass state — as is now generally believed for systems of dimensionality $d = 3$ or greater — many further questions arise. An obvious issue is the value of the transition temperature $T_c(\mathcal{L})$; we will address this specifically in Section 4 et seq.

2. Critical Points for Large Dimensionality

Obtaining explicit expressions for critical points — percolation thresholds, self-avoiding walk limits, or transition temperatures for Ising models — is, in general, a hard task. Exact results are available only in special cases. Thus for the $(d = 2)$-dimensional square lattice the bond percolation threshold is $p_c = \frac{1}{2}$ (Harris 1960, Kesten 1980) and the standard Ising model critical point is given by $\tanh(J/k_{\mathrm{B}}T_c) = \sqrt{2} - 1$ (Onsager 1944). For no other hypercubic lattices are the precise answers known.

For infinite Cayley trees of uniform coordination number q, or Bethe lattices (Domb 1960a), more detailed analytic progress can be made. The branching ratio on such pseudo-lattice structures is $\sigma = q - 1$ and critical points are invariably closely related to σ. For example, for percolation on a Bethe lattice one has $p_c = 1/\sigma$ (Fisher and Essam 1961). Self-avoiding walks are obviously described by $\mu = \sigma$ while for Ising models one finds

$$\tanh(J/k_B T_c) = 1/\sigma \tag{2.1}$$

(Domb 1960a). Note that the inequalities (1.2) and (1.11) became equalities in these cases. Indeed, the departures from these simple results which are found for real lattices are directly related to the existence of closed self-avoiding paths or cycles.

This last assertion can be seen in more quantitative terms in the formal expansions for critical points in inverse powers in the dimensionality, d, constructed by Fisher and Gaunt (1964) for the Ising model and self-avoiding walk problem on hypercubical lattices, $\mathcal{L}_d$. They obtained

$$\frac{T_c^{(d)}}{T_c^0} = 1 - \frac{1}{q} - 1\tfrac{1}{3}\frac{1}{q^2} - 4\tfrac{1}{3}\frac{1}{q^3} - 21\tfrac{34}{45}\frac{1}{q^4} - 133\tfrac{14}{15}\frac{1}{q^5} - \cdots, \tag{2.2}$$

with $T_c^0 = qJ/k_B$ and, as before, $q = 2d = \sigma + 1$, and

$$\frac{\mu(d)}{q} = 1 - \frac{1}{q} - \frac{1}{q^2} - \frac{3}{q^3} - \frac{16}{q^4} - \frac{102}{q^5} - \cdots. \tag{2.3}$$

In terms of σ this last result simplifies to

$$\mu(d) = \sigma\left(1 - \frac{1}{\sigma^2} - \frac{2}{\sigma^3} - \frac{11}{\sigma^4} - \frac{62}{\sigma^5} - \cdots\right), \tag{2.4}$$

which shows that the corrections to the Bethe lattice value for μ are only of order $\sigma^{-2} \sim d^{-2}$; the same is true for the Ising model critical points. The detailed analysis sketched later shows that this correction reflects the fact that through each site on a hypercubic lattice pass precisely $2d(d-1)$ distinct squares constructed of nearest-neighbour bonds (Fisher and Gaunt 1964), as is readily checked.

The question of the convergence of these $1/d$ expansions will be addressed below. It is worth noting here, however, that the first three terms on the righthand sides of (2.2) and (2.3) provide rigorous upper bounds on $T_c(d)$ and $\mu(d)$ correct to order $1/d^3$ when $d \to \infty$; see Fisher (1967).

The Ising model result (2.2) was extended by Gerber and Fisher (1974) to the fixed-length n-vector model yielding

$$\begin{aligned}\frac{T_c(n,d)}{T_c^0} = 1 &- \frac{1}{q} - \frac{1}{q^2}\left(1 + \frac{n}{n+2}\right) - \frac{1}{q^3}\left(3 + \frac{4n}{n+2}\right)\\ &- \frac{1}{q^4}\left(16 + \frac{(21n+32)n}{(n+2)^2} - \frac{2n^2}{(n+2)(n+4)}\right)\\ &- \frac{1}{q^5}\left(102 + \frac{(129n^2+422n+340)n}{(n+2)^3} - \frac{16n^2}{(n+2)(n+4)}\right)\\ &- \cdots.\end{aligned} \tag{2.5}$$

For $n = 1$, which corresponds to the simple Ising case, this reduces correctly to (2.2). In addition, on setting $n = 0$, it evidently reproduces the self-avoiding walk result (2.3)!

Another limit is also of interest here: specifically one may take $n \to \infty$ to obtain

$$\frac{T_c(\infty, d)}{T_c^0} = 1 - \frac{1}{q} - \frac{2}{q^2} - \frac{7}{q^3} - \frac{35}{q^4} - \frac{215}{q^5} - \cdots . \qquad (2.6)$$

Now, as first demonstrated by Stanley (1968b), the limit $n \to \infty$ in the n-vector model yields the *spherical model* devised by Berlin and Kac (see Joyce 1972). This model is exactly soluble in a wide variety of cases. For hypercubic lattices, the critical points, $T_c(d)$, are given by d-fold integrals over the basic lattice generating function which can be reduced to a single integral involving the dimensionality, d, only through a factor $[I_0(x)]^d$, where $I_0(x)$ is the Bessel function of zero order and pure-imaginary argument. Using this fact, Gerber and Fisher (1974) showed that $T_c(d)$ extends naturally into a function of d which is analytic on or near the real axis for $2 < d < \infty$. Furthermore, the inverse dimensionality expansion (2.6) can be checked (and extended indefinitely). The analysis also establishes that the $1/d$ expansion is (for $n = \infty$) asymptotic rather than convergent and suggests that truncation after the term of order $1/d^{l^*}$ with $l^* \simeq 1.62d$ is optimal numerically[2].

The analytic nature of $T_c(n, d)$ and the asymptotic character of the $1/d$ expansion have not been established for general n but it seems likely that both are, in fact, true. Certainly the expansion produces good numerical results even down to $d = 3$ and for $n = 0$ and 1 (see Gerber and Fisher 1974).

The dimensionality expansion can also be carried through for the general, continuous-spin n-vector model (Gerber and Fisher 1975). The expansion coefficients now depend on the reduced noninteracting-spin moments

$$M_{2k}(n) = m_{2k}(n)/m_2^k(n) \equiv \langle|\vec{s}|^{2k}\rangle_0 / \langle|\vec{s}|^2\rangle_0^k, \qquad (2.7)$$

defined in terms of the spin-weighting function via

$$m_{2k}(n) = \int_0^\infty s^{2k+n-1} e^{-w(s^2)} ds \Big/ \int_0^\infty s^{n-1} e^{-w(s^2)} ds. \qquad (2.8)$$

[2]Abe (1976) has proposed a modification of the $1/d$ expansion which is actually convergent in the spherical model limit.

To order $1/d^2$ the calculations yield

$$\frac{T_c(n,d)}{T_c^0(n)} = 1 - \frac{1}{q}\left[1 - \tfrac{1}{2}n(M_4 - 1)\right] - \frac{1}{q^2}\left[2\frac{n+1}{n+2} - \tfrac{1}{2}\frac{n^2}{n+2}(M_4^2 - 1) - \tfrac{1}{8}n^2(M_6 - 2M_4^2 + M_4)\right] - \cdots, \quad \text{with} \quad T_c^0(n) = qJm_2(n)/k_{\mathrm{B}}. \tag{2.9}$$

The term of order $1/d^3$ was also found by Gerber and Fisher (1975) but is not quoted here because of its length. It is interesting to note that the moments M_{2k}, of the spin-weighting factor enters for the first time only in order $1/d^{k-1}$.

Yet another exactly soluble magnetic model is now accessible, namely, the so called Gaussian model characterized by the spin weighting function $w(s^2) = w_0 s^2$. This model was originally studied by Kac and Berlin in connection with the spherical model (Joyce 1972). The spin moments for the Gaussian model are

$$M_{2k}(n) = (n+2)(n+4)\dots(n+2k-2)/n^{k-1}, \tag{2.10}$$

for $k \geq 2$. On substituting this form into the terms of order q^{-1}, q^{-2}, and q^{-3} in (2.9) one finds that all the coefficients vanish identically! This is, in fact, in agreement with the exact Gaussian model result $T_c(n,d) = T_c^0(n)$; clearly the series converges absolutely in this special case!

It has, furthermore, been demonstrated that if the n-vector model is continued to $n = -2$, the Gaussian model is again recaptured (Balian and Toulouse 1973; Fisher 1973) irrespective of the form of $w(s^2)$. This conclusion, which is contingent on $w(s^2)$ remaining bounded as $s^2 \to 0$, can also be checked in (2.9) by analytically continuing the integrals in (2.8) to $n < 0$. One finds $M_{2k}(-2) = 0$ for $k \geq 2$ (see Gerber and Fisher 1975); this value agrees with the Gaussian form (2.10).

Lastly, it is worth quoting the limit for self-interacting walks which follows from (2.9) with (1.13). One obtains

$$\frac{M\{f_r\}}{q} = 1 - (1 - f_2)q^{-1} - (1 - f_3)q^{-2} - (3 + f_2 - 4f_2^2 - 2f_2f_3 + 3f_3^2 - f_4)q^{-3} + \cdots, \tag{2.11}$$

which, of course, reproduces (2.3) when the weight factors, f_r, for r-fold intersections vanish. Conversely, when $f_r = 1$ for all r the series reduces simply to $\mu = q$.

3. Expanding in Inverse Dimensionality

The results summarized above for uniform, nonrandom lattices rest on an analysis of the high-temperature series expansions for the susceptibilities, $\chi_n(T;\mathcal{L}_d)$, for the magnetic systems in question (Fisher and Gaunt 1964). In terms of $K = J/k_B T$ one has

$$\chi_n(T;\mathcal{L}_d) = \sum_{l=0}^{\infty} a_l(n,d)K^l. \tag{3.1}$$

In practice one observes that all the coefficients a_l are nonnegative. Then the nearest singularity of $\chi_n(T)$ lies on the real positive K axis at a K_c which locates the physical critical point, $T_c(d)$. Consequently one can write

$$\ln[k_B T_c(d)/J] = \limsup_{l\to\infty} l^{-1} \ln a_l(n,d). \tag{3.2}$$

The aim is then to calculate $a_l(n,d)$ to leading orders in d for all sufficiently large l.

Now, rather generally, the susceptibility expansion coefficients can be written in graphical form as

$$a_l(n,d) = \sum_{G_l} (G_l,\mathcal{L}_d)A(n;G_l), \tag{3.3}$$

in which G_l represents a graph or multigraph of l lines (following the graph-theoretical terminology set out by Essam and Fisher 1970). The dominant graph in all cases is a chain, C_l, of l lines and $l+1$ vertices. The statistical weights, $A(n;G_l)$, depend on the nature of the interaction, the spin-dimensionality, n, and on the graph G_l but *not* on the lattice $\mathcal{L}_d$. Often the theory is arranged so that $A(n;G_l)$ vanishes if G_l is not a connected graph. In more favorable cases the weights may also vanish if G_l is not multiply connected; the sum in (2.14) may then be restricted to star graphs which is advantageous since they are much less multitudinous.

The lattice-dependence of $a_l(n,d)$ and, hence, the dependence on dimensionality, is isolated in the embedding constants or lattice constants, $(G_l,\mathcal{L}_d)$, which represent the number of ways of embedding the graph, G_l, in the lattice $\mathcal{L}_d$ *per site.* (The rules of embedding may depend on the details of the analysis but the so called *weak lattice constants* — see Essam and Fisher — are usually most convenient.) As stressed by Fisher and Gaunt (1964) the lattice constant, $(G_l,\mathcal{L}_d)$, for a graph of l lines is, on reflexion, easily seen to be a polynomial in d or, equivalently, in q or σ, of degree at most l. Thus for the chain, C_l, the lattice constant is $(2d)^l[1+O(d^{-1})]$; the square has a lattice constant $\frac{1}{2}d(d-1)$; in leading order the lattice constant for a double bond with one or two tails (of all possible lengths totalling $l-2$ lines) varies as $\frac{1}{2}(l-1)(2d)^{l-1}$; and so on.

Finally, then, the coefficient $a_l(n,d)$ is itself a polynomial in d of order l, the coefficient of d^{l-k} being a polynomial in l of order at most k. This structure enables one to remove a factor d^l in (3.2) leading to an overall additive term $\ln l$. Then one may expand the logarithm formally in powers of d^{-1} and take the limit $l \to \infty$ term by term. The desired large-dimensionality expansion or, rather its logarithm, results.

This method adapts readily to other problems when a susceptibility-like function can be identified. Thus Gaunt, Sykes and Ruskin (1976) and Gaunt and Ruskin (1978) considered site and bond percolation problems, respectively. Working with $S(p)$, the mean cluster size function for $p < p_c$, which diverges strongly as $p \to p_c$, they obtained expansions for the percolation threshold. Specifically Gaunt and Ruskin (1978) found

$$p_c(d) = \sigma^{-1}\big(1 + 2\tfrac{1}{2}\sigma^{-2} + 7\tfrac{1}{2}\sigma^{-3} + 57\sigma^{-4} + \cdots\big), \tag{3.4}$$

for bond percolation while, for site percolation one has

$$p_c(d) = \sigma^{-1}\big(1 + 1\tfrac{1}{2}\sigma^{-1} + 3\tfrac{3}{4}\sigma^{-2} + 20\tfrac{3}{4}\sigma^{-3} + \cdots\big), \tag{3.5}$$

(Gaunt, Sykes and Ruskin 1976). It is interesting that site percolation has a leading correction of order σ^{-1} whereas for bond percolation, as for self-avoiding walks, see equation (2.4), this term vanishes.

It should also be mentioned that Harris (1982) has reviewed the use of $1/d$ expansions in a more general context and has developed a method which, in principle, can cast a variety of problems into amenable form. With his methods, he obtained explicit results for the critical fugacity for lattice animals and for the mobility edge in localization theory.

Here we ask the question: "How can similar results be obtained for spin glasses and what do they tell us?" In the following sections we answer these questions. [A brief announcement of our results has been published (Singh and Fisher 1988).]

4. Ising Spin-Glass Susceptibilities

We will study the Ising spin glass with Hamiltonian (1.15) and spins $s_i = \pm 1$ at each site i of $\mathcal{L}_d$. For simplicity we restrict attention to symmetric spin glasses for which

$$[(J_{ij}^{2k+1}]_J = 0 \quad (k = 1, 2, \ldots). \tag{4.1}$$

Such distributions embody the crucial 'frustration' induced by competing interactions which lies at the heart of the spin-glass problem. While retaining symmetry, it is of interest, however, to allow for general distributions of the J_{ij}. In particular one would like a theoretical basis for comparing the predictions of Monte Carlo simulations of spin glass behavior, notably by Bhatt and Young (1985) and Ogielski and Morgenstern (1985) (see also Ogielski 1985), which in some cases have used only a Gaussian distribution, with extensive series expansion studies by Singh and Chakravarty (1986, 1987a,b) which employed the $\pm J$ model.

The basic indicators of order are the spin-spin correlation functions, $\langle s_i s_k \rangle$; see (1.6). For a general Ising model these have a graphical expansion in terms of the auxiliary temperature variables

$$v_{ij} = \tanh(J_{ij}/k_{\mathrm{B}}T), \tag{4.2}$$

which may be written

$$\langle s_i s_k \rangle = Z^{-1} \sum_{G^2(i,k)} \prod_{(j,l)} v_{jl} \qquad (i \neq k), \tag{4.3}$$

where the partition function, Z, has a similar expansion, namely,

$$Z = 1 + \sum_{G^0} \prod_{(j,l)} v_{jl}. \tag{4.4}$$

In these expressions the sums run over all distinct weak embeddings of the graphs $G^2(i,k)$ and G^0 in the lattice $\mathcal{L}$, which is most conveniently regarded here as finite with N sites and periodic boundary conditions. The products run over all lattice bonds, (j,l), covered in the embedding of the graph. The graphs G^0 are generalized polygons: they have no repeated lines and at each site of the lattice an even number, 0, 2, 4, ..., of lines must meet. The prescription for the two-rooted graphs, $G^2(i,k)$ is the same except that an *odd* number of lines must meet at the sites i and k. Note that both Z and the numerator for $\langle s_i s_k \rangle$ are *linear* functions of each v_{jl}.

Having obtained the correlation functions, or any other property, for a given realization, $\{J_{ij}\}$, of the couplings of the spin glass, one must perform the average, $[\cdot]_J$, over the coupling distributions. Because of the linearity of Z and $\langle s_i s_k \rangle Z$ in the v_{jl}, each term in the full expansion of $\langle s_i s_k \rangle$ contains an odd power of at least one bond variable, v_{jl}. Consequently we have

$$[\langle s_i s_k \rangle]_J \equiv 0 \quad (\text{all } i \neq k). \tag{4.5}$$

This result will remain true in the thermodynamic limit, $N \to \infty$, in the disordered, high-temperature region above any transition. As a result, the standard susceptibility of a spin glass, as calculated by averaging the expression in (1.7), reduces simply to a constant, explicitly one has $\chi =$

$[\langle s_0^2\rangle]_J = 1$. Evidently this susceptibility is totally independent of the spin-spin interactions and contains no information about any possible spin-glass transition!

The way around this difficulty is to consider, instead, the generalized susceptibilities

$$\chi_{(q,r)}(T) = N^{-1}\sum_i\sum_k [\langle s_i s_k\rangle^q]_J^r \tag{4.6}$$

(Singh and Chakravarty 1987a). Indeed, the case $q = 2$, $r = 1$ corresponds to the so called spin-glass susceptibility, χ_{SG}, introduced originally by Edwards and Anderson (1975). We will focus on this special susceptibility. [Of course, $\chi_{(q,r)}$ vanishes identically whenever q is odd; the next nontrivial case, $q = 2, r = 2$, yields extra information in studies based on the numerical extrapolation of high-temperature power series (Singh and Chakravarty 1987a).]

The graphical expansion for χ_{SG} follows from (4.3) by squaring, dividing through using (4.4), and averaging over the bond distribution term by term. It is instructive to consider, first, a spin glass on a Bethe lattice. This problem is analytically tractable having been first studied by Japanese workers (Oguchi and Ueno 1976; Katsura, Fujiki and Inawashiro 1979: see also references in Chayes et al. 1986). It has been revisited more recently by Thouless (1986) and coworkers (Chayes, Chayes, Sethna and Thouless 1986), particularly to investigate behavior in an external field.

Now there are no closed polygons on a Bethe lattice so, by (4.4), one has $Z = 1$. Likewise, any two sites, i and k, are connected by a single chain of $l(i,k)$ bonds. Thus, after averaging, the only graphical contribution to $[\langle s_i s_k\rangle^2]_J$ comes from a chain, $C^2(i,k)$, of *doubled bonds* reaching from i to k. If we define the moments of the coupling distribution via

$$w_q(T) = [v_{ij}^{2q}]_J \equiv [\tanh^{2q}(J_{ij}/k_B T)]_J, \tag{4.7}$$

we thus have

$$[\langle s_0 s_k\rangle^2]_J^0 = w_1^{l(0,k)}, \tag{4.8}$$

where the superscript zero indicates the Bethe lattice. To use this, we may formally take the thermodynamic limit in (4.6) by dropping the first summation and the factor N^{-1}; this yields the analogue of (1.7). To perform the remaining sum over the sites k, we note that there are just $q\sigma^{l-1}$ distinct self-avoiding paths of l steps leaving the origin, 0, of a Bethe lattice of coordination number q. Summing on l yields the explicit, high-temperature spin-glass susceptibility for a Bethe lattice, namely,

$$\chi_{SG}^0 = [1 + w_1(T)]/[1 - \sigma w_1(T)]. \tag{4.9}$$

Evidently χ_{SG}^0 diverges at a critical point given by

$$w_{1c}^0 \equiv [\tanh^2(J_{ij}/k_B T_c^0)]_J = 1/\sigma. \tag{4.10}$$

This is a natural analogue of the formula (2.1) for a ferromagnetic Bethe lattice. Note, incidentally, that for a $\pm J$ distribution one has $\Delta J = J$ (> 0) and the moments become $w_q = \tanh^{2q}(J/k_{\mathrm{B}}T)$. Thus (4.10) reduces simply to $\tanh(J/k_{\mathrm{B}}T_c) = 1/\sqrt{\sigma}$. Comparing with (2.1) shows that the critical temperature of the spin glass is much lower than of the corresponding ferromagnet, in accord with the obvious effects of having negative antiferromagnetic bonds competing with positive, ferromagnetic couplings.

5. Expansion for a Hypercubic Spin Glass

We may anticipate that (4.10) will provide the leading large-d behavior for spin glasses on hypercubic lattices with $q = 2d$. To show this, we must allow for the lattice polygons. The first point then is that, $c_l(\mathcal{L}_d)$, the total number of self-avoiding paths or chains, C_l, of l steps leaving the origin, is no longer given by $q\sigma^{l-1}$. Rather this large-d form must be multiplied by a correction factor $\lambda(C_l)$ — the *reduced lattice constant* (Gerber and Fisher 1974). It is instructive to reproduce the calculation of this lattice constant in leading nontrivial order. The dominant correction to $c_l \simeq q\sigma^{l-1}$ comes from the closure of a square, P_4, of four bonds (or steps). This may occur at any one of $(l-4+1) = (l-3)$ positions along the chain. As mentioned in Section 2, through each point in the lattice there pass $2d(d-1) = \frac{1}{2}(\sigma+1)(\sigma-1)$ distinct squares. Each such square may be traced by a chain/walk in two possible senses. The remaining $l-4$ bonds of the chain may, in leading order be regarded as 'free' and so are associated with $q\sigma^{l-5}$ configurations. In total, therefore, one must subtract the term $(l-3)(\sigma^2-1)q\sigma^{l-5}$ from $q\sigma^{l-1}$. Finally, to leading order, the desired correction factor is thus

$$\lambda(C_l) = 1 - (l-3)\sigma^{-2} - \cdots . \tag{5.1}$$

A little reflection shows that allowing for hexagons, P_6, yields a correction of order σ^{-4}; however, a correction of lower order, σ^{-3}, arises from subtracting generic configurations, to be denoted $CP_4^{(1)}$, in which the chain or path overlaps one side of the square so yielding a doubled bond: see the graph labelled [c] in Fig. 1. Such configurations were not eliminated in the leading order calculation. This term and higher order ones up to order σ^{-5} were originally calculated by Fisher and Gaunt (1964). The resulting formula for $\lambda(C_l)$ and for other reduced lattice constants needed here have been listed correct to order σ^{-5} by Gerber and Fisher (1974).

To go further in the calculation of χ_{SG}, one must account for polygons which arise directly in the graphical expansion as products of bond factors v_{ij}. To this end let $\{G\}$ denote the sum of all products of the v_{ij} corresponding to the embeddings in $\mathcal{L}$ of all graphs isomorphic to G, as required in (4.3) and (4.4). Then the expansion of the correlation function may be

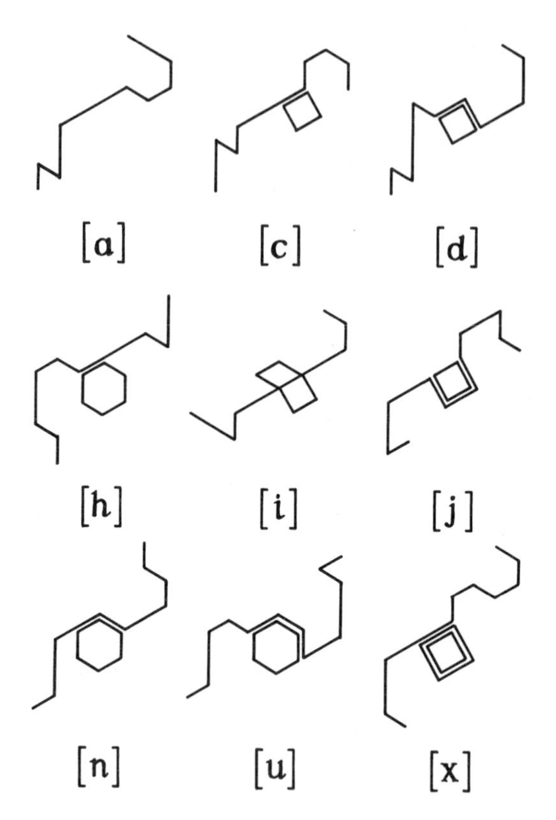

FIG. 1. Generic graphs of l lines which are needed in the calculation of the $1/d$ expansion for a spin glass. Except for $[u]$ and $[x]$, the labelling follows Gerber and Fisher (1974) who give the reduced lattice constants. Note the graphs $[c]$, $[j]$ and $[x]$ have the same skeleton graph of $l' = l-1, l-3$, and $l-5$ lines and thus have simply related lattice constants.

written

$$\langle s_0 s_k \rangle = \sum_l \frac{\{C_l\} + \{C_l, P_4\} + \{C_l, P_6\} + \{C_l, P_4, P_4\} + \cdots}{1 + \{P_4\} + \{P_6\} + \{P_4, P_4\} + \{P_8\} + \cdots}, \tag{5.2}$$

for $k \neq 0$. Here it is understood that each chain of l lines, C_l, is rooted at site 0 and terminates at site k. The polygons of m sides, P_m, may occur singly or, as indicated by a comma, as *disconnected* multiplets, which means, in the present case, that they have *no common bonds* although they may share one or more sites. In dividing out the denominator one obtains products of terms; most, however, cancel to leave only terms with repeated bonds. Thus one has

$$\{C_l, P_m\} - \{C_l\}\{P_m\} = -\sum_{r=1}^{m-1}\{C_lP_m^{(r)}\} - \sum\{C_lP_m^*\}. \tag{5.3}$$

Here, as above, $C_lP_m^{(r)}$ denotes a generic multigraph consisting of a polygon P_m and a chain, C_l, which overlaps it in all possible ways, on r bonds (which are thus doubled); such graphs cannot appear in the numerator of (5.2). The extra terms $\{C_lP_m^*\}$ include contributions from connected graphs in which the chain touches the P_m at an isolated vertex two or more times but does not overlap a bond of P_m; such graphs do arise in the numerator of (5.2) but they appear only once whereas the product in (5.3) generates them more often. Likewise, excess terms having both overlapping bonds and vertex contacts must be subtracted. Leading contributions to the expansion on a hypercubic lattice are then

$$\begin{aligned}\langle s_0s_k\rangle = \sum_l\Big(&\{C_l\} - \{C_lP_4^{(1)}\} - (\{C_lP_4^{(2)}\} + \{C_lP_6^{(1)}\} + \{C_lP_6^{\dagger}\})\\ &- (\{C_lP_4^{(3)}\} + \{C_lP_6^{(2)}\} + \{C_lP_8^{(1)}\} + \cdots)\\ &- (\{C_lP_6^{(3)}\} + \{C_lP_6^{(4)}\} + \{C_lP_6^{(5)}\} + \cdots)\\ &- (-\{C_lP_4^{2(1)}\} + \cdots) - \cdots\Big).\end{aligned} \tag{5.4}$$

The second term, involving $P_4^{(1)}$, generates a contribution of relative order σ^{-3} in the expansion for the critical point of an Ising *ferro*magnet; successive terms in parentheses likewise contribute to terms of order $\sigma^{-4}, \sigma^{-5}, \ldots$. The symbol $C_lP_6^{\dagger}$ in the third term denotes a chain which cuts a hexagon diametrically forming two squares sharing a common bond: see the graph [*i*] in Fig. 1. This will not actually be needed to the order developed here. However, the generic graph $C_lP_4^{2(1)}$, in the last term displayed, will be needed: this is a chain of l single bonds that overlaps a square, P_4, of *doubled* bonds along one side: see graph [x] in Fig. 1.

The first three terms presented in (5.4) actually suffice to generate the expansion (2.2) for a hypercubic Ising ferromagnet correct to order $1/q^4$. For the spin glass, however, it transpires that most of the further graphical terms exhibited are also needed even at order $1/q^3$. To see which matter, we square the expansion for $\langle s_0s_k\rangle$ and perform the spin-glass average, $[\cdot]_J$. Prior to averaging, the expansion will contain multigraphs with bonds of all multiplicities; but, on averaging, any graph containing a bond of odd multiplicity makes a vanishing contribution. On the other hand, each double bond contributes a factor $w_1(T)$, each quadruple bond, a factor $w_2(T)$, and so on. Evidently the square of each term in (5.4) contributes directly so that, for example, $\{C_l P_4^{(1)}\}$ (see [c] in Fig. 1) appears with each bond doubled and weight $w_1^{l+2} w_2$, the factor w_2 arising from the doubling of the original doubled bond. However, cross terms also contribute. Thus the product $\{C_l\}\{C_l P_4^{2(1)}\}$ of the first and last terms displayed in (5.4) appears twice and yields two further terms of the same weight. The product $\{C_l P_4^{(1)}\}\{C_l\}$ yields a contribution with the same skeleton as $C_l P_{\underline{4}}^{(1)}$ but with each bond doubled and hence weight w_1^{l+3}. (In the *skeleton*, $\overline{G}$, of a multigraph G, all multibonds are collapsed to single lines.)

Overall, we obtain the spin-glass susceptibility in the form

$$\chi_{SG}(T;d) \equiv \sum_k [\langle s_0 s_k \rangle^2]_J = 1 + \sum_{l=1} w_1^l b_l(d), \tag{5.5}$$

with coefficients given graphically by

$$b_l = [\overline{C}_l] - 4[\overline{C_{l-3}P_4^{(1)}}] - 2[\overline{C_{l-2}P_4^{(2)}}] - 2\sum_{r=1}^{5}[\overline{C_{l-6+r}P_6^{(r)}}] + 3(w_2/w_1^2)[\overline{C_{l-4}P_4^{(1)}}] + \cdots, \tag{5.6}$$

where the notation $[\overline{G}]$ now denotes the generic lattice constants with one end of the chain rooted at the origin. Note the factor (w_2/w_1^2) which comes from the quadruple bond which arises as explained above. Factors, (w_4/w_1^4), (w_3/w_1^3), etc. appear in higher order terms.

As mentioned, all but one of the required lattice constants in (5.6) have been computed by Gerber and Fisher (1974). It is clear that a given lattice constant depends only on the skeleton graph; however, the precise expressions for the reduced lattice constants, $\lambda(G)$, depend on the total number of lines. With this in mind, the spin-glass susceptibility coefficient may be written, adapting the notation of Gerber and Fisher (1974), as

$$b_l(d) = q\sigma^{l-1}\Big\{\lambda([a]_l) - 4\lambda([c]_{l+1}) - \lambda([d]_{l+2}) - 4\lambda([h]_{l+1}) - 4\lambda([n]_{l+2}) - 4\lambda([u]_{l+3}) - 3(w_2/w_1^2)\lambda([c]_l) + \cdots\Big\}. \tag{5.7}$$

The generic graph $[u]$ corresponds to $C_l P_6^{(3)}$ (see Fig. 1). It was not considered by Gerber and Fisher but its lattice constant is the same, to leading order, as that for $[n]$. Finally, using the data for the $\lambda(G)$ yields

$$b_l(d) = q\sigma^{l-1}\left\{1 - \frac{7l-11}{\sigma^2} - \frac{24l - \delta - 3(l-4)w_2/w_1^2}{\sigma^3} - \cdots\right\}, \tag{5.8}$$

where the integer δ is determined by terms of order l^0 in the lattice constants, which were not retained by Gerber and Fisher (1974); however, the value of δ proves immaterial here.

Now, following the procedure outlined in Section 3, we may finally compute the expansion for $(b_l)^{1/l}$ and take the limit $l \to \infty$ in order to identify the critical value, $(w_{1c})^{-1}$: this yields

$$\frac{1}{w_1(T_c)} = \sigma\left(1 - \frac{7}{\sigma^2} - \frac{24 - 3(w_2/w_1^2)}{\sigma^3} - \cdots\right), \tag{5.9}$$

which is the desired critical point expansion.

In the case of the ferromagnetic models we remarked, in partial justification of the last step, that the known expansion coefficients, a_l, for the susceptibility, χ, are observed to be positive; if true for all l, this means the limit of $(a_l)^{1/l}$ does correctly generate the physical singularity. For the spin-glass susceptibility, χ_{SG}, however, one finds *negative* coefficients, b_l, for $d = 2$ and 3 (Singh and Chakravarty 1986). The known coefficients (with $l \leq 15$) for $d = 4$ are all positive but have a strong alternation and might well become alternating in sign for a larger l. For low d, at least, it thus seems likely that the nearest singularity in the complex $1/T$ plane is *not* the physical singularity. In that case an exact computation of the limit of $[b_l(d)]^{1/l}$ at fixed d would not yield the critical point. We believe, nonetheless, that the procedure we have used will generate the correct asymptotic expansion for the spin-glass critical point.

The factor (w_2/w_1^2) on the right of (5.9) takes the value unity for the $\pm J$ distribution. More generally, however, it must depend on T_c: see equation (4.7). In that case the expansion is really implicit rather than explicit. Furthermore, by examining the higher order terms one sees that factors (w_2/w_1^2) and $(w_2/w_1^2)^2$ appear in order σ^{-4}. The reduced sixth moment of the bond distribution appears first, via a factor (w_3/w_1^3) only in order σ^{-6}. One may, however, generate an explicit expansion for the spin-glass critical temperature for a fixed bond-coupling distribution with reduced moments

$$\rho_q = [(J_{ij})^q]_J/[(J_{ij})^2]_J^{q/2}, \tag{5.10}$$

by expanding $w_1 = [\tanh^2(J_{ij}/k_{\rm B}T)]_J$ in powers of $1/T$, reverting the series and using (5.9). This yields

$$\begin{aligned}\frac{k_B^2 T_c^2}{\Delta J^2} = \sigma\Big[1 - \tfrac{2}{3}\rho_4\frac{1}{\sigma} - \left(7 + \tfrac{4}{9}\rho_4^2 - \tfrac{17}{45}\rho_6\right)\frac{1}{\sigma^2} \\ - \left(24 - 3\rho_4 + \tfrac{62}{315}\rho_8 - \tfrac{34}{45}\rho_4\rho_6 + \tfrac{16}{27}\rho_4^3\right)\frac{1}{\sigma^3} - \cdots\Big].\end{aligned} \tag{5.11}$$

Before discussing the expansion in quantitative terms we present an alternative method of derivation which has enabled us to generate the series for the $\pm J$ models correct to order σ^{-5}.

6. The Inverse Susceptibility Expansion

Singh and Chakravarty (1986, 1987a) discovered that it was possible to expand the free energy and inverse spin-glass susceptibility, $1/\chi_{\text{SG}}$, of an Ising spin glass on a general lattice in a way that only required star graphs (i.e., multiply connected graphs). The significant advantage of such an expansion, which entails a cluster algorithm to generate the appropriate weights, is that for a given number of lines there are far fewer star graphs than the more general graphs required in a direct calculation of χ_{SG}. As a result, for hypercubic lattices Singh and Chakravarty were able to calculate the expansion for the $\pm J$ model to orders w_1^{19}, w_1^{17}, and w_1^{15} for the square, simple cubic, and $(d = 4)$-dimensional hypercubic lattices, respectively. This greatly extended the pioneering work of Fisch and Harris (1977) whose series proved, unfortunately, too short for reliable numerical extrapolation.

Now, as explained in Section 3, the lattice constant for a given star graph is a finite polynomial in d (or q) with, as one easily sees, a vanishing constant term. The most 'open' star graph of l lines is a polygon. Since a polygon of $l = 2p$ lines can explore at most p different spatial dimensions its lattice constant is of order at most d^p (Fisher and Gaunt 1964). Thus each term in the expansion for $1/\chi_{\text{SG}}$ can be written as a polynomial in d. Furthermore, since the star graphs of l lines enter no earlier than in the term of order w_1^l, the expansion to order w_1^{2p} entails no powers of d higher than d^p.

Given this information it is actually possible to obtain the polynomials representing the expansion coefficients of $1/\chi_{\text{SG}}$ for a fixed bond distribution knowing only their numerical values for various dimensionalities. Specifically, from the numerical expansions in dimensions $d = 1$ [for which the Bethe lattice form (4.9) is exact], and $d = 2, 3,$ and 4 (Singh and Chakravarty 1986) one can compute the first nine polynomials. However, using the lattice constants of Fisher and Gaunt, which include all stars of 10 lines, one can go to order w_1^{10} and use the numerical results as a cross check. Writing $w_1 \equiv w$, the result for the $\pm J$ model is found to be

$$\begin{aligned}\chi_{\text{SG}}^{-1}(T) = 1 &- qw + qw^2 - qw^3 + (7q^2 - 13q)w^4 \\ &- (30q^2 - 59q)w^5 + (44q^3 - 169q^2 + 163q)w^6 \\ &- (352q^3 - 1712q^2 + 2017q)w^7 \\ &+ (405q^4 - 3026q^3 + 8503q^2 - 8141q)w^8 \\ &- (3968\tfrac{2}{3}q^4 - 35266\tfrac{2}{3}q^3 + 107011\tfrac{1}{3}q^2 - 104704\tfrac{1}{3}q)w^9 \\ &+ (4712q^5 - 61157q^4 + 336356q^3 - 846314q^2 + 761069q)w^{10} \\ &- \cdots . \end{aligned} \tag{6.1}$$

Now the critical point is determined by the divergence of χ_{SG} or the vanishing of χ_{SG}^{-1}. If we set $qw = y$ in the result (6.1), we can write the critical equation $\chi_{\rm SG}^{-1}(y) = 0$ as

$$y = 1 + y^2 q^{-1} + (7y^4 - y^3)q^{-2} + (44y^6 - 30y^5 - 13y^4)q^{-3} \\ + (405y^8 - \cdots + 59y^5)q^{-4} + (4712y^{10} - \cdots + 163y^6)q^{-5} \\ + O(q^{-6}). \tag{6.2}$$

This is readily solved by reversion which finally yields the expansion

$$\frac{1}{w_1(T_c)} = q\left(1 - \frac{1}{q} - \frac{7}{q^2} - \frac{28}{q^3} - \frac{219}{q^4} - \frac{1905\frac{1}{3}}{q^5} - \cdots\right). \tag{6.3}$$

Of course, this result for the $\pm J$ model can be checked to order $1/q^3$ against the original $1/\sigma$ expansion (5.9). However, the two further terms prove helpful in using the expansion numerically. For convenience we also quote

$$\frac{1}{\tanh(J/k_{\rm B}T_c)} = \sqrt{\sigma}\left(1 - \frac{3\frac{1}{2}}{\sigma^2} - \frac{10\frac{1}{2}}{\sigma^3} - \frac{91\frac{1}{8}}{\sigma^4} - \frac{699\frac{5}{12}}{\sigma^5} - \cdots\right). \tag{6.4}$$

d	$1/\sigma$ series w_c	order	High-T series w_c	Biassed Padé w_c
3	0.5036	4	0.48 ± 0.04	0.48 (± 0.04)
4	0.2133	5	0.21 ± 0.01	0.21 (± 0.01)
5	0.1322	5	0.139 ± 0.002	0.133 ± 0.003
6	0.1002	5	0.102 ± 0.002	0.1005 ± 0.0006
7	0.0818	5	0.083 ± 0.001	0.0819 ± 0.0002
8	0.0696	5	0.070 ± 0.001	0.06964 ± 0.00005

TABLE 1. Estimates of critical temperatures for the $\pm J$ Ising spin glass on a d-dimensional hypercubic lattice: values of $w_c = \tanh^2(J/k_{\rm B}T_c)$ are listed.

7. Spin-Glass Critical Temperatures

Having obtained the expansions (5.9), (6.3) and (6.4) for the critical points, $T_c(d)$, of Ising spin-glass models on hypercubic lattices, let us examine the numerical aspects. Note, first, that the ratios of successive coefficients in (6.3) are increasing rapidly; the pattern is somewhat erratic but suggests an approximately linear increase with order. Thus it seems likely, as in other

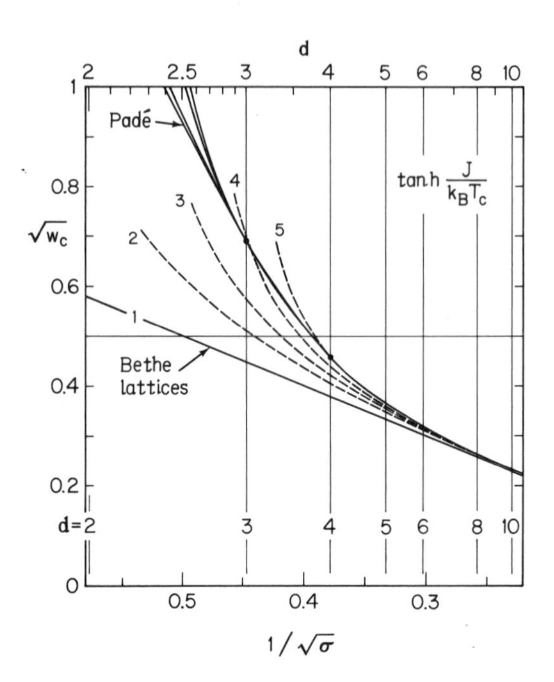

FIG. 2. Plots of $\sqrt{w_c} = \tanh(J/k_\mathrm{B}T_c)$ vs. $1/\sqrt{\sigma}$ with $\sigma = 2d - 1$ for the critical temperature of the $\pm J$ Ising spin-glass model on a d-dimensional hypercubic lattice. The numerals $1, \ldots, 5$ label the order of truncation of the $1/d$ expansion for $1/w_c$. The first-order truncation is exact for Bethe lattices of coordination number $q = \sigma + 1$. The curves marked 'Padé' represent approximants to the $1/\sigma$ expansion, of equation (6.4), biassed to reproduce the favored estimates for $d = 3$ and 4, marked by solid circles. (These estimates result from high-temperature series analysis.)

cases, that the series for $T_c(d)$ is no better than asymptotic as $d \to \infty$. For numerical estimation, truncation of the series close to the smallest term, which is roughly that of order $1/q^{d+1}$, thus seems reasonable. Table 1 shows values of $w_c = \tanh^2(J/k_\mathrm{B}T_c)$ for the $\pm J$ model calculated this way from (6.4) for dimensions $d = 3, 4, \ldots, 8$. Also shown are corresponding estimates for w_c based on the high-temperature series extrapolation analysis of Singh and Chakravarty (1986, 1987). The agreement is rather encouraging.

A graphic portrayal of the large d series is presented in Fig. 2 which plots $\sqrt{w_c} = \tanh(J/k_\mathrm{B}T_c)$ vs. $1/\sqrt{\sigma}$. In such a plot the values for a Bethe lattice of coordination number $q = \sigma + 1$ lie on a straight line. The partial sums to order $1/q^k$ of the series (6.3) are shown; the corresponding sums of the $1/\sigma^k$ series, (6.4), are quantitatively very similar.

In order to improve the numerical performance of the large-d expansion we have accepted the central values of the high-temperature series estimates for $d = 3$ and 4: see Table 1. Then one may generate Padé approximants to the (truncated) series (6.4) *biassed* to ensure that the preferred values are reproduced for $d = 3$ and 4. The four near-diagonal biassed approximants, [2/5], [3/4], [4/3], and [5/2] are displayed: they agree very closely for $d > 4$. Indeed we believe that these approximants are rather reliable for $d \gtrsim 3$: their predictions are listed in the last column of Table 1. The uncertainties assigned there take into consideration the uncertainties in the biassing points at $d = 3$ and 4.

An interesting theoretical point may be addressed using the Padé approximants for $T_c(d)$; this concerns the *lower critical* (or borderline) *dimensionality*, $d_<$, for Ising spin glasses. The lower critical dimensionality is defined for systems with Hamiltonians that belong to a given universality class, in the usual renormalization-group or critical-phenomena sense (see e.g. Fisher 1983), as the dimensionality below which the critical point, $T_c(d)$, vanishes. For Ising-like ferromagnets etc., (with $n = 1$) one has $d_< = 1$; however, for n-vector ferromagnets with $n \geq 2$ one finds $d_< = 2$. As regards Ising spin glasses, the Monte Carlo simulations and the high-T series analyses strongly suggest $2 < d_< < 3$ (Bhatt and Young 1985; Ogielski and Morgenstern 1985; Singh and Chakravarty 1986). If one notes that $w_c = 1$ implies $T_c = 0$, it is evident that the intersections of the plots of the biassed Padé approximants for $T_c(d)$ with the frame in Fig. 2 provide explicit estimates for $d_<$. It is reasonable to conclude

$$d_< \simeq 2.5_0 \pm 0.01_5. \tag{7.1}$$

However, if, as is not implausible, $T_c(d)$ departs from zero like $(d - d_<)^\omega$ with $\omega > 1$, this estimate could prove somewhat too high since such behavior cannot be accounted for in the high-d approximants. Nevertheless, the value $d_< \simeq 2.5$ agrees quite well with various approximate real-space renormalization-group calculations (see, e.g., Bray and Moore 1984).

Finally, it is of interest to gain some feel for the effect of the actual distribution of couplings on the spin-glass critical temperature (Singh and Fisher 1988). By (5.9) we can write the deviation of $T_c(d)$ from that for the $\pm J$ model in the form

$$\frac{w_c - w_c^{(\pm J)}}{w_c^{(\pm J)}} = \frac{3[1 - (w_2/w_1^2)]}{\sigma^3} + O\left(\frac{1}{\sigma^4}\right), \tag{7.2}$$

while to the same order one might replace w_2/w_1^2 by ρ_4: see (5.10). The Gaussian distribution of the J_{ij} is usually taken proportional to

$\exp(-\frac{1}{2}J_{ij}^2/J^2)$. From the value of $w_c \equiv w_1(T_c)$ for $d = 3$ in Table 1 we find $k_\mathrm{B}T_c^{(\pm J)}/J = 1.2 \pm 0.1$; for the Gaussian model, however, this value of $w_1(T_c)$ implies $k_\mathrm{B}T_c^{(G)}/J \simeq 0.79 \pm 0.09$. The correction to this implied by (7.2) increases $k_\mathrm{B}T_c^{(G)}/J$ by about 0.02 . Although small, the change does serve to bring the series-based estimate for $T_c^{(G)}$ somewhat closer to the central estimate of $k_\mathrm{B}T_c^{(G))}/J = 0.9 \pm 0.1$ obtained by Bhatt and Young (1988) in their Monte Carlo simulations. For $d = 4$ the value of w_c in Table 1 yields $k_\mathrm{B}T_c^{(\pm J)}/J = 2.02 \pm 0.06$; the corresponding, uncorrected Gaussian estimate is $k_\mathrm{B}T_c^{(G)}/J \simeq 1.74 \pm 0.07$. The result (7.2) again yields only a small increase, to about 1.76; however, that compares well with the Monte Carlo estimate $k_\mathrm{B}T_c^{(G)}/J \simeq 1.8$ (Bhatt and Young 1988).

In summary, the large-dimensionality expansions for the critical points of Ising spin glasses prove effective and informative. Of course, the remaining theoretical challenge, which seems likely to prove hard, is to provide some better basis for the expansions (5.9) and (6.3) than the heuristic calculations we have expounded here. Mathematical progress along such lines could add significant insight into the statistical mechanics of random systems, a subject in which Hammersley's pioneering contributions remain a striking landmark.

Acknowledgments

It is a pleasure to contribute this article in honor of John M. Hammersley and to take the opportunity to thank him, on behalf of Michael E. Fisher, for past kindnesses, instruction and stimulation. The researches reported here have enjoyed the support of the U.S. National Science Foundation through the Condensed Matter Theory Program of the Division of Materials Research. The interest of Daniel S. Fisher in this work has been appreciated.

References

Abe, R. (1976). A modified $1/d$ expansion for critical temperature of spherical model on hypercubic lattice. *Progress of Theoretical Physics* 56, 494–497.

Balian, R. and Toulouse, G. (1973). Critical exponents for transitions with $n = -1$ components of the order parameter. *Physical Review Letters* 30, 544–546.

Bhatt, R.N. and Young, A.P. (1985). Search for a transition in the three-dimensional $\pm J$ Ising spin glass. *Physical Review Letters* 54, 924–928.

——— (1988). Numerical studies of Ising spin glasses in two, three and four dimensions. *Physical Review B* 37, 5606–5614.

Binder, K. and Young, A.P. (1986). Spin glasses: experimental facts, theoretical concepts, and open questions. *Reviews of Modern Physics* 58, 801–976.

Bowers, R.G. and McKerrell, A. (1973). An exact relation between the classical n-vector model ferromagnet and the self-avoiding walk problem. *Journal of Physics C: Solid State Physics* 6, 2721–2732.

Bray, A.J. and Moore, M.A. (1984). Lower critical dimension of Ising spin glasses: a numerical study. *Journal of Physics C: Solid State Physics* 17, L463–L468.

Broadbent, S.R. and Hammersley, J.M. (1957). Percolation processes I. Crystals and mazes. *Proceedings of the Cambridge Philosophical Society* 53, 629–641.

Chayes, J.T., Chayes, L., Sethna, J.P., and Thouless, D.J. (1986). A mean field spin glass with short-range interactions. *Communications in Mathematical Physics* 106, 41–89.

Domb, C. (1960a). On the theory of cooperative phenomena in crystals. *Advances in Physics* 9, 149–361, Sects. 4.3 and 4.6.

——— (1960b). *Loc. cit.*, Sects. 3.4.1 and 3.6.2.

Edwards, S.F. and Anderson, P.W. (1975). Theory of spin glasses. *Journal of Physics F: Metal Physics* 5, 965–974.

Essam, J.W. and Fisher, M.E. (1970). Some basic definitions in graph theory. *Reviews of Modern Physics* 42, 271–288.

Fisch, R. and Harris, A.B. (1977). Series study of a spin-glass model in continuous dimensionality. *Physical Review Letters* 38, 785–787.

Fisher, D.S., Grinstein, G.M., and Khurana, A. (1988). Theory of random magnets. *Physics Today* 41 (12), 56–67.

Fisher, M.E. (1966). The shape of a self-avoiding walk or polymer chain. *Journal of Chemical Physics* 44, 616–622.

——— (1967). Critical temperatures of anisotropic Ising lattices II. Upper bounds. *Physical Review* 162, 480–485.

——— (1973). Classical n-component spin systems or fields with negative even integral n. *Physical Review Letters* 30, 679–681.

——— (1983). Scaling, universality, and renormalization group theory. In *Critical Phenomena*, ed. F.J.W. Hahne, Lecture Notes in Physics No. 186, 1–139, Springer-Verlag, Berlin.

Fisher, M.E. and Essam, J.W. (1961). Some cluster size and percolation problems. *Journal of Mathematical Physics* 2, 609–619.

Fisher, M.E. and Gaunt, D.S. (1964). Ising model and self-avoiding walks on hypercubical lattices. *Physical Review* 133, A224–A239.

Fisher, M.E. and Sykes, M.F. (1959). Excluded volume problem and the Ising model of ferromagnetism. *Physical Review* 114, 45–58.

Gaunt, D.S. and Ruskin, H. (1978). Bond percolation processes in d dimensions. *Journal of Physics A: Mathematical and General* 11, 1369–1380.

Gaunt, D.S., Sykes, M.F., and Ruskin, H. (1976). Percolation processes in d dimensions. *Journal of Physics A: Mathematical and General* 9, 1899–1911.

Gennes, P.G. de (1972). Exponents for the excluded volume problem as derived by the Wilson method. *Physics Letters A* 38, 339–340.

Gerber, P.R. and Fisher, M.E. (1974). Critical temperatures of classical n-vector models on hypercubic lattices. *Physical Review* 10, 4697–4703.

——— (1975). Critical temperatures of continuous spin models and the free energy of a polymer. *Journal of Chemical Physics* 63, 4941–4946.

Hammersley, J.M. (1957a). Percolation processes II. The connective constant. *Proceedings of the Cambridge Philosophical Society* 53, 642–645.

——— (1957b). Percolation processes: lower bounds for the critical probability. *Annals of Mathematical Statistics* 28, 790–795.

Hammersley, J.M. and Morton, K.W. (1954). Poor man's Monte Carlo. *Journal of the Royal Statistical Society B* 16, 23–38.

Harris, A.B. (1982). Renormalized $(1/\sigma)$ expansion for lattice animals and localization. *Physical Review B* 26, 337–366.

Harris, T.E. (1960). A lower bound for the critical probability in a certain percolation process. *Proceedings of the Cambridge Philosophical Society* 56, 13–20.

Jasnow, D. and Fisher, M.E. (1976). Self-interacting walks, random spin systems, and the zero-component limit. *Physical Review B* 13, 1112–1118.

Joyce, G.S. (1972). Critical properties of the spherical model. In *Phase Transitions and Critical Phenomena*, Vol. 2, ed. C. Domb and M.S. Green, Academic Press, New York, 375–442.

Katsura, S., Fujiki, S., and Inawashiro, S. (1979). Spin-glass phase in the site Ising model. *Journal of Physics C: Solid State Physics* 12, 2839–2846.

Kesten, H. (1980). The critical probability of bond percolation on the square lattice equals $\frac{1}{2}$. *Communications in Mathematical Physics* 74, 41–59.

Ogielski, A.T. (1985). Dynamics of three-dimensional Ising spin-glasses in thermal equilibrium. *Physical Review B* 32, 7384–7398.

Ogielski, A.T. and Morgenstern, I. (1985). Critical behavior of three-dimensional Ising spin-glass model. *Physical Review Letters* 54, 928–932.

Oguchi, T. and Ueno, Y. (1976). Statistical theory of the random ordered phase in quenched bond mixtures. *Journal of the Physical Society of Japan* 41, 1123–1128.

Onsager, L. (1944). Crystal statistics I. A two-dimensional model with an order-disorder transition. *Physical Review* 65, 117–149.

Singh, R.R.P. and Chakravarty, S. (1986). Critical behavior of an Ising spin-glass. *Physical Review Letters* 57, 245–248.

——— (1987a). Critical exponents for Ising spin-glasses through high temperature series analysis. *Journal of Applied Physics* 61, 4095–4096.

——— (1987b). High temperature series expansion for spin glasses. I: Derivation of the series. II: Analysis of the series. *Physical Review B* 36, 546–558 and 559–566.

Singh, R.R.P. and Fisher, M.E. (1988). Short-range Ising spin-glasses in general dimensions. *Journal of Applied Physics* 63, 3994–3996.

Stanley, H.E. (1968a). Dependence of critical properties on dimensionality of spins. *Physical Review Letters* 20, 589–592.

——— (1968b). Spherical model as the limit of infinite spin dimensionality. *Physical Review* 176, 718–722.

——— (1969). Exact solution for a linear chain of isotropically interacting classical spins of arbitrary dimensionality. *Physical Review* 179, 570–577.

Thouless, D.J. (1986). Spin-glass on a Bethe lattice. *Physical Review Letters* 56, 1082–1085.

Wilson, K.G. and Fisher, M.E. (1972). Critical exponents in 3.99 dimensions. *Physical Review Letters* 28, 240–243.

Institute for Physical Science and Technology
The University of Maryland
College Park
Maryland 20742.

R.R.P.S. is now at:
AT&T Bell Laboratories
Murray Hill
New Jersey 07974.

Bistability in Communication Networks

R.J. Gibbens[1], P.J. Hunt[1], and F.P. Kelly

1. Introduction

Advances in the technology of modern telecommunications networks have led to considerable interest in schemes which can dynamically control the routing of calls within a network. The aim of such schemes is to adjust routing patterns within the network in accordance with varying and uncertain offered traffics, to make better use of spare capacity, and to provide extra flexibility and robustness to respond to failures or overloads. However, unless care is taken a dynamic routing strategy that appears to be beneficial may under some circumstances be detrimental.

One of the simplest dynamic routing strategies is Random Alternative Routing, which operates as follows. Every call type that can arrive at the network has a fixed first choice route and a set of possible second choice routes. If possible a call will be carried on its first choice route. If not, then an alternative route is selected at random from the set of possible second choice routes. The call is carried on this route if possible and otherwise it is lost.

At first sight the existence of second choice routes appears beneficial, as it gives each call more ways of being accepted. However if second choice routes require more network resource (hold more circuits or possibly hold circuits for longer) then the network performance, as measured for example by its overall loss probability, may be worse than if a call has access to just its first choice route. Further, Random Alternative Routing (and several other dynamic routing schemes) can lead to instability and hysteresis: several modes of behaviour are possible, the initial conditions of the network determining which is obtained. In this paper we use a combination of analytical, numerical and simulation approaches to investigate these phenomena, in the simplest case of a symmetric fully connected network.

The possibility of bistable behaviour was first noted by Nakagome and Mori (1973), using a simple analytical fixed point approximation for the equilibrium behaviour of a network. Using a development of this approximation Krupp (1982) showed that a network's performance could be improved by using a simple priority technique, known as trunk reservation.

[1] Supported by the SERC.

Akinpelu (1984) and Ackerley (1987) have presented simulation results illustrating hysteresis, and Schwartz (1987) gives a review of this area. In this paper we extend this earlier work by establishing limit theorems for sequences of approximating processes and by obtaining systems of integral equations to describe transient behaviour. We confirm the power of these analytical approaches by comparing numerical solutions to the systems of integral equations with simulated sample paths.

The organisation of this paper is as follows. In Sections 2 and 3 we define formally the network we consider, and obtain a functional law of large numbers for a sequence of approximating processes, using results from Whitt (1985) and Ethier and Kurtz (1986). Fixed point approximations of the form considered by Nakagome and Mori (1973) and, amongst others, Krupp (1982) and Kelly (1986), emerge naturally as fixed points of the integral equations we obtain. Section 4 illustrates the integral equations with a simple example. In Section 5 we deal with systems involving multiple alternative routes and trunk reservation. In Section 6 we introduce a one-dimensional diffusion approximation and use the approximation to elucidate bistable and tunnelling behaviour. Nelson (1986) describes how diffusions can be used to illuminate various types of catastrophic behaviour in performance models of computer systems, and our approximation can be viewed within his framework as an example of a stochastic cusp catastrophe.

Although in this paper we consider just a symmetric fully connected network operating under simple random routing schemes, many of the insights carry over to more general network structures and routing strategies. For example, the insights into trunk reservation obtained from the fixed point approximation of Section 5 were important in the development of Dynamic Alternative Routing (Stacey and Songhurst 1987, Gibbens 1988, Gibbens, Kelly and Key 1988), the dynamic routing strategy currently being implemented by British Telecom in the UK main digital trunk network.

2. A Simple Model

The symmetric fully connected network that we wish to study is as follows. There are a total of N nodes and every pair of nodes is connected by a link of capacity C, giving a total of $K = N(N-1)/2$ links. For all $\alpha \neq \beta$, calls between node α and node β arrive as a Poisson process of rate ν, all arrival streams being independent. If there is free capacity on the direct link between α and β then the call is routed along this path. If not, we try to route the call along two links via a randomly chosen third node $\gamma \neq \alpha, \beta$. If there is free capacity on both these links then the call is routed. Otherwise the call is lost. A call that has been successfully routed holds one circuit from each link on its path for the holding period of the call. The holding period is independent of earlier arrival times and holding periods, and is exponentially distributed with unit mean.

The network described above can be treated as a finite state space Markov process and we can derive equations for the equilibrium distribution. However the full state space is rather complicated, involving the graph structure of the network. It is difficult to analyse the process, even in equilibrium.

So we will consider a simplified model for this network, defined as follows. There are K links, each link comprising C circuits. Calls requesting link k as their first choice arrive as a Poisson process of rate ν. If a call is blocked on its first choice link it tries two other links chosen at random from the $K-1$ remaining links, with each pair of links having equal probability of being chosen. If neither of the links in the chosen pair is full the call is set up along these two links. Otherwise the call is lost. When a circuit is used by a call, the circuit is held for an exponential time, mean 1. All circuit holding times are independent of one another and of earlier arrival times. In particular, a call that requires two links holds each link independently for an exponential length of time, and so these circuits will become free at different times. Thus the simplified model differs in two ways from the original network: circuit holding times are independent, and the graph structure relationship between links has been lost.

The simplified model is an approximation that we would expect to be good for large K. The approximation is much simpler to analyse than the original and can be described by the following Markov process. Let $n_j^K(t)$ be the number of links with j circuits in use at time t, $j = 0, 1, \ldots, C$. Let

$$x_j^K(t) = \frac{n_j^K(t)}{K}, \quad \mathbf{x}^K(t) = (x_j^K(t))_j. \tag{1}$$

So $\sum_{j=0}^C x_j^K(t) = 1$ for all t. For $i \neq j$, $0 \leq i, j \leq C$ let T_{ij} be an operator defined on $\mathbf{x}^K$ given by

$$T_{ij}\mathbf{x}^K = \mathbf{x}^K + K^{-1}(\mathbf{e}_j - \mathbf{e}_i)$$

where $\mathbf{e}_i$ is the unit vector in the ith direction. Then $\mathbf{x}^K$ is a Markov process with transition rates

$$\mathbf{x}^K \to T_{j,j+1}\mathbf{x}^K \qquad \text{at rate} \quad \nu x_j^K K, \qquad j = 0, 1, \ldots, C-1$$

$$\mathbf{x}^K \to T_{j,j-1}\mathbf{x}^K \qquad \text{at rate} \quad j x_j^K K, \qquad j = 1, 2, \ldots, C$$

$$\mathbf{x}^K \to T_{i,i+1}T_{j,j+1}\mathbf{x}^K \qquad \text{at rate} \quad 2\left(\frac{K}{K-1}\right)\nu K x_C^K x_i^K x_j^K, \qquad i > j,\ i, j = 0, 1, \ldots, C-1$$

$$\mathbf{x}^K \to T^2_{j,j+1}\mathbf{x}^K \qquad \text{at rate} \quad \left(\frac{K}{K-1}\right)\nu K x_C^K x_j^K \left(x_j^K - \frac{1}{K}\right),$$
$$j = 0, 1, \ldots, C-1.$$

With this process we can prove results that we also expect to hold for the original fully connected network.

3. Weak Convergence

We now prove a functional law of large numbers for the $\mathbf{x}^K$ process defined in Section 2. Note that $\mathbf{x}^K$ lies in the simplex

$$\Delta = \{\, \mathbf{x}^K \in \mathbb{R}_+^{C+1} : \sum_{i=0}^{C} x_i^K = 1 \,\}.$$

Let $\Rightarrow$ denote convergence in distribution as $K \to \infty$ of random elements in the state space Δ or the space of all sample paths $D_\Delta[0,\infty)$; for background see Billingsley 1968, Lindvall 1973, Whitt 1980, Ethier and Kurtz 1986.

LEMMA 1. *The sequence $\mathbf{x}^K$ is relatively compact in $D_\Delta[0,\infty)$ and the limit of any convergent subsequence has continuous sample paths.*

PROOF: This result follows from a minor modification of Lemma 1 of Whitt (1985, p. 1843). ■

THEOREM 2. *If $\mathbf{x}^K(0) \Rightarrow \mathbf{x}(0)$ then $\mathbf{x}^K(\cdot) \Rightarrow \mathbf{x}(\cdot)$ where $\mathbf{x}(\cdot)$ is the unique solution to the equations*

$$x_0(t) = x_0(0) + \int_0^t \{x_1(u) - (\nu + \lambda(u))x_0(u)\}\, du \tag{2}$$

$$x_j(t) = x_j(0) + \int_0^t \{(\nu + \lambda(u))x_{j-1}(u) - (\nu + \lambda(u) + j)x_j(u) + (j+1)x_{j+1}(u)\}\, du \qquad j \neq 0, C \tag{3}$$

$$x_C(t) = x_C(0) + \int_0^t \{(\nu + \lambda(u))x_{C-1}(u) - Cx_C(u)\}\, du \tag{4}$$

and

$$\lambda(t) = 2\nu x_C(t)(1 - x_C(t)). \tag{5}$$

PROOF: Let

$$v(\mathbf{x}^K(t)) = \lim_{h \downarrow 0} E\left[\frac{\mathbf{x}^K(t+h) - \mathbf{x}^K(t)}{h}\middle| \mathbf{x}^K(t)\right]$$
$$D^K(t) = \int_0^t v(\mathbf{x}^K(u))\, du$$
$$\mathbf{M}^K(t) = \mathbf{x}^K(t) - \mathbf{x}^K(0) - D^K(t).$$

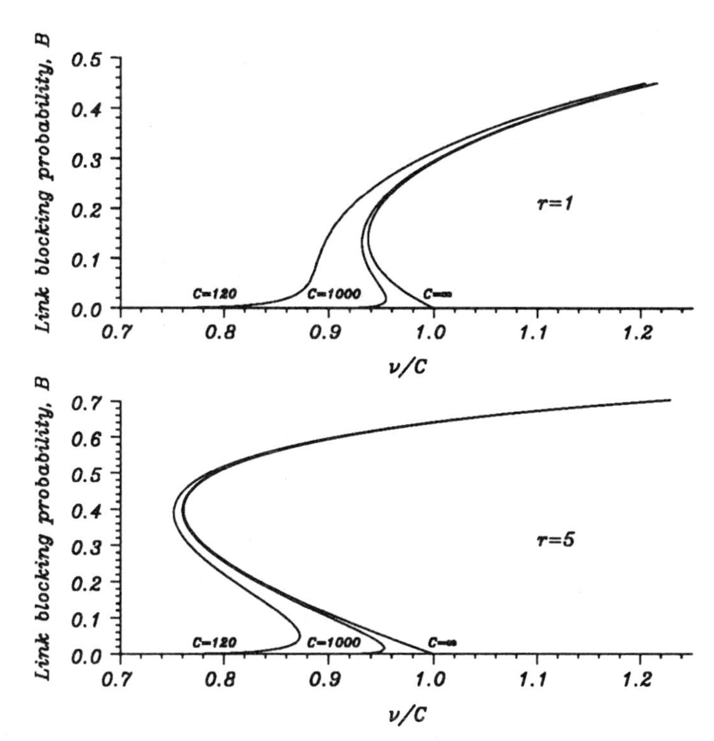

FIG. 1. Instability of blocking probability:
(i) with one retry (ii) with five retries.

Then $\mathbf{M}^K$ is an $\{\mathcal{F}_t^{\mathbf{x}^K}\}$-martingale. It is now easy to check the conditions in Ethier and Kurtz (1986, Theorem 1.4, p. 339). Hence, since $[M_i^K, M_j^K](t) \to 0$ as $K \to \infty$, we have that $\mathbf{M}^K \Rightarrow \mathbf{0}$.

Now along any convergent subsequence of $\{\mathbf{x}^K\}$ we can use the continuous mapping theorem (see, for example, Whitt 1980) to show that $\mathbf{M}^K \Rightarrow \mathbf{M}$ for some $\mathbf{M}$. But by the above result we know that $\mathbf{M} = \mathbf{0}$ and thus we have (2)–(5) satisfied by the limit of a convergent subsequence. But the result now follows since (2)–(5) have a unique solution. (See Arnold 1973, pp. 50, 57.) ■

From equations (2)–(5)

$$\sum_{i=0}^{j} x_i(t) = \sum_{i=0}^{j} x_i(0) + \int_0^t \{(j+1)x_{j+1}(u) - (\nu + \lambda(u))x_j(u)\}\, du$$

$$j = 0, 1, \ldots, C-1$$

Thus $\mathbf{x} = (x_0, x_1, \ldots, x_C) \in \Delta$ is a fixed point of the system of equations

(2)–(5) if and only if

$$(j+1)x_{j+1} = (\nu + \lambda)x_j \qquad j = 0, 1, \ldots, C-1$$

where

$$\lambda = 2\nu x_C(1 - x_C).$$

A fixed point $\mathbf{x}$ is thus of the form

$$x_j = \frac{\xi^j}{j!}\left(\sum_{i=0}^{C} \frac{\xi^i}{i!}\right)^{-1} \qquad j = 0, 1, \ldots, C$$

where ξ solves

$$\xi = \nu + 2\nu E(\xi, C)\,(1 - E(\xi, C))\,. \tag{6}$$

Here

$$E(\xi, C) = \frac{\xi^C}{C!}\left(\sum_{i=0}^{C} \frac{\xi^i}{i!}\right)^{-1} \tag{7}$$

is Erlang's formula for the loss probability of a single link offered Poisson traffic at rate ξ. The equation (6) for ξ is equivalent to the equation

$$B = E\big(\nu + 2\nu B(1 - B), C\big) \tag{8}$$

for B, under the transformation $B = E(\xi, C)$. The parameter B corresponds to the *link blocking probability*, x_C. Equation (8) is usually derived from an approximation that links block independently: see, for example, Kelly 1986. Under such an approximation the probability that a call overflows is B, and the probability it can be accepted at the other link of a two-link path is $1 - B$; the arrival rate of overflowing calls at a link is then $2\nu B(1 - B)$. The locus of points satisfying equation (8) is illustrated in Figure 1(i) for $C = 120, 1000$ and infinity. Observe the possibility of multiple solutions for B, for C large enough and for a narrow range of the ratio ν/C. The upper and lower solutions correspond to stable fixed points for the system of equations (2)–(5), while the middle solution corresponds to an unstable fixed point. We discuss the possibility of multiple fixed points further in Section 6.

4. An Illustration

The integral equations of Theorem 2 apply to the limit process obtained from the simplified model. It is natural to ask how well they model the behaviour of the fully connected network. Figure 2 shows that the model is in fact very good. If $\overline{x} = \sum_{i=0}^{C} i x_i$, then Figure 2 shows the projection of the path given by the integral equation for several initial points. Also

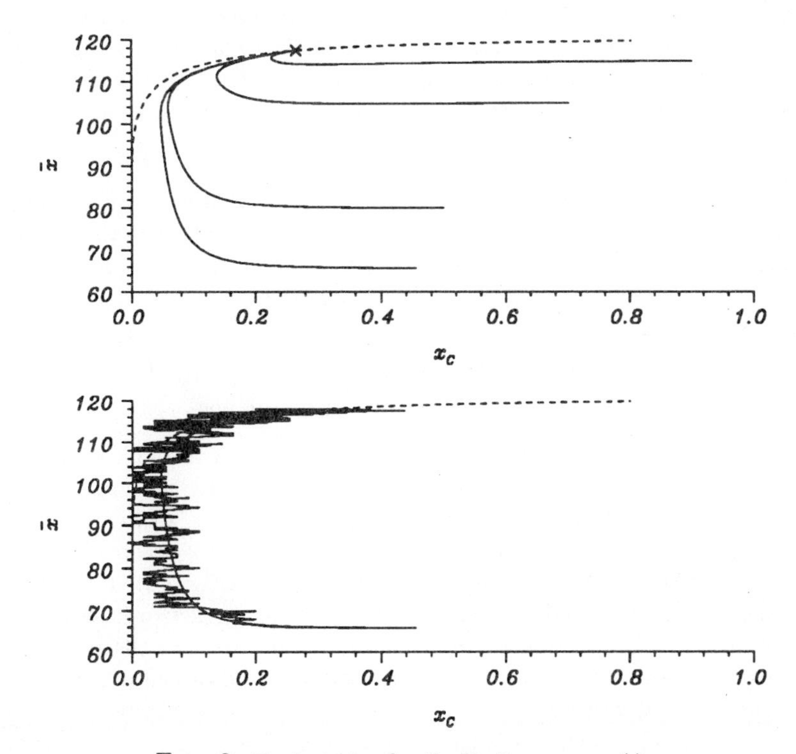

FIG. 2. Trajectories for the limit process $\mathbf{x}(\cdot)$.

shown is the sample path for a fully connected network starting with the same initial configuration as one of the points. The parameters used to obtain these simulation results were $\nu = 115$, $C = 120$ and the number of nodes $N = 11$.

Let

$$\Xi = \left\{ \mathbf{x} : x_i = \frac{\xi^i}{i!} \left(\sum_{j=0}^{C} \frac{\xi^j}{j!} \right)^{-1} ; \xi \in (0, \infty) \right\},$$

a one-dimensional submanifold of the space Δ. The submanifold Ξ is a natural space to consider: if $\lambda(t)$ is held fixed at a value λ then the solution to the integral equations (2)–(4) will move exponentially quickly to the submanifold Ξ, to the point parametrised by $\xi = \nu + \lambda$. The submanifold Ξ is not closed under the integral equations (2)–(5), but notice the way in which trajectories head rapidly towards the projection of Ξ (shown as a dashed curve), and more slowly towards the fixed point. We exploit this observation later in Section 6.

The two dimensions shown in Figure 2 are natural choices since x_C controls the rates of the process and $\overline{x}$ measures the total network utilization. Calculations of the trajectories for various different initial vectors $\mathbf{x}$ with the same projection have given very similar trajectories supporting the belief that this projection is natural and sufficient to summarise the process $\mathbf{x}(t)$. The overall network loss probability is given by

$$L = x_C[1-(1-x_C)^2]$$

and for the example simulated this value is 0.12 . If alternative routing is not allowed, so that a call blocked on its direct link is lost, then the network loss probability is given by Erlang's formula (7) to be 0.05. Observe that allowing a blocked call to attempt a two-link alternative actually *increases* the loss probability of the network.

5. Trunk Reservation and Multiple Alternatives

We have seen that allowing a blocked call to attempt a two-link alternative route may *increase* the loss probability of a network, and we might expect this effect to become even more pronounced if a blocked call can attempt a sequence of alternative routes. Observe that if a link accepts an alternatively routed call it may later have to block a directly routed call which will then attempt to find two circuits elsewhere in the network. A natural response is to allow a link to reject alternatively routed calls if the link occupancy is above a certain level. Suppose then that a call attempting a two-link alternative route is only accepted if on each of the two links the number of circuits occupied is less than $C-s$. This method of giving priority at a link is known as *trunk reservation*, and the constant s is known as the trunk reservation parameter for the link.

The above model for a fully connected network of N nodes is difficult to analyse, and so instead we suppose there are $K = N(N-1)/2$ links, and that a call blocked on its first choice link tries two other links chosen at random from amongst the $K-1$ remaining links. If the number of circuits occupied on each of the two links is less than $C-s$ then the call is routed via that pair of links. If not the call can try another pair of links chosen at random from amongst the $K-3$ remaining links. On each link a trunk reservation parameter of s acts against alternatively routed calls, and a call is lost after it has tried r pairs. As in Section 2 we suppose that all circuit holding times are independent, even the holding times of two circuits used by an alternatively routed call. Let $n_j^K(t)$ be the number of links with j circuits in use at time t, and define $\mathbf{x}^K(t)$ by (1). Then the following result can be established by the methods used to prove Theorem 2.

THEOREM 3. *If* $\mathbf{x}^K(0) \Rightarrow \mathbf{x}(0)$ *in* Δ *then* $\mathbf{x}^K(\cdot) \Rightarrow \mathbf{x}(\cdot)$ *in* $D_\Delta[0,\infty)$ *where* $\mathbf{x}(\cdot)$ *is the unique solution to the equations*

$$x_0(t) = x_0(0) + \int_0^t \{x_1(u) - [\nu + \lambda(u)]x_0(u)\}\, du \tag{9}$$

$$x_j(t) = x_j(0) + \int_0^t \{[\nu + \lambda(u)](x_{j-1}(u) - x_j(u)) + (j+1)x_{j+1}(u) - jx_j(u)\}\, du \qquad (10)$$
$$j = 1, 2, \ldots, C-s-1$$

$$x_{C-s}(t) = x_{C-s}(0) + \int_0^t \{[\nu + \lambda(u)](x_{C-s-1}(u) - x_{C-s}(u)) + (C-s-1)x_{C-s-1}(u) - (C-s)x_{C-s}(u)\}\, du \qquad (11)$$

$$x_j(t) = x_j(0) + \int_0^t \{\nu(x_{j-1}(u) - x_j(u)) + (j+1)x_{j+1}(u) - jx_j(u)\}\, du \qquad (12)$$
$$j = C-s+1, \ldots, C-1$$

$$x_C(t) = x_C(0) + \int_0^t \{\nu x_{C-1}(u) - Cx_C(u)\}\, du \qquad (13)$$

and

$$\lambda(t) = 2\nu x_C(t)\left(\sum_{m=0}^{C-s-1} x_m(t)\right)^{-1}\left\{1 - \left[1 - \left(\sum_{m=0}^{C-s-1} x_m(t)\right)^2\right]^r\right\}. \qquad (14)$$

A fixed point $\mathbf{x} = (x_0, x_1, \ldots, x_C) \in \Delta$ of the system of equations (9)–(14) satisfies

$$(j+1)x_{j+1} = (\nu + \lambda)x_j \qquad j = 0, 1, \ldots, C-s-1 \qquad (15)$$
$$(j+1)x_{j+1} = \nu x_j \qquad j = C-s, \ldots, C-1 \qquad (16)$$

where

$$\lambda = 2\nu B_1(1-B_2)^{-1}\left\{1 - \left[1 - (1-B_2)^2\right]^r\right\} \qquad (17)$$

$$B_1 = x_C, \quad B_2 = \sum_{i=C-s}^{C} x_i. \qquad (18)$$

The network loss probability corresponding to a solution to the fixed point equations (15)–(18) is

$$L = B_1\left[1 - (1-B_2)^2\right]^r.$$

We can interpret this form as follows: a call is lost if it is blocked on its first choice route, which happens with probability B_1, and if it is then blocked on each of r alternatives. It is blocked on an alternative route with

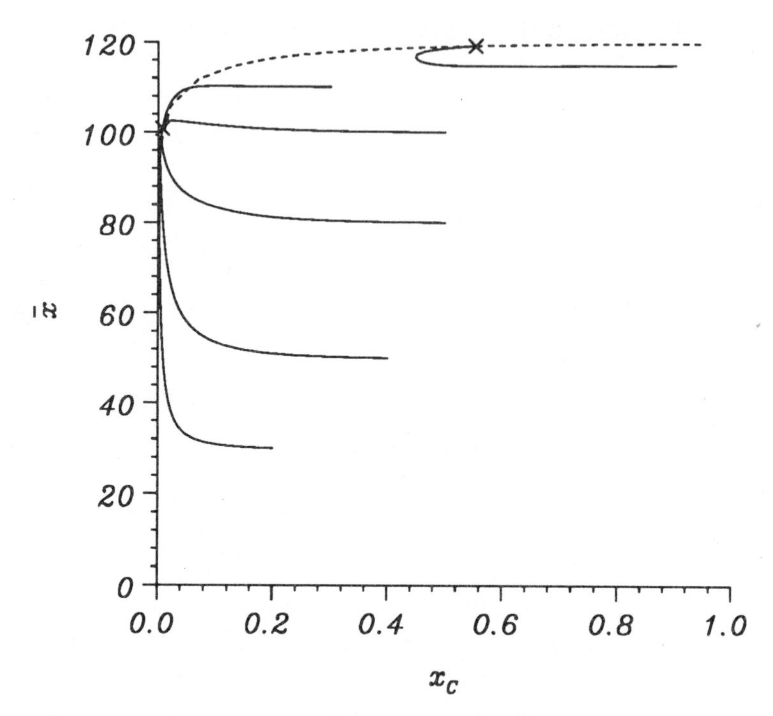

FIG. 3. Trajectories for a network with two stable fixed points.

probability $1-(1-B_2)^2$, where B_2 is the probability a link is occupied above its trunk reservation parameter.

We illustrate the integral equations of Theorem 3 in Figure 3 for a network in which $N = 11, C = 120, \nu = 100, r = 5$ and $s = 0$. For this network $B_1 = B_2 = B$, say, and Figure 1(ii) shows the locus of points satisfying relations (15)–(18). The chosen values of ν and C lead to two stable fixed points; these correspond to the points marked by crosses in Figure 3.

6. A One-Dimensional Approximation

In Section 3 we developed a functional law of large numbers for the simplified model, and in Section 4 we saw that the resulting integral equations can provide a reasonable approximation for the exact network. In Section 3 we observed that there may be multiple fixed points for the integral equations. Of course a finite network corresponds to an irreducible Markov process, with a unique equilibrium distribution. However the unique distribution may be multi-modal, and the time taken for the process to move from one mode to another may be long. How can we investigate this analytically?

One approach would be to use a diffusion approximation to the simplified Markov process to obtain results about the process of interest. However we are dealing with a C-dimensional process where C may be quite large, and it is difficult to obtain useful analytical results and computationally expensive to obtain numerical results. So instead we will reduce the process to one important dimension, as follows.

Consider a fully connected network with r retries and no trunk reservation. Let $n(t) \in \{0, 1, \ldots, CK\}$. We will use $n(t)$ to represent the number of circuits in use at time t. Let $n(\cdot)$ have transition rates

$$\begin{array}{lll} n \to n+1 & \text{at rate} & \nu K(1 - B_n) \\ n \to n+2 & \text{at rate} & \nu K B_n \left\{1 - \left(1 - (1 - B_n)^2\right)^r\right\} \\ n \to n-1 & \text{at rate} & n \end{array}$$

where B_n solves

$$B_n = E(\rho, C), \quad n = \rho(1 - B_n).$$

Thus $n(\cdot)$ has the same transition rates as the total number of circuits in use in the earlier model of Section 2 provided that the number of links full in the earlier model is KB_n. The process $n(\cdot)$ should thus approximate the earlier model close to the one-dimensional submanifold Ξ. (We note that a more refined model would take into account that two circuits may sometimes be freed simultaneously.) Next approximate $n(\cdot)/CK$ by a diffusion $Z(\cdot)$ on the interval $[0,1]$ with drift $\mu(z)$ and infinitesimal variance $\sigma^2(z)$ given by

$$\mu(z) = \frac{\nu}{C}\left[(1 - B_z) + 2B_z\left\{1 - (1 - (1 - B_z)^2)^r\right\}\right] - z$$

$$\sigma^2(z) = \frac{\nu}{C^2 K}\left[(1 - B_z) + 4B_z\left\{1 - (1 - (1 - B_z)^2)^r\right\}\right] + \frac{z}{CK}$$

where

$$zC = \rho(1 - E(\rho, C))$$

and with reflecting barriers at 0 and 1. Thus $\mu(z)$ and $\sigma^2(z)$ are the natural extensions of the drift and infinitesimal variance of the discrete process $n(\cdot)/CK$ to $[0,1]$. We can now use the powerful results for one-dimensional diffusions to gain insight into the network's behaviour.

As an example consider the equilibrium density $\psi(z)$ for the diffusion $Z(\cdot)$. This is given by

$$\psi(z) = A\,\frac{\exp\left(\int_0^z 2\mu(y)/\sigma^2(y)\,dy\right)}{\sigma^2(z)}$$

for some constant A (cf. Karlin and Taylor 1981, Kent 1978, Nelson 1986). Figure 4(i)–(iv) shows this equilibrium density for a network with parameters $N = 11$, $C = 120$, $r = 5$ and $s = 0$ as ν varies. Observe that the high

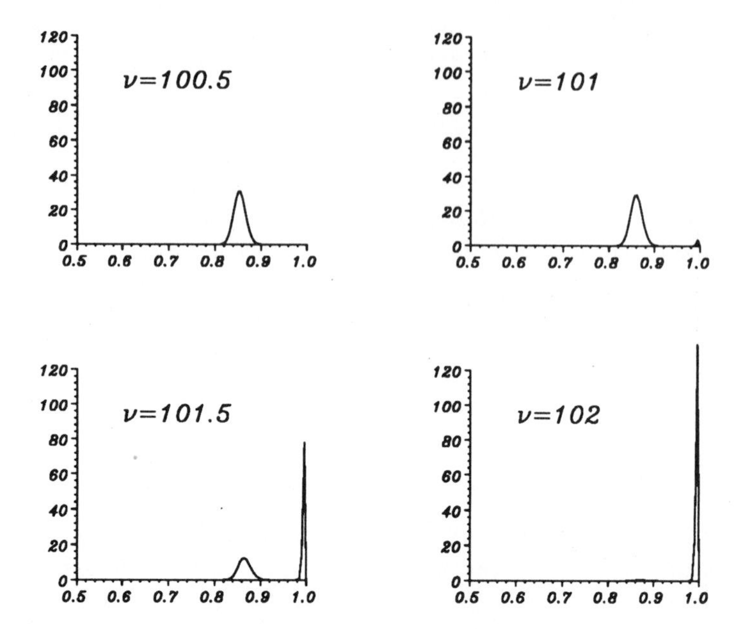

FIG. 4. Equilibrium density for the diffusion $Z(\cdot)$.

blocking state is a lot less stable than the low blocking state for smaller values of ν but becomes more stable as ν increases until finally there is only one stable point. In the region of ν/C for which there are two stable fixed points, illustrated in Figure 1(ii), we expect to see tunnelling. Figure 5 illustrates a sample path for the same network with $\nu = 100.5$ where this tunnelling has occurred. We see that the sample path heads rapidly to the dashed curve; it then moves towards the upper fixed point, about which it wanders for a period before tunnelling to the region of the lower fixed point.

A natural question that arises is how stable are the two fixed points. That is, how long do we expect to wait until we tunnel from one to the other? Again we can use the one-dimensional approximation.

Let $T(x;y)$ be the first time that the diffusion hits y given that it starts at x. Let $f(x;y) = E\,[T(x;y)]$. Then (Karlin and Taylor 1981, p. 193) $f(x;y)$ satisfies

$$\frac{1}{2}\sigma^2(x)\frac{\partial^2 f}{\partial x^2} + \mu(x)\frac{\partial f}{\partial x} = -1$$

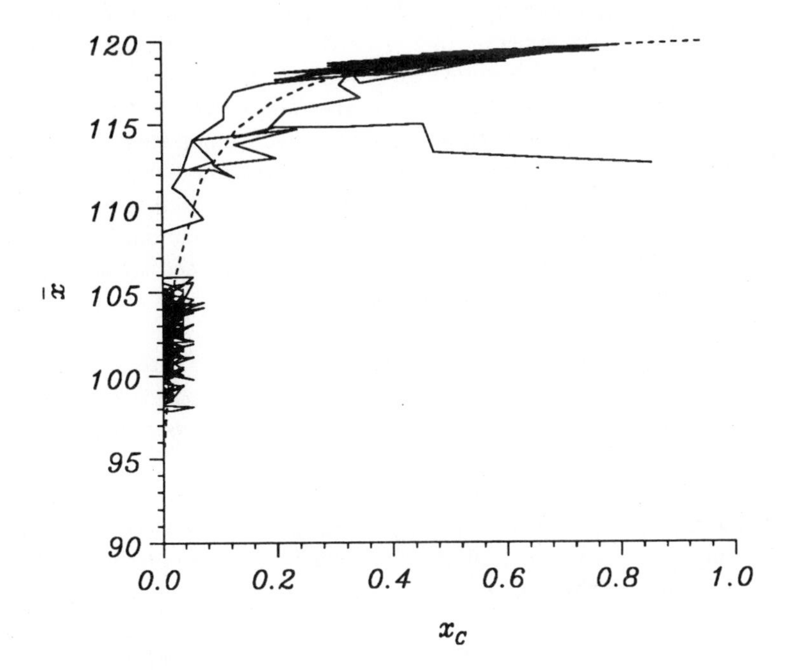

FIG. 5. Tunnelling between stable fixed points.

with boundary conditions

$$\begin{aligned} f(y;y) &= 0 \\ \frac{\partial f(0;y)}{\partial x} &= 0 \qquad x > y \\ \frac{\partial f(1;y)}{\partial x} &= 0 \qquad y > x. \end{aligned}$$

So if $x_1 < x_2 < x_3$ are the three fixed points then we can assess stability from $f(x_1;x_2)$ and $f(x_3;x_2)$.

Remark. For $r = 1$ we find that for some A_1, A_2, A_3, A_4

$$\frac{e^{A_1CK}}{CK} \leq f(x_1;x_2) \leq \frac{e^{A_2CK}}{CK} \tag{19}$$

$$\frac{e^{A_3K}}{CK} \leq f(x_3;x_2) \leq \frac{e^{A_4K}}{CK} \tag{20}$$

as $C, K \to \infty$.

Equations (19) and (20) show that the low blocking state becomes more stable very rapidly as C and K increase. However the high blocking state becomes stable rapidly with K but more *unstable* as C increases. This is as one would expect. In the high blocking state the number of free circuits is $O(1)$ as C becomes large and in the low blocking state it is $O(C-\nu)$. To tunnel from high to low the number of free circuits needs to be unusually large for a time, an $O(1)$ effect, since then more single link calls are routed and the network falls into the low blocking state. To tunnel from low to high blocking the number of free circuits must change by $C-\nu$, an $O(C)$ effect. This accounts for the exponential terms in the expressions. The $1/C$ part comes from the fact that the transition rates increase linearly in C and hence the time taken between events behaves like $1/C$.

The corresponding model with trunk reservation is a two-dimensional process (x_h, x_l), where x_h and x_l are the amounts of, respectively, high and low priority traffic. We do not develop this here: trunk reservation removes the bistability that has been a focus of this paper.

Note Added in Proof: Marbukh (1983), starting from an independent blocking assumption, has derived differential equations corresponding to the integral equations of this paper.

References

Ackerley, R.G. (1987). Hysteresis-type behaviour in networks with extensive overflow. *British Telecom Technology Journal* 5.

Akinpelu, J.M. (1984). The overload performance of engineered networks with non-hierarchical routing. *Bell System Technical Journal*, September, 1261–1281.

Arnold, V.I. (1973). *Ordinary Differential Equations.* The Colonial Press Inc., MIT.

Billingsley, P. (1968). *Convergence of Probability Measures.* John Wiley and Sons, New York.

Ethier, S.N and Kurtz, T.G. (1986). *Markov Processes: Characterization and Convergence.* John Wiley and Sons, New York.

Gibbens, R.J. (1988). *Dynamic Routing In Circuit-Switched Networks: The Dynamic Alternative Routing Strategy.* Ph.D. Thesis, University of Cambridge.

Gibbens, R.J., Kelly, F.P., and Key, P.B. (1988). Dynamic alternative routing — modelling and behaviour. *Twelfth International Teletraffic Congress*, Paper 3.4A.3.

Karlin, S. and Taylor, H.M. (1981). *A Second Course in Stochastic Processes.* Academic Press, London.

Kelly, F.P. (1986). Blocking probabilities in large circuit-switched networks. *Advances in Applied Probability* 18, 473–505.

Kent, J. (1978). Time-reversible diffusions. *Advances in Applied Probability* 10, 819–835.

Krupp, R.S. (1982). Stabilization of alternate routing networks. *IEEE International Communications Conference*, Paper 31.2.

Lindvall, T. (1973). Weak convergence of probability measures and random functions in the function space $D[0, \infty)$. *Journal of Applied Probability* 1, 109–121.

Marbukh, V.V. (1983). Investigation of a fully connected channel switching network with many nodes and alternative routes. *Avtomatika i Telemekhanika* (in Russian) 12, 86–94.

Nakagome, Y. and Mori, H. (1973). Flexible routing in the global communication network. *Seventh International Teletraffic Congress*, Stockholm.

Nelson, R. (1986). How to control those unruly tape measures; or, computer performance modeling using stochastic catastrophe theory. *Mathematical Intelligencer* 8, 50–56.

Schwartz, M. (1987). *Telecommunication Networks.* Addison-Wesley, Reading, Massachusetts.

Stacey, R.R. and Songhurst, D.J. (1987). Dynamic alternative routing in the British Telecom trunk network. *International Switching Symposium*, Phoenix, Arizona.

Whitt, W. (1980). Some useful functions for functional limit theorems. *Mathematics of Operations Research* 5, 67–85.

——— (1985). Blocking when service is required from several facilities simultaneously. *AT&T Technical Journal* 64, 1807–1856 .

Statistical Laboratory
University of Cambridge
16 Mill Lane
Cambridge CB2 1SB.

A Quantal Hypothesis for Hadrons and the Judging of Physical Numerology

I.J. Good

Introduction

Readers who know little or no physics should not be deterred by the many mentions of physics in this chapter, for most of the required physics is spelt out in Appendix A. The chapter has two parts. Part 1 is an updating of Good (1988b) in which a quantal hypothesis is discussed concerning the rest masses of 'elementary particles'. The discussion depends, perhaps necessarily, on subjective probabilities. The hypothesis is largely *numerological* in the non-occult sense to be described more fully in Part 2. That part deals with the difficult topic of judging more general numerological assertions. Both parts might shed some light on how we choose between scientific theories in general. The chapter is somewhat speculative and this is appropriate in a *Festschrift* for John Hammersley. Moreover some of his early work (Hammersley 1950, 1954) dealt in part with a quantal hypothesis. Quantal hypotheses have occurred in physics, chemistry, genetics, and archaeology.

Warning. This chapter contains subjectively oriented material.

Part 1. Numerology for the Masses of Hadrons

A new edition of the Review of Particle Properties was issued in mid-February, 1989: see Particle Data Group (1988/89), abbreviated here as PDG89. Using this edition, and Cohen and Taylor (1987), together with information from Cohen (1989), who also drew my attention to Kinoshita (1989), I have recomputed some of the numbers, based on PDG86, published in Good (1988b); see also Good (1989c). In the present account, which is self-contained, I report the revised implications. The concept of the *relativistic fine structure constant* might have independent interest.

The first formula was

$$R(p,n) = \frac{m(n) - m(p)}{m(p)} \approx \frac{136\alpha}{720} = \left[\binom{4}{2}^{-1} + \binom{10}{2}^{-1}\right]\alpha \qquad (1.1)$$

where $R(p,n)$ may be described as a 'proportional bulge', $m(n)$ and $m(p)$ denote the rest masses of the neutron and proton, and α denotes the fine structure constant (Sommerfeld 1916, p. 91; PDG89, p. 51; Cohen and Taylor 1987, p. 1139; updated by Kinoshita 1989),

$$\alpha = e^2/(\hbar c) = 1/137.0359914(1 \pm 8.1 \times 10^{-9}).$$

(For the sake of a simple and familiar formula for α, I have assumed electrostatic units, in which the permittivity of empty space is unity, but α has the same numerical value, whatever units are used, because it is a dimensionless constant.) Here e denotes the charge on the electron, c denotes the velocity of light, and $\hbar = h/(2\pi)$ where h denotes Planck's constant. (The expressions h and $\hbar$ are equally simple because energy $= h \times$ frequency $= \hbar \times$ angular frequency, while frequency and angular frequency are equally natural concepts. In other words, angles could be measured in *circumfians* instead of radians.) The constant α is generally regarded as measuring the strength of the electromagnetic forces. Kinoshita (1989) argues that α 'may be regarded as the most fundamental parameter of the physical world'.

A very similar conjecture is

$$R(p,n) = \frac{m(n) - m(p)}{m(p)} \approx \frac{136\alpha'}{720} \qquad (1.1')$$

where

$$\alpha' = \tanh^{-1}(\alpha) = 1/137.0335589(1 \pm 8.1 \times 10^{-9})$$

which may be regarded as the *relativistic fine-structure constant.* The idea of this minor adjustment to α is that whereas (i) $\alpha = v/c$ where v is the velocity of the electron in the lowest Bohr orbit (for example, Allen 1928, p. 56, or Whittaker 1953, p. 120), and (ii) if we replace the ratio v/c by its *rapidity* in the sense of A.A. Robb (Eddington 1930, p. 22), namely $\tanh^{-1}(v/c)$, then α is replaced by α'. Unlike two velocities, in the same direction, rapidities are strictly additive (in the Special Theory of Relativity). It does not seem too ad hoc to regard α' as a fundamental constant of nature, perhaps as fundamental as α athough the familiar formula for α, mentioned above, is simpler than that for α'.

The expression in (1.1) containing binomial coefficients ('triangulations') is included partly because 4 and 10 are two of the prominent integers in Eddington's *Fundamental Theory*, namely 4, 6, 10, 16, 120, 136, and 256. Moreover 4 is the number of dimensions of ordinary space-time whereas 10 is the number of dimensions in the currently most fashionable form of string theory when the six coiled up dimensions are included (see, for example, Schwarz 1988, p. 72). But the binomial expression in (1.1) will not be mentioned again in this chapter. Observe that the Eddingtonian integers are closely related to one another; for example, 6 and 10 are the lower and upper triangulations of 4, while 120 and 136 bear the same relationships to 16 and the reader will see other even more obvious relationships. It might not be necessary to rely on Eddington's judgement because, for example, a 16-dimensional torus occurs prominently in 'heterotic' string theory (Gross et al. 1985, p. 260). The 16 dimensions are described as 'internal'.

I am going to argue that (1.1) or (1.1′), and some allied assertions, are very probably 'correct'. Of course a convincing physical explanation would be much better. The meaning of *correctness* will be discussed in Part 2.

When trying to estimate the prior probability of (1.1) or (1.1′) it is appropriate to take a little physics into account; because the formulae are not *purely* numerological. The numerator $m(n) - m(p)$ on the left depends only on electromagnetic forces: see, for example, Rowlatt (1966 p. viii). It is therefore natural to have α or α' in the numerator on the right. Since $\alpha^{-1} = 137.0359914(8)$, and since 136 is so prominent in Eddington (1946), and is the closest nice integer to α^{-1} (see the ranking in Part 2), it is natural to introduce the number $\beta' = 1/(136\alpha)$ in preference to Eddington's $\beta = 137/136$ (Bond's factor).

Eddington even had names for 136 and 120. He called 136 'the basal multiplicity' and called 120 'the number of dormant components in the extended energy tensor'. In his theory (Eddington 1946, p. 30) 136 is expressed as $10^2 + 6^2$, and 120 occurs as $2 \times 6 \times 10$.

According to Slater (1957 p. 5), 136 is the number of mechanical degrees of freedom of the hydrogen atom and presumably of any similar two-particle system. We can also think of 136 and 120 as the numbers of real and imaginary components of a 16 by 16 Hermitian matrix or equivalently as the upper and lower triangulations of 16. These ways of expressing 136 and 120 come to much the same thing as Eddington's expressions, from a numerical point of view, because, for all n,

$$\binom{n}{2}^2 + \binom{n+1}{2}^2 = \binom{n^2+1}{2} \quad \text{and} \quad 2\binom{n}{2}\binom{n+1}{2} = \binom{n^2}{2}. \qquad (1.2)$$

See also Eddington (1946, pp. 30 and 111) to obtain a further impression of why he liked the number 136. Of course $120 = 5!$ but I don't think Eddington was concerned with this fact.

If a formula contains both 120 and 136 it should not on that account be given independent complexity scores (or independent probabilities), one for 120 and one for 136. (See equation (2.7) below.) There should be an 'interaction term' subtracted from the total complexity (or divided into the product of the probabilities) to allow for the close relationship of these two numbers.

If there is any sense whatever in Eddington (1946), then the number 136 is very probably 'fundamental' (prominent in a good theory: see Part 2), more so than is suggested by its good 'ranking' in Table 2 in Part 2. I think the prior probability that it is physically fundamental is not more than $\frac{1}{4}$, because the testable predictions of Eddington's *Fundamental Theory* have been refuted, but I think the probability is at least $\frac{1}{10}$ because his intuition had been outstanding in other problems of physics, and also because of the current interest in the 16-dimensional torus mentioned above. (My subjective probabilities are my estimates of logical probabilities. There are scientists and statisticians who believe they do not use subjective probabilities. We can ask them for their subjective probabilities of these beliefs.) If 136 deserves to be called fundamental then 136α or $136\alpha'$ very probably deserves to be regarded as a fundamental physical constant because α occurs in Sommerfeld's theory of the hydrogen atom and so does 136 in Eddington's theory.

Again, $[m(n) - m(p)]/m(p)$ (which is of course dimensionless) seems like a reasonable measure of the ratio of the electromagnetic forces to the strong forces although *a priori* a denominator of $m(n)$ or $\frac{1}{2}[m(p) + m(n)]$ would be about as good as $m(p)$. We should therefore pay a factor of 3 (or a little less because the proton is the 'ground state') for 'special selection' of $m(p)$. Conditional on 136α or $136\alpha'$ making sense, we need to decide how impressed we should be by the denominator 720 on the right of (1.1). I think it is the simplest integer in the range of say $[600, 800]$, 625 and 729 being 'runners up'. In accordance with the comment (viii) to Table 2 (in Part 2), I assume that the first stage of information is that the denominator lies in this interval. This 'forces' about $\log_{10}(700/100) = 0.8$ correct significant digits. But the following Bayesian argument makes no use of the number of correct significant digits so it is fair to count 0.8 neither as a penalty nor as a reward.

There are only nine integers other than 720 in the range $[600, 800]$ that are of the simple form $2^a 3^b 5^c$, namely 600, 625, 640, 648, 675, 729, 750, 768, and 800. Moreover $720 = 6!$, it is a 'highly composite number' in the sense of Ramanujan (1915) (that is, it has more factors than any smaller number), and is also the product, 6×120, of two Eddingtonian integers, and one of the two is the 'twin' of 136. Of course 6! is the order of the symmetric group of degree six and the theory of finite groups is already basic to the theory of elementary particles. (See Appendix E for a distinctive property of this group.) So a physical explanation of the number 720 might depend on a theory entirely different from Eddington's. Perhaps it is relevant that

480 bosons occur in the heterotic string theory (Gross et al. 1985, p. 265) and we can think of 720 as sesqui-480. Moreover, 720 occurs prominently in Green et al. (1985, pp. 339, 340, 344) and in Candelas et al. (1985, p. 1123).

We need to judge the prior probability (say between $\frac{1}{20}$ and $\frac{1}{10}$) that there is an unknown *reason* why the denominator on the right of (1.1) or (1.1′) is an integer (or very close) and the prior probability (say between $\frac{1}{20}$ and $\frac{1}{5}$) that it is 720 given that it is an integer between 600 and 800. (Readers should make their own judgements.) With my judgements, the prior probability that (1.1) or (1.1′) is 'correct' lies between $1/(3 \times 10 \times 20 \times 20)$ and $1/(3 \times 4 \times 10 \times 5)$, that is, between 1/12000 and 1/600. Estimates should be made, together with some informal reasoning, by several particle physicists, but I have given my estimates to indicate a subjectivistic Bayesian way of thinking about the problem. For this application the approach in Part 2 pays too little attention to the physical background. Perhaps a reader can suggest another approach. Of course, as I said before, it would be better to find a convincing explanation (which by definition must be lucid) instead of just a probability estimate.

The conjectures or hypotheses that (1.1) or (1.1′) is exact, or at least appreciably more accurate than the experiments have proved, will be called H_0 or H_0' respectively. A different but related hypothesis, say H_0'', is that neither is exact but that there is an unknown physical reason why they are very good approximations. Such 'smudging of the null hypothesis' occurs in science more often than not because absolutely precise null hypotheses are rare. To save words, scientists and statisticians often omit explicit mention of this smudging and I shall follow this fashion and usually leave it to the reader to hold H_0'' in mind.

The values, based on PDG89, Cohen and Taylor (1987 p. 1142), and Kinoshita (1989), are

$$\frac{136\alpha}{720} = 0.00137838890 \pm 1 \times 10^{-11}, \tag{1.3}$$

$$\frac{136\alpha'}{720} = 0.00137841336 \pm 1 \times 10^{-11}, \tag{1.3$'$}$$

and

$$R(p, n) = 0.001378404 \pm 9 \times 10^{-9}, \tag{1.4}$$

but Cohen (1989) updates (1.4) by

$$R(p, n) = 0.001378416 \pm 6 \times 10^{-9}, \tag{1.5}$$

although he believes the uncertainty might well be as large as 8×10^{-9} or as small as 3×10^{-9}. To exaggerate the accuracy of H_0' one could say that the numerological estimate of $m(n)/m(p)$ is 1.0013784134 as compared with the current best experimental value 1.001378416(6). But it is fairer to subtract 1 from both sides when considering the proportional accuracy. (See Appendix C.)

We may infer that

$$\frac{136\alpha}{720R(p,n)} = 0.9999806 \pm \sigma \tag{1.6}$$

and

$$\frac{136\alpha'}{720R(p,n)} = 0.9999981 \pm \sigma, \tag{1.6'}$$

where the 'best' estimate of σ is 0.0000044. Thus the fate of the *exactness* of conjecture H_0 depends critically on whether the value of σ is appreciably larger than its nominal value. On the other hand H_0' would be a good fit, even if the uncertainty in (1.5) were, for example, only 3×10^{-9}. If this uncertainty has its nominal value of 6×10^{-9}, so that $\sigma = 0.0000044$ in (1.6′), then H_0' is correct to one part in at least 160,000 (in accordance with the natural formula (2.5) of Part 2 which allows both for discrepancies and uncertainties), or one part in at least about 10^8 if we 'add 1 to exaggerate'. Another way to present the argument is to start with the simple observation that

$$R(p,n)/\alpha' = 0.1888892 \pm 0.0000008, \tag{1.6''}$$

and any schoolgirl would conjecture that $0.1\dot{8} = 17/90$ is exact. But an objective test for 'closeness to rationality' (Good 1969, p. 38, with N there taken as 90 or more) leads to an unimpressive P-value of between 0.078 and 0.098. To be impressed we must write $0.1\dot{8}$ in a more interesting way, for example, as 136/6!. We might want to judge too what fraction of rational numbers, in their lowest terms and with denominators 'subceeding' say 100, and not too distant from $R(p,n)/\alpha'$, can be written in at least as interesting a manner.

Now let X and Y be any pair of hadrons differing in having a u quark in X where there is a d quark in Y or vice versa, and let Y be the heavier. (The 'vice versa' applies only to the pair Λ, Σ^+.) Consider the experimental values of $720R(X,Y)/(136\alpha')$ shown in Table 1, calculated from the data in PDG89 combined with the latest estimate of $m(n)/m(p)$ (see equation (1.5)). As in Good (1989c) the Bayes factors (defined in Appendix B) listed in the last column refer to the hypotheses that each of the ratios is an integer. The method of calculating the factors is described in Appendix D. The product of these factors is the overall Bayes factor in favour of the hypothesis H_1 that all the ratios are integers (at least to an extremely good approximation). If we exclude the pair (B^+, B^0), which is the only pair involving the bottom (or beauty) quarks, the product is 17,500,000. For H_0' alone the factor is 83,000. That H_0' can be extended to H_1 is an example of 'consilience of induction' (Whewell 1847/1967). The overall Bayes factor of 17,500,000 does not allow for the fact that the numbers 48 etc. are all factors of 480 and 720 and are of the simple form 2^a3^b. To support my judgement that such numbers are attractively simple, note that Hardy (1940, p. 69) discusses Ramanujan's interest in numbers of this form. The numbers 48 etc. are orders of subgroups of the symmetric group of degree 6, and this fact might be relevant in an explanation.

I estimated the prior probability of H_0 as between 1/12000 and 1/600. For H_0' I am inclined to lean over backwards and to reduce these lower and upper probabilities to 1/36000 and 1/1800. Thus my (subjective) posterior odds of H_0', not allowing for the other evidence in Table 1, are between 2 and 46 (with a geometric mean of about 10). I shall be interested to know the reader's honest estimates.

The initial probability of H_1 is I think not much less than that of H_0', say by a factor of 5. (Don't forget that H_0' is a part of H_1.) The remaining Bayes factor in favour of H_1, from Table 1, is 211 not allowing for (B^+, B^0). Thus, not allowing for (B^+, B^0), my posterior odds that H_1 has a 'physical meaning' would be between 100 and 2000.

Quark compositions	X	Y	$\frac{R(X,Y)}{136\alpha'/720}$	Close integer	Bayes factor
(uud, udd)	p	n	0.9999981 $\pm$0.0000044	1	83,000
(uds, uus)	Λ	Σ^+	47.95 ± 0.055	48	4.798
(uus, uds)	Σ^+	Σ^0	1.94 ± 0.07	2	3.947
(uds, dds)	Σ^0	Σ^-	2.974 ± 0.048	3	7.177
$(u\bar{s}, d\bar{s})$	K^+	K^0	5.914 ± 0.046	6	1.511
(uss, dss)	Ξ^0	Ξ^-	3.54 ± 0.033	3 or 4	0.775
$(u\bar{c}, d\bar{c})$	$\overline{D}^0$	D^-	1.844 ± 0.11	2	1.327
$(u\bar{b}, d\bar{b})$	B^+	B^0	0.26 ± 0.15	-	-

TABLE 1. Experimental values of $R(X, Y)/(136\alpha'/720)$, based on PDG89 and Cohen (1989).

The result for (B^+, B^0) is somewhat of a setback. The closest integer to 0.26 ± 0.15 is of course zero, but it seems physically unlikely that B^+ and B^0 have the same rest mass (Blecher 1989). (Moreover 0 is of course not a factor of the two numbers 720 and 480 mentioned above.) According to PDG89 (pp. 20 and 218), the mass difference is 1.9 ± 1.1 MeV/c^2. This estimate was based on only one experiment (performed by 85 experimenters), so the estimated standard error might not be wholly reliable, but it now seems most unlikely that the pair (B^+, B^0) can be included in our conjecture, H_1.

If we are to retain H_1 we must assume either that

(i) the masses of B^+ and B^0 are equal; or

(ii) the standard error for the (B^+, B^0) mass difference is much larger than 1.1 MeV$/c^2$ (to allow $R(B^+, B^0)$ to be equal to $R(p, n)$); or

(iii) the hypothesis applies only to particles made of light quarks (u, d, s and c quarks: see PDG89, p. 102, col. i). Then the pair (B^+, B^0) should be excluded. In PDG89 (p. 6, col. ii) the light quarks are defined as the u, d, and s quarks (the c quark not being included), and, with this definition, I would have to exclude the pair $(\overline{D}^0, D^-)$ also, but this would lose a Bayes factor of only 1.327. See Appendix A for information about the masses of the quarks. As in my previous work I have not allowed here for the elegance of the integers 48 etc. although this elegance clearly supports H_1.

I think it will be generally agreed that (iii) is much more reasonable than (i) or (ii). The restriction to the light quarks is not a big restriction because the hypothesis was already restricted to pairs of particles differing only in the interchange of a down quark with an up quark, and these are by far the lightest of the quarks.

The bottom quark is so much heavier than u, d, and s that the removal of the pair (B^+, B^0) is only a small loss of beauty for H_1. As a good analogy, consider the following argument that might have been used against the Copernican system. The moon always shows the same face towards the earth (as if the moon were embodied in a sphere that rotates around the earth) whereas the earth shows a variable face towards the sun. Copernicus could have replied, without much adhockery, that the two pairs (Earth, Moon) and (Sun, Earth) are quantitatively so different that a qualitative difference is not surprising.

We should not forget that the hypothesis H_0' is 'logically' independent of the rest of H_1; but probabilistically (in the epistemic sense) they support one another because they imply that $R(X, Y)$ has a fundamental interpretation, unknown at present but presumably related to the relative strengths of the strong and the electromagnetic forces.

The numerical values of $R(X, Y)/(136\alpha'/720)$ are changed negligibly if the denominator is replaced by $R(p, n)$, so, even if (1.1′) or (1.1) is coincidental, we still have evidence that *the proportional bulges* $R(X, Y)$ *bear a simple rational relationship to one another* when heavy quarks are not involved.

Further Discussion and a Modification of H_1

The restriction to the light quarks suggests that H_1 might be only a good approximation (because if heavy quarks damage the numerology why shouldn't the light ones do a little damage?). Thus the approximations to integers might not be as exact as I hope, but I believe they are at least *close enough to demand an explanation.* For the time being, I define the 'null hypothesis' H_1 in terms of exact integers.

Note that the experimental values of $R(X,Y)/(136\alpha'/720)$ in Table 1 all fall short of the integers 1, 48, etc., though this is ambiguous for the pair (Ξ^0, Ξ^-), and the 'short-fall' is statistically significant only for (B^+, B^0). If we combine the short-falls for the first seven pairs (excluding (B^+, B^0)), each divided by its 'uncertainty' (regarded as a standard deviation), we get 7.42. Since $7.42/\sqrt{7} = 2.80$, the null hypothesis H_1 that all the seven ratios are integers might be rejected with a P-value of 0.0052 (the double-tail) if the non-null hypothesis asserts that the short-falls are all positive or all negative. One would then wish to consider a modified hypothesis H_2 that the true values of the ratios $R(X,Y)/(136\alpha'/720)$ are the integers 1, 48, 2, 3, 6, 4, 2 (and 1?) minus small quantities that are monotonically related to the masses of the corresponding quarks even if the pair (B^+, B^0) is included. Among those pairs of particles for which the strangeness (which is minus the number of s quarks) is 0, ± 1, or ± 2, the best result is for strangeness 0 (the Bayes factor for H_1 being 83000) and the results are good for the four pairs of particles with strangeness ± 1 (Bayes factor 205). For the pair (Ξ^0, Ξ^-), where the strangeness is -2, the hypothesis H_1 loses a little ground (the Bayes factor being 0.775). For giving H_2 a more precise formulation we require improved experimental values for the mass differences, especially for $m(\Xi^-) - m(\Xi^0)$.

I believe that the numerology is good enough to be taken very seriously. How can the numbers 136, 720, 48, etc. be explained by a coherent and intelligible physical theory?

The pair (Λ, Σ^+) stands out in Table 1 in that the value of $R(\Lambda, \Sigma^+)$ is much larger than for the other pairs. This pair is also exceptional in that (i) the 'isospins' I of Λ and Σ^+ are unequal, in fact $I(\Lambda) = 0$ and $I(\Sigma^+) = 1$; and (ii) it is the only 'vice versa' pair, as mentioned a few sentences below equation (1.6″). We could avoid this 'vice versa' property by changing the sign of the numerator of $R(\Sigma^+, \Lambda)$, and then 48 would be replaced by -48 as a 'Pontryagin number' in Green et al. (1985, p. 338).

The pair (Λ, Σ^0) does not qualify for Table 1 because Λ and Σ^0 have the *same* quark composition, *uds*. Moreover the value of $R(\Lambda, \Sigma^0)$ had to be close to $R(\Lambda, \Sigma^+) + R(\Sigma^+, \Sigma^0)$ because the masses of Λ, Σ^+ and Σ^0 don't differ much. In fact $R(\Lambda, \Sigma^0)/(136\alpha'/720) = 50.02 \pm 0.065$.

Philosophical Discussion

Eddington (1946) and Einstein (1949, p. 63) believed that the fundamental constants of physics could be calculated from qualitative assumptions just as π can be calculated from the assumptions of Euclidean geometry. Eddington's main speculations along these lines, though stimulating, seem to have been fairly unsuccessful. There is now a theoretical argument, though an unconvincing one, that it is impossible to attain his Pythagorean goal and I shall mention that argument. First note, however, that if his dream is unattainable then the prior probability that (1.1) or (1.1′) is exact is somewhat reduced. It is not *much* reduced because those equations determine only one computable constraint on the fundamental constants.

The theoretical argument for believing that Eddington's goal cannot be achieved is related to one interpretation of the so-called anthropic principle, better called the *biotic* principle. This interpretation of the principle asserts that the fact that carbon-based life exists implies (but of course does not cause) severe constraints on the fundamental constants; see, for a review, Barrow & Tipler (1986). (The name *anthropic principle* is misleading because it is anthropomorphic to base the deduction on the fact that *humans* exist, and 'astronomomorphic' on the fact that *astronomers* exist: see Barrow & Tipler (1986, p. 15) who say wittily 'certain properties of the universe are necessary if it is to contain carbonaceous astronomers like ourselves.') An early and unconvincing example of the biotic principle can be read into an argument by Boltzmann. He said that the low entropy in the neighbourhood of the earth is *a priori* exceedingly improbable but in an infinite universe everything that is possible occurs somewhere. (See, for example, Porter 1986, pp. 215 and 216.) Boltzmann could have reversed the argument and said that the existence of life on earth is evidence that the universe is infinite or perhaps that there are an infinite number of universes. Another hypothesis is the one of which Laplace 'had no need', the existence of God. A third hypothesis is that Boltzmann's argument is simply wrong. For an extensive discussion of relevant matters see Prigogine and Stengers (1984). To put the matter in general terms: if an explanation requires an amazing coincidence, whether in a legal or a scientific context, then we have probably overlooked something. (Compare the discussion of Sherlock Holmes's law in Good 1950, page 67; and the seeming occurrence of two nearly independent murders in the same house: see, for example, *The Times*, London, 19 September 1970, p. 19.)

As another example, theories of the origin of the solar system based on the close encounter of a second star were proposed because there were difficulties in the theories of a nebular origin. The unlikelihood of a close encounter encouraged astronomers to remove the difficulties in the nebular theories. (See Nieto 1972.) Similarly, the isotropy of the universe would be explained by the 'chaotic cosmological principle' (Barrow & Tipler 1986, §6.11). According to this principle the present isotropic condition of the universe does not require isotropy in the initial conditions, but the principle runs into difficulties pointed out by Collins and Hawking. If, however, the chaotic principle is abandoned the isotropy seems to be an amazing coincidence, and one way to begin to explain the coincidence is to use the biotic principle. But another explanation of the coincidence would be that the chaotic cosmological principle will probably be reinstated by means of a theoretical correction. An example of such a loophole is indeed mentioned by Barrow & Tipler (1986 p. 425).

There is nothing wrong with the biotic principle as such, but a form of it that requires an amazing coincidence needs to be extremely watertight. If such an argument is valid then we are faced with a metaphysical option; very probably either God exists (a 'design argument') or there are myriads of universes (or both). In this case, to believe in Eddington's dream would be analogous to supposing that the first several hundred digits of π would, under some simple encoding into letters, spell out a sonnet by Shakespeare. (To believe that hypothesis you'd have to be a numerologist in the occult sense or perhaps a monkey.) In short, the attaining of Eddington's dream, (1.1′) being a step in that direction, would be effectively incompatible with the 'metaphysical option'.

Part 2. The Judging of Numerological Assertions

The quantal hypothesis in Part 1 was not entirely numerological for it made use of some physical theory though not convincingly. In the present part we consider how one might try to judge numerology with less explicit reference to physical theory.

At one time *numerology* meant divination by numbers, but during the last few decades it has been used in a sense that has nothing to do with the occult and is more fully called *physical numerology.* The expression *numerology* has been applied to one or more proposed formulae of the form (a 'null hypothesis')

$$x = y \text{ or } x \approx y$$

where x and y are numbers that might involve physical constants. The formula is regarded as numerological by a person who thinks it has not been explained. There will also be people who know that an explanation has been proposed, but who have not understood the explanation. These people either accept the judgements of the understanders or they might treat a formula *as if* it were numerological and judge it partly by its simplicity (or elegance) and its accuracy, or they might adopt a compromise position.

Numerological activity can be regarded as the search for patterns in collections of numbers so it is a kind of exploratory data analysis, though not necessarily of a Tukeyesque kind. The ultimate aim is to help in the formulation of scientific theories.

A statement can be partly numerological and partly scientific. For example, in 1815, William Prout suggested that all atomic weights are multiples of that of hydrogen and, as an inference (which contains a little truth), that all elements are composed of hydrogen (see Ihde 1964, p. 154). The evidence at the time was weak but the estimates of 1960 (Ihde, p. 142) are impressive even if we don't take isotopes into account. Of 49 atomic weights given in Ihde's table, 27 are within 0.1 of an integer (apart from oxygen whose atomic weight was taken there as 16 exactly, by definition), whereas only $49/5 = 9.8$ would be expected if Prout's hypothesis were entirely wrong. That Prout's hypothesis is false is obvious from the table, but the discrepancy between 9.8 expected and 27 observed shows that there is enough in the hypothesis to *demand an explanation*. This anachronistic example verifies that a hypothesis can be clearly wrong and yet clearly partly right at the same time. To say it was right would be absurd whereas to say simply that it was wrong would be extremely misleading. Such examples are not exceptional.

Even allowing for isotopes, the atomic weights are not exact integers because of the so-called 'packing fractions' of special relativity, and because the masses of protons and neutrons are not exactly equal. Prout's hypothesis is a good illustration of the need to take approximate laws seriously. Unfortunately such laws cause difficulties for statisticians. For the sake of simplicity, statisticians often assume sharp null hypotheses.

A *sharp hypothesis* is another name for a simple statistical hypothesis. We often test sharp hypotheses although we know they are most unlikely to be exactly true. This activity can often be justified on grounds of simplicity: if the available evidence is not sufficient to reject a sharp hypothesis we might find it convenient to regard it as true to an adequate approximation. But if we *reject* a sharp hypothesis we should be more careful. We ought sometimes to consider smudging, smearing, or desharpening the null hypothesis (Laplace 1774; and independently, but somewhat later, Good 1950, pp. 90–94). If we don't desharpen there'll be a risk of rejecting a hypothesis that points in the general direction of the truth. It is unfortunate that the terminology of 'rejection' is entrenched in statistical jargon, for it causes us, by the 'tyranny of words' (Chase 1938; Good 1969, p. 62), to be too ready to ignore hypotheses that are (probably) 'wrong' but might be suggestive of something better. Examples are Prout's law and the Titius-Bode law. (Good 1969, especially p. 62; Good 1971; Efron 1971; and Nieto 1972.) Perhaps the term *rejected* when applied to hypotheses or theories should often be replaced by *smudged* or *shown to be inexact*.

It is only in the non-occult senses that the word *numerology* is used in this chapter. An example (Lenz 1951) is

$$m(p)/m(e) \approx 6\pi^5 \tag{2.1}$$

where $m(p)$ and $m(e)$ denote the rest masses of the proton and electron. This formula is correct to one part in 50,000 but is now known to be inexact. (See Table 3.) An example from pure mathematics that will seem to some to be numerological, but which has a known explanation (Weber, c. 1908,

§125), is

$$\exp\left(\frac{\pi}{3}\sqrt{67}\right) \approx \text{number of feet in a mile,} \tag{2.2}$$

which is correct to one part in 300,000,000. (Naturally Weber doesn't mention feet or miles.) Of course a mere numerical computation doesn't explain *why* the left side is so close to an integer.

There have been a few examples of numerology that have led to theories that transformed society: see the mention of Kirchhoff and Balmer in Good (1962, p. 316) and in Barrow & Tipler (1986, p. 219ff) and one can well include Kepler on account of his third law. It would be fair enough to say that numerology was the origin of the theories of electromagnetism, quantum mechanics, gravitation, and quantitative chemistry (by Proust's law and Prout's hypothesis). So I intend no disparagement when I describe a formula as numerological.

There is, however, much bad numerology, and we shall discuss methods for attacking the difficult problem of distinguishing the good from the bad when the distinction is not obvious. This project is a special case of the even more difficult and more familiar one of distinguishing between good and bad scientific theories or hypotheses, and the consideration of the special case might provide some insights for the more general project which is of course a main problem in the philosophy of science. I believe that part of the solution should depend on the concept of explicativity (Good 1977) but this approach cannot be carried out without first thinking along somewhat Bayesian lines as in the present work. My purpose here is to contribute ideas that might lead to a more satisfactory solution than has been attained so far. See also Cover (1973).

When a numerological formula is proposed, then we may ask whether it is *correct*. The notion of *exact correctness* has a clear meaning when the formula is purely mathematical, but otherwise some clarification is required. I think an appropriate definition of *correctness* is that the formula has a good explanation, in a Platonic sense, that is, the explanation could be based on a good theory that is not yet known but 'exists' in the universe of possible reasonable ideas. A good but undiscovered theory is like a work of art waiting to be chiselled out of a block of marble.

A formula might be *partly correct* in the sense that some reasonable theory (possibly unknown) might explain why it is a good approximation. Such a theory is *known* for the mathematical formula (1.2) (with 5280 on the right of course). Lenz's formula (2.1) is so simple, and so nearly true, that I would not be surprised if it turned out to be partly correct. It will be discussed again below. Leaving such approximations aside, a correct numerological formula might sometimes be used for predicting new decimal places of observations before an explanation is found. A formula might be partly numerological in the sense that it can already be partly supported by rational arguments, although there is not yet a good explanation. Part 1 provides an example. There is no precise demarcation between numerology and a scientific theory.

We naturally ask the following questions (which are not *demands*) of a numerological formula:

(a) Was it 'consistent' with the experimental observations when it was published? (See below for the meaning of *consistency*.)
(b) Is it consistent with the latest experimental observations?
(c) Has its accuracy improved or deteriorated since it was first suggested?
(d) To how many significant digits (for example, in radix 10) is it correct according to the latest experimental results?
(e) How complex or simple is it? Here, as in (f), subjectivity can hardly be avoided.
(f) What, in some sense, was its prior probability of being correct (without allowing for the experimental results)?
(g) If it is consistent with the latest experimental results, is it likely to remain consistent with future experiments? This question can be attacked by Bayes's theorem, though still with difficulty, if an answer to (f) can be accepted.
(h) Is is part of a set of similar formulae, in other words does it satisfy, in a numerological sense, the desideratum of 'consilience of induction'? (Whewell 1847/1967, Vol. 2, pp. 77–78.)

We now elaborate on aspects (a) to (g).

(a) and (b). *Consistency with experiment.* Let x and y have estimated standard errors σ and τ, and estimated correlation coefficient ρ. (For the sake of simple notation I am not writing $\hat{\sigma}$, $\hat{\tau}$, and $\hat{\rho}$.) Let

$$s = \frac{|x-y|}{(\sigma^2+\tau^2-2\rho\sigma\tau)^{1/2}} \tag{2.3}$$

and call s the *sigmage* (to rhyme with *porridge*) of $|x-y|$. (In the special case of a purely mathematical piece of numerology we have $\sigma = \tau = 0$, and the sigmage is infinite unless the piece of numerology is exact.) We should hold in mind that, historically, standard errors of physical constants have tended to be too small (Henrion & Fischhoff 1986). Also ρ is seldom presented, and we might have to assume that $\rho = 0$. We might say that the equation $x = y$ is *inconsistent* with present experiments if $s > 3$ (compare Jeffreys 1937, p. 83, or 1957, p. 72) and *consistent* if $s \leq 2$. Like most things, consistency is a matter of degree, for example $s = 1$ is appreciably more consistent with experiment than is $s = 2$. In fact if $s \leq 1$ it would hardly be misleading to say that the agreement with experiment is *perfect* and in many Bayesian models such good agreement would give a little support to the null hypothesis, the *Bayes factor* (see Appendix B) being roughly proportional to $\exp(-s^2/2)$. A rough rule of thumb for the weight of evidence (logarithm of a Bayes factor), against the null hypothesis, would be $\frac{1}{2}(s^2-1)$ 'natural bans' or $0.217(s^2-1)$ bans (Turing's name for decimal units). This takes the 'cut even' sigmage as unity. As recalled by Good (1989a), if it is possible for the outcome of an experiment to undermine a hypothesis, then it is also possible for the outcome to support the hypothesis provided that all outcomes are observable. A proof of this little theorem is obvious, once the theorem is stated, but the result might surprise some Popperians and non-Bayesian statisticians. An example of a hypothesis tested by an experiment for which not all possible outcomes are observable is that there is life after death.

(c) and (d). *Number of correct significant digits (n.c.s.d.).* People sometimes use an integer to measure the number of correct significant digits of an approximation, but this constraint is of course unnecessary and ties one too much to radix 10. It is more informative to measure the number of correct significant digits by means of a real positive number and to take the standard errors into account. In Good (1988a) I suggested two related definitions but made an error (plus some verbal ones). Of the two, I think most statisticians will prefer *the smaller of the two numbers*

$$\left|\log_{10}\left|\left\{\left|\log_e\left(\frac{x}{y}\right)\right| \pm \left(\frac{\sigma^2}{x^2}+\frac{\tau^2}{y^2}-\frac{2\rho\sigma\tau}{xy}\right)^{1/2}\right\}\right|\right|, \qquad (2.4)$$

where σ, τ, and ρ denote respectively the 'uncertainties' or standard errors of x and y and the estimated correlation between the measurements. The proportional accuracy is then defined as

$$1 \text{ in } 10^{\text{n.c.s.d.}}, \qquad (2.5)$$

where one may well add the words 'or better' or 'at least' when $|x-y| \leq 2(\sigma^2+\tau^2-2\rho\sigma\tau)^{1/2}$. The proportional accuracy (2.5) does not depend on the radix 10.

To allow both for the n.c.s.d. and the sigmage s, the rule of thumb mentioned above for the weight of evidence in bans in favour of the null hypothesis (to be added to the prior log-odds to get the posterior log-odds) is

$$\text{n.c.s.d.} - 0.217(s^2-1), \qquad (2.6)$$

the corresponding Bayes factor being

$$10^{\text{n.c.s.d.}} e^{-(s^2-1)/2}. \qquad (2.6\text{A})$$

But the most difficult and controversial problem is the estimation of the initial odds of the null hypothesis. The initial odds, the n.c.s.d., and the sigmage are all relevant for an evaluation.

(e) *How complex or simple is the numerology?* Although no one has found an entirely satisfactory definition of simplicity or complexity, it is obviously easier to *force* several significant digits of accuracy by inventing complex formulae. Certainly simplicity has something to do with brevity as hardly anyone would deny apart from some Humpty Dumpty philosophers who change the meanings of words completely. In Good (1968) I suggested that the complexity $\kappa(H)$ of a proposition H could be defined by a measure of information,

$$\kappa(H) = -\log P(H), \tag{2.7}$$

where $P(H)$ denotes the prior probability of H. This definition, which links headings (e) and (f), leads to some difficulties when applied to arbitrary propositions (for example, $H \vee K$ is less simple but more probable than H), and a modification was discussed by Good (1975, pp. 46–48 = 1983, pp. 154–156) where negations and logical disjunctions are avoided. For the computable numbers that I shall discuss I believe formula (2.7) is reliable enough. The base of the logarithms merely determines the unit of complexity and I shall assume base 10 for convenience. My article on surprise, Good (1984/88), contains some further discussion of this matter.

A natural axiom for complexity is

$$\kappa(H\&H') = \kappa(H) + \kappa(H' \mid H) \tag{2.8}$$

in which the second term is a *conditional complexity*. This axiom is an immediate deduction from (2.7) when (2.7) is applicable. Also the axiom forces (2.7) if $\kappa(H)$ is assumed to depend only on $P(H)$, just as in information theory (Good 1950, p. 75).

When considering the complexity of an *arbitrary-looking* real positive number based on measurement (and hence not 'computable') such as 4357.073 we should hold in mind that this almost certainly is just an abbreviated way of denoting an estimate with a standard error which is *sometimes* roughly equal to 0.0005 (or this divided by $\sqrt{3}$). The definition of the complexity of the number ought to depend on how large it is as compared with its standard error, say σ (Rissanen 1983, p. 419). An approximate measure of the complexity is $\log_{10}(4357.073/\sigma)$ in decimal units. This definition is consistent with (2.7) if we assume the Jeffreys-Haldane improper prior density $1/u$ for a random positive number u. The definition does not depend on the use of the decimal system any more than the theories of information and weight of evidence depend on the units 'bits', 'decibans', etc.

(f) and (g). *What was the prior (epistemic) probability?* We shall be discussing, in two different senses, the prior probability of a computable number x. Sometimes the prior probability refers to the probability that a computable number chosen at random from some context will be x, and sometimes it refers to the prior probability that $x = y$ where y is some physical quantity. These two senses are very different and to distinguish between them I shall call the prior probability in the former sense the *preprior* probability. It is the probability of coming across x in some context and is not at all the same as the prior probability that $x = y$. But sometimes the context is influenced by that in which y arises. The complexity of x can be defined as minus the logarithm of its preprior probability. Thus the complexity of x might be relative to a context. This comment is in the spirit of the more general concept of complexity discussed in Good (1975), as referenced here just below equation (2.7). The complexity of the equation $x = y$ is equal to that of y plus that of x in the contexts in which y appears.

The preprior probability of x is especially pertinent when a piece of numerology is consistent with experimental results. It is more fundamental than the concept of simplicity (or complexity) but the concept of simplicity helps one to judge the preprior probability. The preprior probability can be expressed as the product of that of the functional form or algorithm assumed in the piece of numerology, and that of the choice of computable numbers that are the arguments of this functional form. (More precisely, this probability would be computed for the simplest algorithm that gives the same answer. Still more precisely, the probability should be estimated for all the algorithms, the maximum of these probabilities being selected.) In this article I have mainly in mind the functional form of $a_1a_2\dots a_m/(b_1b_2\dots b_n)$ (a *pseudorational number* so to speak), where the a's and b's are positive integers or π or e or one of a few operations such as square-rooting, but what I say about the probabilities of numbers of this form might well apply to most functional forms.

We have mentioned the Jeffreys-Haldane prior for measured quantities but for computable numbers it is much more difficult to suggest reasonable prior probabilities. In principle, this can be done consistently because computable numbers (like statable hypotheses) form only a countable set. This problem of assigning probabilities to computable numbers was propounded but not attacked by Good (1950 p. 55n). Here I shall try to make some small steps towards a solution. See also Rissanen (1983). In Part 1 I gave an example in which the computable number is 136/720. This example shows that in a special case one might need to invoke special arguments.

When judging any prior probability, such as the occurrence of an integer, it is necessary to take the context into account. Consider, for example, M.H.A. Newman's bus problem (Jeffreys 1938, p. 186): We know that a town has N buses, numbered 1 to N, but we don't know N, and we have little idea of the population of the town. We see a bus numbered k. What is the posterior probability distribution of N? Jeffreys obtained a reasonable solution by assuming an (improper) prior roughly proportional to $1/N$ by analogy with the Jeffreys-Haldane prior density $1/x$ for a positive random real number x (which has the property that powers of x have the same distribution as x has). This can also be called suggestively the log-uniform prior. Some proper distributions that 'approximate' the Jeffreys-Haldane distribution are mentioned by Good (1989b).

When the density $1/x$ is 'approximated' by a proper density $f(x)$ the definition of the complexity $\kappa(x)$ of a real number x with standard deviation σ needs to be modified. A natural definition is $\kappa(x) = -\log_{10} f(x/\sigma)$.

For general physical numerology one needs to put the positive integers and a few operations in some rank order, and a distribution somewhat resembling $1/N$ should be applied to the *ranks* and not to the original integers. For example, we would tend to favour composite integers over primes of about the same size because many formulae in physics, and in mathematics, consist of the products of various quantities. It would be possible to sample computable numbers in many texts, but so far my sample is too small. It will be described below. When sampling we must of course avoid including numbers, like 100, when they occur as approximations merely because of the use of radix 10. Also angles should be measured in radians or 'circumfians' and not in degrees.

Apart from the large amount of work required to obtain an adequate sample, there are further sampling difficulties such as

(i) A sample of individual numbers ignores relationships between pairs or larger groups of numbers; for example, the two 'triangulations' of n, namely $\frac{1}{2}n(n-1)$ and $\frac{1}{2}n(n+1)$, are logically related so it would be unfair to multiply their preprior probabilities when they both occur in the same piece of numerology. This case occurred in Part 1.

(ii) The field of theoretical physics is not homogeneous, and which population is appropriate for a given piece of numerology might be difficult to judge.

For these and other reasons a large reliance on subjective judgement is necessary. This necessity was exemplified in Part 1 and will be further exemplified below. Science seems objective only when it is compared with non-science. Fortunately, an exact ordering is unimportant.

My sample consisted of all the integers and π, 2π, and e occurring in Ramond (1985) and Goldman & Haber (1985), although e had zero frequency. I chose these two articles because my examples relate to elementary particles and because these two articles contain a lot of integers. I included also the operations $\times$, $\div$, $+$, $-$, squaring, powering other than squaring, ! (factorial), and triangulating up or down. Let us think first of numerological formulae of the form $x = y$ where x is of the form $a_1 a_2 \ldots a_m/(b_1 b_2 \ldots b_n)$ in which the a's and b's all belong to the set consisting of positive integers, etc., and their square roots, squares and other integer powers. I assume further that ζ and $1/\zeta$ have equal complexities and preprior probabilities, where ζ is any computable number, so at first we need to consider only products $a_1 a_2 \dots a_n$. For this limited project the preprior probabilities of $\times$, $\div$, $+$, $-$, and 1 are irrelevant and are mentioned only in the *heading* of Table 2 and not in the body of that table. The table consists of the integers etc. in a ranking order that I have chosen mainly subjectively because my samples are too small to give very much information about the ranking.

A principle of this ranking is to begin by thinking of the (unique) prime factorization of integers, where small primes are preferred to larger primes. Note that unique factorization also applies to rational (or pseudorational) numbers expressed in their lowest terms, even if π and e are regarded as pseudoprimes, for presumably π is not a rational or pseudorational power of e. (That looks like a difficult problem in the theory of numbers.) A simple measure of complexity of a rational number

$$2^{n_1} 3^{n_2} 5^{n_3} 7^{n_4} \dots, \tag{2.9}$$

where the n_i are integers, positive, negative or zero, would be

$$\lambda_1 |n_1| + \lambda_2 |n_2| + \lambda_3 |n_3| + \cdots, \tag{2.10}$$

where λ_r is $-\log$(relative frequency of the rth prime). The frequencies of 2 and 3, as prime factors, in the two samples combined, were respectively 1237 and 269 (out of 1665), so the ratio of the frequencies of 2 and 3, regarded as *prime factors*, differs substantially from the ratio when they are regarded just as integers. This discrepancy suggests that the simple rule (2.10) can be only a crude approximation. If used, the λ's could instead be estimated by means of monotonic regression, where the 'dependent variable' is someone's subjective measure of complexity.

The samples suffer from the same disadvantage as the use of continuous patches of prose for sampling the frequencies of English words — that a specific number can occur many times in a single context and yet can be rare in a more general population. I have listed the two samples separately to show how much they differ. This difference shows the desirability of a very large sample of samples. Having only two samples is like estimating the time by consulting two bad clocks, and averaging the result. But I shall regard the sum of the two samples as adequate, in this early discussion, for estimating a smooth fit to the relative frequencies of the *ranks*.

A fairly good smoothing is given by

$$p_r = \frac{2}{(r+1)(r+2)} \quad (r = 1, 2, 3, \dots), \tag{2.11}$$

where p_r denotes the probability of the *rank* r. (For example, the rank of 2 is 1.) Compare (2.11) to H_9 of Good (1953, p. 249) where, for greater flexibility, a factor of ξ^r occurs in the numerator, ξ being close to and less than 1. I have found that H_9 often fits biological data. As an example note that the frequencies of 2, 3, and *squaring* according to (2.11), would be 342, 171, and 103, while the observations were 327, 138, and 104. The fit p_r is good for the ranges $1 \leq r \leq 39$ and $r \geq 60$ but not when $40 \leq r \leq 59$. The fit would be improved by giving the number 56 a much smaller rank or else by assuming that the relative frequency of 56 will be greatly reduced when the sample of samples is much larger. Because p_r gives a better fit than the other distributions mentioned in Good (1989b), namely the log-Cauchy and the distribution of Rissanen (1983), let us accept p_r for the present. The fit leads to results that are harsher on numerology than the less formal method used by Good (1984/88, p. 107).

TABLE 2. *Frequency counts of positive integers etc. in two articles on physics, and a subjective ranking.* Because I was aiming only to evaluate products (and quotients), the following pairs of counts were excluded from the table: $1 : (273, 43)$; $\times : (145, 126)$; $\div : (60, 84)$; $+ : (133, 41)$; $- : (40, 34)$. Although the ranking is mainly subjective it is influenced somewhat by the two samples.

Rank, r	Number	Sample 1	Sample 2	Total
1	2	286	41	327
2	3	93	45	138
3	squaring	7	97	104
4	4	56	14	70
5	8	58	26	84
6	6	26	3	29
7	powering	1	20	21
8	$\sqrt{}$	-	17	17
9	$\pi, 2\pi$	-	14	14
10	16	8	4	12
11	24	1	-	1
12	e	-	-	-
13	5	11	14	25
14	10	13	4	17
15	!	1	-	1
16	32	2	4	6
17	9	11	4	15
18	12	4	1	5
19	21	-	-	-
20	7	16	-	16
21	48	-	2	2

22	720	-	-	-
23	120	-	-	-
24	28	6	-	6
25	64	1	-	1
26	triang.	3	-	3
27	27	1	9	10
28	36	-	-	-
29	60	-	-	-
30	72	-	-	-
31	96	-	-	-
32	15	3	-	3
33	20	7	1	8
34	136	-	-	-
35	144	-	-	-
36	216	3	1	4
37	81	-	-	-
38	105	-	-	-
39	7!	-	-	-
40	25	1	-	1
41	50	1	-	1
42	18	-	1	1
43	56	10	10	20
44	45	-	-	-
45	55	1	-	1
46	256	3	1	4
47	1024	-	-	-
48	512	-	-	-
49	128	8	1	9
50	576	-	1	1
51	125	-	-	-
52	49	-	-	-
53	480	2	-	2
54	35	6	-	6
55	42	1	-	1
56	65536	1	-	1
57	32768	3	2	5
58	40	3	-	3
59	30	3	-	3

The remaining numbers occurred but are not *ranked* here. (The multiples of 10 are exact integers, not mere roundings.)

	22	1	-	1
	26	1	-	1
	31	1	-	1

44	1	-	1
82	2	-	2
84	2	-	2
110	2	-	2
126	1	-	1
133	2	-	2
560	1	-	1
672	1	-	1
1120	2	-	2
1160	1	-	1
1520	3	-	3
2040	1	-	1
2640	2	-	2
3200	1	-	1
3696	1	-	1
8800	1	-	1
Totals	688	337	1025

Table 2 should be considered together with the following notes. (i) The numbers beyond rank 59 in this table have not been ranked but occurred in the first sample. (ii) Composite numbers can be represented in more than one way, for example $576 = 24^2$. In this example we might well regard the prior probability of 576 as the larger of $\frac{1}{2}p_{50}$ and $\frac{1}{4}p_{11}^2p_3$ where 50, 11, and 3 are the ranks of 576, 24 and *squaring* as listed and where the quotients 2 and 4 are explained under note (vii). (iii) Similarly, the probability of $v = a_1a_2\ldots a_m/(b_1b_2\ldots b_n)$ should strictly be computed as the maximum of the probabilities of all the different ways of expressing v. This leads to a difficult unsolved mathematical problem and will not be explicitly taken into account. When v is a rational number it can be expressed *uniquely* as a product of prime powers (or, more generally, one could include π and e as 'pseudoprimes') where the powers can be negative. The primes and pseudoprimes could be ranked in order, such as $2, 3, \pi$ or $2\pi, e, 5, 7, 11, \ldots$, and their probabilities estimated from samples. (Allowance for square roots, factorials etc. is still necessary.) Some of this idea is of course implicit in the ranking in Table 2. (iv) The subjective ranking would be modified if the number of samples of samples were greatly increased but p_r might still be adequately estimated by formula (2.11). (v) The number 120 deserves a good ranking because it is both a factorial and a triangulation of a nice number, 16. See also Part 1. (vi) When estimating the probability of a product (or quotient) it is necessary to take into consideration whether the factors are in some manner logically related to each other. (vii) For each parameter one should pay a Bayes factor of 2 because the parameter could have occurred as its reciprocal. I shall call this the *binary factor*. Instead, we could pay a factor of 2 for each multiplication or division. (viii) It helps the judgement to imagine that the information about the experimental observations arrives in two stages. In the first stage we are told only that the unknown number lies in some wide (but not excessively wide) interval (y_1, y_2), with a uniform prior, or slightly more accurately a log-uniform prior, for the non-null hypothesis, within that interval. For example, the interval might be $(y/\sqrt{2}, y\sqrt{2})$. The full information (in the form $y \pm \tau$) arrives at the second stage. The first stage gives so little information that it is reasonable to 'condition' on it. This device might help you to replace the preprior by a prior. The conditioning on $(y_1 < x < y_2)$ provides a method for multiplying by a *ballpark factor* to obtain the prior probability. The ballpark factor is a 'reward' for x's being in the right ballpark, and, when the ballpark interval is $(y/\sqrt{2}, y\sqrt{2})$ I take this factor as $2y$ in accordance with the following argument.

Let us condition on $x \neq 1$, assume a probability of $\frac{1}{2}$ that a positive computable number exceeds 1 (since a rational number and its reciprocal are assumed to be equally probable), and that the probabilities have the geometric distribution

$$P(2^n < x \leq 2^{n+1}) = 2^{-(n+2)} \quad (n = 0, 1, 2, \dots). \tag{2.12}$$

(This is more consistent with (2.11) than it looks.) Then, if $y > 1$, we have

$$P(y/\sqrt{2} < x \leq y\sqrt{2}) \approx \frac{1}{2^{3/2}y}. \tag{2.12A}$$

By conditioning on x's lying in this interval we force a proportional accuracy of $1/\sqrt{2}$ so the n.c.s.d. of the observed x should strictly be adjusted to allow for this. Instead, I shall absorb this small adjustment into the ballpark factor which is therefore taken as $2y$. (ix) $65536 = 2^{2^{2^2}}$ is the number of possible functionals of two binary variables (Good 1985) whereas 16 is the number of functions. (x) From a geometrical point of view π and 2π are equally simple for an obvious reason. (xi) I hope to carry out a small survey to see to what extent judgement of the ranking varies from one judge to another.

I now apply my ranking of computable numbers, such as it is, to several examples related to $m(p)/m(e)$.

The Mass Ratio of Proton to Electron

I have assembled eight pieces of numerology for $m(p)/m(e)$, the ratio of the rest masses of the proton and the electron. (All but one were previously assembled in Good, 1987.) They are not all of the form $a_1 a_2 \dots a_m/(b_1 b_2 \dots b_n)$ which was mentioned above. They are shown in Table 3 together with the observational value and I here use them as examples for a method of evaluation. Although it is obvious, I emphasize that these evaluations are largely based on my personal judgement, but the kind of reasoning might help others to make their own judgements about numerological assertions. The sources were

(i) Cohen and Taylor (1987, pp. 1126 and 1139), the observational value.
(ii)* Eddington (1946, pp. 38 and 58).
(iii)* Lenz (1951).
(iv) Worrall (1960, p. 602).
(v)* Good (1960, 1962).
(vi) Good (1962, p. 318).
(vii) Sirag (1977).
(viii)* Parker-Rhodes (1981, p. 185).
(ix) A modification of (iii) proposed here.

The four items marked with an asterisk were 'within experimental error' when they were announced but, in common with the other four items, they are now 'contradicted' (shown not to be exact) by the observed value given in row (i). The sigmages are given in the third column of the table. The numbers of correct significant digits (n.c.s.d.) are shown in the fourth column. In the fifth column I give a rough subjective estimate of the complexity κ_0 of the numerology measured in decimal digits and adjusted by the binary factor and ballpark factor of notes (vii) and (viii) to Table 2. These estimates are based largely on formulae (2.7) and (2.11) where r is the *rank* of an integer or symbol as shown in Table 2. The adjusted complexities are denoted by κ_0 to distinguish them from the κ used above and in Good (1988c) where no allowance was made for the binary and ballpark factors. Because I am ignoring note (iii) to Table 2, apart from allowing for the obvious permutations, the measures of adjusted complexity might be somewhat too harsh. The last column is discussed at the end of the article. The approach leaves much to be desired, but I don't know a better one, and the results seem to me to make overall approximate sense.

Elaborate arguments were provided by the authors of items (ii) and (viii) but I have not yet understood their arguments so I here treat these items *as if* they were purely numerological. This treatment might not be excessively unfair because the formulae seem to have been empirically refuted. Note, however, that item (viii) is stated only as an approximation on page 475 of Bastin et al. (1979). Those who have understood the arguments of the authors won't need to adopt the numerological approach for items (ii) and (viii) but they provide examples for my discussion.

The following notes describe how I arrived at the rough estimates of the adjusted complexity measures κ_0. In each case the ballpark factor is taken as $2 \times 1836 = 3672$. The paragraph numbers (ii) to (viii) correspond to those in Table 3.

		'sigmage'	n.c.s.d.	κ_0 (adjusted complexity)	n.c.s.d.–κ_0 (score)
(i)	1836.152701 ± 0.000037	0	7.6	-	-
(ii)*	1836.34, the ratio of the roots of $10x^2 - 136x + \beta^{5/6} = 0$ where $\beta = 137/136$	5000	4.0	7.5	−3.5
(iii)*	$1836.118 = 6\pi^5$	900	4.7	3.8	0.9
(iv)	$1839.39 = 10^4/(2e)$	90000	2.8	4.1	−1.3
(v)*	$1836.10 = 137^2/(10\gamma)$ where $\gamma = 46/45$	1400	4.5	8.9	−4.4
(vi)	$1836 = 1728 + 108$	4000	4.1	4.7	−0.6
(vii)	$1836 = 136 \times 135/10$	4000	4.1	4.4	−0.3
(viii)*	$1836.15150 = \frac{137\pi \times (5/4)}{\frac{3}{14}\left(1+\frac{2}{7}+\frac{4}{49}\right)}$	32	6.2	11.0	−4.8
(ix)	$1836.15070 = 6\pi^5\alpha'/\alpha$	54	6.0	-	-

TABLE 3. Numerology for $m(p)/m(e)$.

(i) This is the best available experimental value.

(ii) I judge that the prior probability that $m(p)/m(e)$ is equal to the ratio of the roots of a quadratic, with interesting and ultimately explicable coefficients, is less than 1/200, but I'll use this value to be generous. (The use of this judgement shows that my prior probability in this example is not entirely 'preprior'.) Conditioning on that assumption we have to measure the complexities of 10, −136, 137/136, powering, 5 and 6. The numbers 6, 10, and 136 are somewhat related to one another, as said in Part 1, so it would not be fair to multiply all of the values of p_r $(= 2/[(r+1)(r+2)])$ for the ranks. To allow for the relationship just mentioned, and out of respect for Eddington, Bastin, and others, I shall generously allow no penalty for 10 and 137. We recover a factor of 2 because the quadratic could have been written in reverse order. The adjusted complexity (in decimal units) then comes to at least

$$\kappa_0 = -\log_{10}\left(\frac{2}{200}\cdot\frac{2}{35\cdot 36}\cdot\frac{2}{14\cdot 15}\cdot\frac{2}{7\cdot 8}\cdot\frac{2}{8\cdot 9}\cdot\frac{3672}{16}\right) = 7.5.$$

(iii) Lenz's paper is probably the shortest physics paper on record. He did not try, at least not there, to dress up his formula with a theory, whether lucid or obscure. The values of r for 6, π, powering and 5, are 6,

9, 7, and 13. We gain a factor of 2 because the product can be written as $\pi^5 6$. This gives $\kappa_0 = 3.8$. This is slightly less than the number of correct significant digits, so I judge that Lenz's formula was better than an evens bet at the time it was first proposed. It is interesting to write the formula as

$$m(p)/m(e) \approx 6!\, V_{10}/C_{10} \tag{2.13}$$

where V_{10} is the volume of a ten-dimensional ball of radius say k and C_{10} is that of a ten-dimensional cube of side k (compare Good 1970). The occurrence of 6! again is somewhat striking.

(iv) The ranks for 10, powering, 4, 2, and e are 14, 7, 4, 1, and 12, and we recover a factor of $3! = 6$ for permuting the factors, so we get $\kappa_0 = 4.1$. This item should not have been published because it was extremely inaccurate when first suggested. Moreover the author tried to explain the factor e^{-1} as 'exponential decay'. The correct factor representing exponential decay is of the form $e^{-\lambda t}$ and it is hardly conceivable that there can be any reason for λt to be equal to 1.

(v) The constant $\gamma = 46/45$ was introduced by analogy with 137/136 which Eddington denoted by β. Whereas $136 = 10^2 + 6^2$ (or the number of independent elements in a 16×16 symmetric matrix), we have $45 = 6^2 + 3^2$ (as for a 9×9 matrix) where 3, 6, and 10 are three consecutive triangular numbers. To allow for the relationships I count 10 only once, having regarded 45 as a triangulation of 10 (while losing a factor of 2 because there are always two ways to triangulate a number). I think of 46 as 45 + 1 and 137 as 136 + 1. Take the ranks of 'squaring', 10, and triangulation $(\binom{10}{2} = 45)$ (ranks 3, 14, and 26), and judge a rank of 12 for +1. We recover a factor 4! for permutations and get

$$\kappa_0 = -\log_{10}\left(\tfrac{2}{35\cdot 36}\cdot\tfrac{2}{13\cdot 14}\cdot\tfrac{2}{4\cdot 5}\cdot\tfrac{2}{15\cdot 16}\cdot\tfrac{2}{27\cdot 28}\cdot\tfrac{2}{13\cdot 14}\cdot\tfrac{24\cdot 3672}{16\cdot 2}\right) = 8.9.$$

(vi) The numbers 1728 and 108 were originally given credit for arising in Klein's theories of groups connected with the icosahedron and dodecahedron. If (on grounds of self-criticism) we harshly credit nothing for these reasons we can write the formula as $12^3(1 + \frac{1}{16})$, and use ranks for 12, powering, 3, 16, and addition of 1, namely 18, 7, 2, 10 and say 12. The expression is unchanged under four permutations so $\kappa_0 = 4.7$.

(vii) The formula can be written as $\binom{136}{2}/5$. The ranks for 136, triangulation and 5 are 34, 26, and 13. We should pay a factor of 2 for the choice of the lower triangulation and recover a factor of 2 for 'permutation'. We get $\kappa_0 = 4.4$.

(viii) Write the formula as

$$\frac{137\pi(5/4)}{\frac{3}{2\times 7}\left(1 + \frac{2}{7} + \frac{4}{7^2}\right)}$$

and take ranks for 136, +1, π, 5, 4, 3, 2, 7, +1, 2, and squaring, namely 34, 12, 9, 13, 4, 2, 1, 20, 12, 1, and 3. It seems roughly right to count the 7 only once and allow for only one plus sign in the denominator. We recover 8! 2! for 'permutations' of the factors 137, π, 5, 4, 3, 2, 7 and $1 + \frac{2}{7} + \frac{4}{49}$ and for reversal of the terms in the last factor. The binary factor is 512. The outcome is $\kappa_0 = 11.0$. It is possible that Parker-Rhodes constructed his obscure explanation of (viii) after noticing that, according to the best value of $m(p)/m(e)$ known at the time,

$$\frac{1}{137\pi} \cdot \frac{m(p)}{m(e)} \approx \frac{1715}{402} = \frac{5 \times 7^3}{6 \times 67}$$

which would have been readily obtained by using continued fractions. He then might have rewritten this in various ways while constructing his explanation. (For example, $67 = 7^2 + 2 \times 7 + 2^2$.) If only he were still alive he could confirm or deny this conjecture. The fundamental correctness of his explanation would have been far more convincing (to those who do not understand it) if he had produced it without first knowing the observed value. The same statement applies to Eddington's explanations which I surmise were largely numerological though nominally based on a theory.

(ix) This modification of Lenz's formula is correct to one part in a million. I have found it too difficult to estimate the adjusted complexity, for I cannot decide how much to 'pay' for the factor α'/α. This factor seems to require a self-contradictory explanation. It might be better to replace it by $1 + \frac{1}{3}\alpha^2$ or $1 + \frac{1}{3}\alpha'^2$ or $\exp(\frac{1}{3}\alpha^2)$ etc.

The last column of the table gives the difference n.c.s.d. $-\kappa_0$ and is a rough measure of how good each piece of numerology is when the sigmage s is ignored or equivalently is assumed to be equal to 1. The difference is suggestive of the posterior log-odds of the corresponding piece of numerology, at the time it was proposed, if it was consistent with the observations at that time as were items (ii), (iii), (v), and (viii). I say 'suggestive of' rather than 'roughly equal to' because my arguments are not rigorous enough to justify the latter expression. It is safe enough to describe the expression n.c.s.d. $-\kappa_0$ as a *score* in its ordinary English sense. It gives an indication of whether the numerology (even if only an approximation) might point towards the truth when its sigmage is not taken into account by means of the factor $\exp\left[-\frac{1}{2}(s^2 - 1)\right]$ of formula (2.6A).

Lenz's formula is the only one, among formulae (ii) to (viii), having a positive score.

Concluding Comments

The methods used in Parts 1 and 2 differ considerably. This is because the examples in Part 2 are treated as almost purely numerological apart from my judgement of the rankings of numbers and operations in Table 2. But the two parts share the 'two stages of information' as described in note (viii) to Table 2. There are precedents for two stages of information. For example, in statistical consulting a client might suggest a null hypothesis, and then the statistician might take this hypothesis seriously out of respect for the client's scientific judgement. Similarly a medical or legal investigation often begins with 'presenting symptoms' or a 'prima facie case'.

Part 1 supports the hypothesis of *the relative rationality of proportional bulges (of hadron masses)*, at least to a good approximation when the heavy quarks are not involved. The numbers 720, 48 etc. suggest that an explanation based on finite groups might be found, the symmetric group of degree 6 being a candidate, or the 'heterotic string theory' might be relevant. Part 2 argues that the judgement of physical numerology does not need to be made only in a gestalt manner, but can be largely analyzed in terms of judgements concerning the complexities of integers, familiar constants, and familiar mathematical operations. At present these judgements are subjective and depend on what mathematical language, or calculator, one believes to be appropriate for a specific application. (Compare, for example, Good 1977, pp. 326–327.)

The methods of Part 2 are exemplified by various numerological expressions for $m(p)/m(e)$. One conclusion was that Lenz's formula $6\pi^5$ was seemingly 'odds on' when it was suggested though it is now known not to be accurate. Equation (2.13) expresses it in a form that might lead to a geometrical interpretation in ten dimensions.

Acknowledgements

This work was supported in part by a grant from the U.S. National Institute of Health. I am also much indebted to E. Richard Cohen for a correction and for some important information concerning the latest observational values of physical constants. My thanks are also due to Leslie Pendleton Miller who typed the original draft of this long paper.

Appendix A. Physics

Many of the 'elementary particles', namely the *hadrons*, are believed to be mainly made up of *quarks* of which there are various kinds. The quarks are often described as up, down, strange, charmed, bottom (or beauty) and top (or truth). These can be regarded as nicknames or mnemonics for the official names, u, d, s, c, b, and t (Cohen and Giacomo 1987, p. 12). Corresponding antiquarks are denoted by $\overline{u}$, $\overline{d}$, etc. The charges on u, d, s, c, b, and t are respectively $-\frac{1}{3}$, $\frac{2}{3}$, $-\frac{1}{3}$, $\frac{2}{3}$, $-\frac{1}{3}$ and $\frac{2}{3}$ where the unit is the charge on the electron, while the antiparticles have the signs of the charges reversed. The compositions, in terms of quarks, of the particles mentioned in Table 1, are shown in the first column of that table. The particles containing three quarks are baryons while those containing one quark and one antiquark are mesons.

The standard notation for the mass of the proton, for example, is m_p, but I have used the notation $m(p)$ to make the production of the document a little easier. The 'mass' $m(X)$ of a particle X means its rest mass. The relativistic mass of X moving with velocity v, relative to a specified frame of reference, is equal to $m(X)(1 - v^2/c^2)^{-1/2}$.

The masses of the quarks. PDG89 (p. 102, col. i) gives the masses of the quarks, in units of MeV/c^2, as $m(u) = 5.6 \pm 1.1$, $m(d) = 9.9 \pm 1.1$, $m(s) = 199 \pm 33$, $m(c) = 1350 \pm 50$, $m(b) \approx 5000$, and $m(t) > 50,000$. These are described as 'running masses evaluated at 1 GeV'. Perhaps the charmed quark should be regarded as of intermediate mass, neither light nor heavy. Much of the mass of a quark is converted into 'packing energy' so a particle can be lighter than the sum of the masses of the quarks that lie within it.

From Balmer to Bohr. The well-respected textbook Messiah (1961, p. 38n) is historically somewhat inaccurate when it says 'The quantization of circular orbits led Bohr to find the Balmer formula ...'. For Bohr postulated this quantization to *explain*, not to 'predict', Balmer's formula when that formula was shown to him by Hans Marius Hansen (Barrow & Tipler 1986, p. 222).

Appendix B. Odds and Bayes Factors

If an event or proposition has (possibly conditional) probability p, then its odds are defined as $p/(1-p)$. (Odds of, for example, 3.5 are also expressed as 7 to 2 on.) If the result of an experiment is denoted by E, then the prior odds $O(H)$ of a hypothesis H are multiplied by $B(H : E)$ to obtain the posterior odds of H, where $B(H : E)$ is called the *Bayes factor* in favour of H provided by E, and is given by

$$B(H : E) = \frac{O(H|E)}{O(H)} = \frac{P(E|H)}{P(E|\overline{H})} \tag{B1}$$

where $\overline{H}$ denotes the negation of H. This odds form of Bayes's theorem was stated by Wrinch and Jeffreys (1921), although they did not use the terminology of odds. One can think of $O(H|E)/O(H)$ as the definition of the Bayes factor, and the right side of (B1) as the method usually used for its calculation or estimation. When H and $\overline{H}$ are simple statistical hypotheses the right side is an uncontroversial simple likelihood ratio, otherwise it is undefined in non-Bayesian statistics. In Bayesian statistics, in this latter case, some judgement is needed to estimate the right side. Sometimes much depends on how the negation of H is interpreted.

Some writers use the term 'odds-ratio' which can mean, ambiguously, odds or the ratio of odds. Hence the expression *Bayes factor* is linguistically better as well as being historically earlier.

Appendix C. The One-Plus Exaggeration

The one-plus exaggeration, which was mentioned in the text, has occurred in relation to quantum electrodynamics (QED). While mentioning this I have no wish to question that highly successful theory.

The experimental value of the magnetic moment of the electron, in units of $e\hbar/(2cm(e))$ (where e in the numerator denotes the charge of an electron), according to PDG89 (p. 24) or Cohen and Taylor (1987, p. 1141), was 1.001 159 652 193(10), while the value given by QED was 1.001 159 652 46(20) (see Feynman 1985, pp. 6 and 7). In an interesting book, Watkins (1986, p. 46), said that the accuracy was better than one part in a million million. He confirmed in correspondence that he was referring to the magnetic moment of the electron. If we take the observed and theoretical values as the x and y of our formula, we find that n.c.s.d. was only (at least) 8.3 which is an accuracy of one part in at least two hundred million. But, according to Dirac's previous theory, which did not allow for the interaction of electrons with light, the theoretical value would be 1. If we are evaluating the *further* advance of QED, for this observation, it seems to me that we should consider that

$$x/y = (115965246 \pm 20)/(115965219 \pm 1)$$

and this reduces the n.c.s.d. to at least 6.6 or one part in at least 4,000,000 (instead of one part in a million million). Thus this crowning achievement of QED was 24 times as accurate as my piece of numerology H_0' (and both had small sigmages and therefore had 'room for improvement'). Of course a numerical success based on an attractive and otherwise successful theory is very much more convincing than even an equally accurate largely numerological result. This is because the result based on a successful theory has the higher prior probability as judged by most of the people who are paid to do physics.

The latest experimental and theoretical values for the magnetic moment of the electron are (Kinoshita 1989) 1.001159652164(7) and 1.001159652188(4) with an accuracy of 1 in 30 billion, or 1 in 40 million if we don't 'add one to exaggerate'. The 'official' sigmage is now 3.0 *so the theory has possible reached the limit of its accuracy.*

Feynman (1985, p. 9) says that nobody understands QED and then proceeds to explain it brilliantly! Similarly Bohr said that any one who is not shocked by quantum mechanics hasn't understood it. This can be reworded, paradoxically: Anyone who understands quantum mechanics knows that she has not understood it. And this is at least as true for QED. The 'instrumentalism' of modern physical theories detracts somewhat from their stati as explanations and in this respect they have a soupçon of numerology.

Appendix D. Quantal Hypotheses

Suppose that a hypothesis or theory H states that a certain physical constant is exactly equal to an unspecified integer n, while the experimental estimate is $N(x, \sigma^2)$. For the sake of elegance I allow n to be positive, negative, or zero, and I assume that σ is known precisely.

Suppose that, before the measurement was made, we had little idea about what value x would have. Then I claim that the Bayes factor in favour of H is approximately

$$\frac{1}{\sigma\sqrt{2\pi}} \sum_{n=-\infty}^{\infty} \exp\left(-\frac{(n-x)^2}{2\sigma^2}\right) \tag{D1}$$

$$= 1 + 2\sum_{n=1}^{\infty} e^{-2\pi^2\sigma^2 n^2} \cos(2\pi n x) \tag{D2}$$

$$= \vartheta_3(\pi x \mid 2\pi i \sigma^2). \tag{D3}$$

The equivalence of (D1) and (D2) is a special case of Poisson's summation formula given as (5.13) of Good (1986) where further details and applications of formula (D2), as well as historical comments, can be found.

Formula (D1) is based on the idea that, given H, the prior distribution of the relevant integer is nearly uniform over a wide range of integers, while, given the negation of H, the corresponding real number has a prior that is nearly uniform, as a real number, over much the same range. In fact I am regarding this as the *definition* of the negation of H. It is important not to forget that this assumption has been made because, for example, the result would be very different if the negation of H stated that the real number is equal to half an odd integer. The Bayes factor would be expressible as ϑ_3/ϑ_4.

If $\sigma < \frac{1}{2}$, formula (D1) can be well approximated by just a few terms of the series, and if σ is small a single term is adequate. If $\sigma > \frac{1}{2}$, the Bayes factor is close to 1, as can be readily seen both intuitively from the meaning of the statistical problem and also from formula (D2). It is also interesting to note the check that, if x is an integer, expression (D2) exceeds unity, as it should, while if x is half an odd integer the expression 'subceeds' unity, and this again makes perfect intuitive sense. This last fact follows at once from Jacobi's infinite product for ϑ_4.

If σ is regarded as having a prior distribution we could multiply formula (D1), (D2), or (D3) by that distribution and integrate to get an improved value for the overall Bayes factor. But I shall not try to carry out this refinement.

Example. In Part 1 the hypothesis H_1 states that seven independent physical constants are all equal to integers. The corresponding observations are 0.9999981 ± 0.0000044, 47.95 ± 0.055, etc. The seven corresponding Bayes factors are therefore approximately

$$\frac{1}{\sigma_1\sqrt{2\pi}}\exp\left(-\frac{0.0000019^2}{2\sigma_1^2}\right) \quad (\sigma_1 = 0.0000044),\ \text{etc.},$$

$$\frac{1}{\sigma_6\sqrt{2\pi}}\left[\exp\left(-\frac{0.46^2}{2\sigma_6^2}\right) + \exp\left(-\frac{0.54^2}{2\sigma_6^2}\right)\right] \quad (\sigma_6 = 0.33),$$

and

$$\frac{1}{\sigma_7\sqrt{2\pi}}\exp\left(-\frac{0.156^2}{2\sigma_7^2}\right) \quad (\sigma_7 = 0.11).$$

The seven Bayes factors are respectively as shown in Table 1.

Discussion. The topic of this appendix is closely related to that of 'Quantum hunting' which is surveyed by Kendall (1986). In quantum hunting one searches for a quantity q such that all observations are multiples of q 'within experimental error', where the experimental error (standard deviation σ) is assumed to be the same for all observations. Our problem is the case where q has a specified value, but where σ varies from one observation to another and has an approximately known value for each observation. As far as I know, the published work on quantum hunting has all been non-Bayesian but it could be tackled by a Bayesian approach. Even without assuming a prior distribution $F(q)$ for q it would be of interest to draw a graph of $B(q)$ where $B(q)$ is the Bayes factor (or *a* Bayes factor) in favour of the quantum hypothesis (say H_q) that q has a specified value, the rival hypothesis being that *no* value of q exists. Clearly $B(q) \to 1$ as $q \to 0$. By definition H_q is supposed to denote the hypothesis that q is the largest quantum. This definition makes the various hypotheses H_q mutually exclusive. Without this constraint, H_q would *imply* $H_{q/2}$, $H_{q/3}$, etc.

The integral of $B(q)dF(q)$ would be the overall Bayes factor in favor of the quantal hypothesis without specifying a value for q.

Another quantal problem was treated by Hammersley (1950), that of estimating a parameter when it is known in advance of sampling that the parameter certainly belongs to a specified set of numbers, such as the set of integers, whereas in this appendix we have been concerned with *testing* this hypothesis. In the example, the problem of estimation is not entirely absent because the favoured integers seem all to be of the form $2^a 3^b$, but I have not taken this nice-looking feature into account when calculating the overall Bayes factor.

Appendix E. The Symmetric Group of Degree 6

Burnside (1911 or 1955, p. 209) states the following theorem which gives a distinctive property of the symmetric group of degree 6:

> The symmetric group of degree n ($n \neq 6$) contains n and only n sub-groups of order $(n-1)!$ .... The symmetric group of degree 6 contains 12 sub-groups of order 5!, which are simply isomorphic with one another and form two conjugate sets of 6 each.

It is tempting to conjecture that the two conjugate sets correspond to the six quarks and six antiquarks.

References

Allen, H.S. (1928). *The Quantum and its Interpretation.* Methuen, London.

Barrow, J.D. and Tipler, F.J. (1986). *The Anthropic Cosmological Principle.* Clarendon Press, Oxford.

Bastin, Ted, Noyes, H.P., Amson, J., and Kilmister, C.W. (1979). On the physical interpretation and the mathematical structure of the combinatorial hierarchy. *International Journal of Theoretical Physics* 18, 445–488.

Blecher, Marvin (1989). Private communication.

Burnside, W. (1911). *Theory of Groups of Finite Order.* 2nd edition. Cambridge University Press, Cambridge. Reprinted by Dover Publications, New York, 1955.

Candelas, P., Horowitz, G.T., Strominger, A., and Witten, E. (1985). Vacuum configurations for superstrings. In Schwarz 1985, 1107–1133.

Chase, Stuart (1938). *The Tyranny of Words.* Harcourt, Brace, and Co., New York.

Cohen, E. Richard (1989). A letter dated March 7 based on recent information obtained from Robert S. Van Dyck Jr., G. Audi, and A.H. Wapstra.

Cohen, E. Richard and Giacomo, Pierre (1987). Symbols, units, nomenclature and fundamental constants in physics. *Physica* 146A, 1–68.

Cohen, E. Richard and Taylor, Barry N. (1987). The 1986 adjustment of the fundamental physical constants. *Reviews of Modern Physics* 59, 1121–1148.

Cover, T. (1973). On determining the irrationality of the mean of a random variable. *Annals of Statistics* 1, 862–871.

Davies, P.C.W. and Brown, Julian (1988). *Superstrings: A Theory of Everything?* Editors, Cambridge University Press, Cambridge.

Eddington, A.S. (1930). *The Mathematical Theory of Relativity.* Cambridge University Press, Cambridge.

——— (1946). *Fundamental Theory.* Cambridge University Press, Cambridge.

Efron, B. (1971). Does an observed sequence of numbers follow a simple rule? (Another look at Bode's law.) (with discussion). *Journal of the American Statistical Association* 66, 552–568.

Einstein, A. (1949). Autobiographical notes. Translated from the German by P.A. Schilpp, written in 1946. In *Albert Einstein: Philosopher-Scientist,* ed. P.A. Schilpp, Tudor, New York, 3–95.

Feynman, R.P. (1985). *QED: The Strange Theory of Light and Matter.* Princeton University Press, Princeton, N. J.

Goldman, T. and Haber, H.E. (1985). Gluinonium: the hydrogen atom of supersymmetry. *Physica* 15D, 181–196. Reprinted in Kostelecký and Campbell (1985).

Good, I.J. (1950). *Probability and the Weighing of Evidence.* Charles Griffin, London, and Hafners, New York.

——— (1953). The population frequencies of species and the estimation of population parameters. *Biometrika* 40, 237–264.

——— (1960). Some numerology concerning the elementary particles or things. *Journal of the Royal Naval Scientific Service* 15, 213.

——— (1962). Physical numerology. In *The Scientist Speculates,* ed. I.J. Good, A.J. Mayne, and J. Maynard Smith, Heinemann, London, 315–319.

——— (1968). Corroboration, explanation, evolving probability, simplicity, and a sharpened razor. *British Journal of Philosophical Science* 19, 123–143.

——— (1969). A subjective evaluation of Bode's law and an 'objective' test for approximate numerical rationality (with discussion). *Journal of the American Statistical Association* 64, 23–66. Reprinted in part in Good 1983.

——— (1970). The proton and neutron masses and a conjecture for the gravitational constant. *Physics Letters* 33A, 383–384.

——— (1971). The evolving explanation of a 'numerological' law, being an invited 'rebuttal' to Efron (1971). *Journal of the American Statistical Association* 66, 559–562.

——— (1975). Explicativity, corroboration, and the relative odds of hypotheses. *Synthèse* 30, 39–73. Reprinted in Good 1983.

——— (1977). Explicativity: a mathematical theory of explanation with statistical applications. *Proceedings of the Royal Society (London) A* 354, 303–330, and 377 (1981), 504. Reprinted in part in Good 1983.

——— (1983). *Good Thinking: The Foundations of Probability and its Applications.* University of Minnesota Press.

——— (1984/88). Surprise index. *Encylopedia of Statistical Sciences* 9, 104–109.

——— (1985). The number of functionals: a combinatorial curiosity. C228 in *Journal of Statistical Computing and Simulation* 21, 90.

——— (1986). Some statistical applications of Poisson's work. *Statistical Science* 1, 157–170.

——— (1987). Good and bad scientific speculations. An invited lecture at the meeting of the British Association for the Advancement of Science, Belfast.

——— (1988a). The number of correct significant digits in a piece of physical numerology. C305 in *Journal of Statistical Computing and Simulation* 29, 358–359.

——— (1988b). What are the masses of the elementary particles? *Nature* 332, 495–496 .

——— (1988c). Physical numerology. Technical Report No. 88–26, Department of Statistics, Virginia Polytechnic Institute and State University (December 30), 44 pp.

——— (1989a). The theorem of corroboration and undermining, and Popper's demarcation rule. C317 in *Journal of Statistical Computing and Simulation* 31, 119–120.

——— (1989b). On the judgement of "numerological" assertions. C333 in *Journal of Statistical Computing and Simulation*, in press. [Or see Good 1988c.]

——— (1989c). On the neutron and proton masses: a numerological case study. C334 in *Journal of Statistical Computing and Simulation*, in press. [Or see Good 1988c.]

Green, M.B., Schwarz, J.H., and West, P.C. (1985). Anomaly-free chiral theories in six dimensions. *Nuclear Physics* B254, 327–348. Reprinted in Schwarz 1985.

Gross, D.J., Harvey, J.A., Martinec, E., and Rohm, R. (The 'Princeton String Quartet') (1985). Heterotic string theory, I. *Nuclear Physics* B256, 253–284. Reprinted in Schwarz (1985).

Hammersley, J.M. (1950). On estimating restricted parameters (with discussion). *Journal of the Royal Statistical Society, Series B* 12, 192–240.

Hammersley, J.M. and Morton, K.W. (1954). Poor man's Monte Carlo. *Journal of the Royal Statistical Society, Series B* 16, 23–38, with discussion on pp. 61–75.

Hardy, G.H. (1940). *Ramanujan*. Cambridge University Press, Cambridge.

Henrion, M. and Fischhoff, B. (1986). Assessing uncertainty in physical constants. *American Journal of Physics* 54 (9), 791–798.

Ihde, A.J. (1964). *The Development of Modern Chemistry*. Harper & Row, New York.

Jeffreys, H. (1937/57). *Scientific Inference*. 1st and 2nd editions. Cambridge University Press, Cambridge.

——— (1938). *Theory of Probability*. Clarendon Press, Oxford.

Kendall, D.G. (1986). Quantum hunting. In *Encyclopedia of Statistical Sciences*, Volume 7, ed. S. Kotz, N.L. Johnson, and G.B. Read, John Wiley & Sons, New York, 435–439.

Kinoshita, T. (1989). Accuracy of the fine structure constant. *I.E.E.E. I.M. [Instrumentation and Measurement]* 38, 172–174.

Kostelecký, V.A. and Campbell, D.K. (1985). *Supersymmetry in Physics*. Proceedings of a conference on supersymmetry in physics held at the Center for Nonlinear Studies, Los Alamos, NM 87545, USA, December 15–20, 1983.

Laplace, P.S. (1774/1986). Mémoire sur les probabilités des causes par les événements. *Mémoires de Mathématique et de Physique, Académie Royale des Sciences* 6, 621–656. Reprinted in Laplace's *Oeuvres Complètes* 8, 27–65.

English translation by S.M. Stigler, *Statistical Science* 1 (1986), 359–378, with an introduction by Stigler.

Lenz, P. (1951). The ratio of proton and electron masses. *Physical Review* 82, 554.

Messiah, A. (1961). *Quantum Mechanics*, Volume 1. John Wiley& Sons, New York.

Nieto, M.M. (1972). *The Titius-Bode Law of Planetary Distances: Its History and Theory*. Pergamon Press, New York.

Parker-Rhodes, A.F. (1981). *The Theory of Indistinguishables*. Reidel, Dordrecht.

Particle Data Group (1986) ['PDG86']. Review of particle properties. *Physics Letters* 170B, 1–350.

——— (1988/89) ['PDG89']. Review of Particle Properties, *Physics Letters* 204B, 1–486. (Imprinted April 1988, but distributed in February 1989. Covers data up to December 1987.)

Porter, T.M. (1986). *The Rise of Statistical Thinking, 1820–1900*. Princeton University Press, Princeton, N.J.

Prigogine, I. and Stengers, I. (1984). *Order Out of Chaos*. Bantam, New York.

Ramanujan, S. (1915). Highly composite numbers. *Proceedings of the London Mathematical Society* 14, 347–409. Reprinted in *Collected Papers of Srinivasa Ramanujan*, ed. G.H. Hardy, P.V.S. Aiyar, and B.M. Wilson; Cambridge University Press, Cambridge, 1927, and Dover Publications, 1962.

Ramond, P. (1985). Supersymmetry in physics: an algebraic overview. *Physica* 15D, 25–41. Reprinted in Kostelecký and Campbell (1985).

Rissanen, J. (1983). A universal prior for integers and estimation by minimum description-length. *Annals of Statistics* 11, 416–431.

Rowlatt, P.A. (1966). *Group Theory and Elementary Particles*. Longmans, London. (Elementary particles are hardly mentioned except in the Introduction.)

Schwarz, J.H. (1985). *Superstrings*, Volume 2. Editor, World Scientific, Singapore.

——— (1988). Chapter 2 of Davies and Brown (1988).

Sirag, S.P. (1977). A combination [combinatorial] derivation of the proton-electron mass ratio. *Nature* 268, 294.

Slater, N.B. (1957). *Eddington's Fundamental Theory*. Cambridge University Press, Cambridge.

Sommerfeld, A. (1916). Zur Quantentheorie der Spektrallinien. *Annalen der Physik* 51, 1–94.

Watkins, P. (1986). *Story of the W and Z*. Cambridge University Press, Cambridge.

Weber, H. (c. 1908). *Lehrbuch der Algebra*, Volume 3. Chelsea reprint, New York, undated.

Whewell, W. (1847/1967). *The Philosophy of the Inductive Sciences Founded Upon Their History*, two volumes (with an introduction by John Herival). Johnston Reprint Corporation, New York and London.

Whittaker, E.T. (1953). *A History of the Theories of Aether and Electricity: The Modern Theories, 1906–1926.*

Worrall, R.L. (1960). Ratio of neutron and electron mass. *Nature* 185 (Feb. 27), 602.

Wrinch, D. and Jeffreys, H. (1921). On certain fundamental principles of scientific inquiry. *Philosophical Magazine, Series 6* 42, 369–390.

Department of Statistics
Virginia Polytechnic Institute and State University
Blacksburg
Virginia 24061.

Percolation in $\infty + 1$ Dimensions

G.R. Grimmett and C.M. Newman

Abstract

We investigate percolation on the graph of the direct product $\mathbb{T} \times \mathbb{Z}$ of a regular tree $\mathbb{T}$ and the line $\mathbb{Z}$, in which each 'tree' edge is open with probability τ and each 'line' edge with probability λ. There are three non-trivial phases, corresponding to the existence of 0, ∞, and 1 infinite open clusters. Such results may be obtained also for the graph $\mathbb{T} \times \mathbb{Z}^d$ where $d \geq 2$.

1. Introduction

The mathematical theory of percolation was conceived by Simon Broadbent and John Hammersley three decades or so ago as a stochastic model for the flow of material through a porous medium (see Broadbent and Hammersley 1957). In more recent years it has been the subject of much attention from mathematicians and physicists, and progress has been great. Substantial advances have been made in the last ten years, during which time percolation theory has become established as a fundamental tool in modelling random media.

The two phases of percolation are now understood reasonably well. For bond percolation (say) on $\mathbb{Z}^d$, there is a critical density p_c of open edges with the property that if the actual density p satisfies $p < p_c$ then all open clusters are (a.s.) finite, whereas if $p > p_c$ then there exists (a.s.) a *unique* infinite open cluster. The majority of the main unanswered questions about percolation relate to the behaviour of the process when p is close or equal to p_c. The picture is somewhat different for the easier case of bond percolation on a regular tree. For such a graph (which is regarded as 'infinite-dimensional' by physicists), we learn from the theory of branching processes that there exists a critical density p_c ($= k^{-1}$, where $k + 1$ is the common degree of the vertices) such that for $p \leq p_c$ all open clusters are (a.s.) finite whereas if $p_c < p < 1$ then there exists (a.s.) *infinitely* many infinite open clusters. Thus for both lattice and tree there exist two phases; however, the corresponding supercritical phases differ qualitatively in the *number* of infinite clusters (*one* for the lattice, and *infinitely many* for the tree). One reason for this dichotomy lies in the fact that the growth function (i.e. the volume of the n-ball, or the number of vertices within distance n of the origin) grows polynomially (like n^d) for $\mathbb{Z}^d$ but exponentially for the tree (this is one of the reasons trees are sometimes thought of as infinite-dimensional). It is not difficult to see that, for a large class of graphs with periodic structures including all lattices and regular trees, the number N of infinite open clusters satisfies exactly one of $P_p(N=0)=1$, $P_p(N=1)=1$, $P_p(N=\infty)=1$, for any given value of p (see Newman and Schulman 1981). The existing proofs of the uniqueness of the infinite open cluster (Aizenman, Kesten, and Newman 1987; Gandolfi, Grimmett, and Russo 1988; Burton and Keane 1989) may be adapted to all 'periodic' graphs having the property that the surface-to-volume ratio of the n-ball tends to 0 as $n \to \infty$, and this covers all periodic graphs with sub-exponential growth functions; an interesting class of such graphs is discussed implicitly by Grigorchuk (1983).

Lattices and trees have two distinct phases. It is our purpose in this paper to explore the phase diagram of a graph which possesses (at least) three distinct phases, in which the number of infinite clusters is (a.s.) 0, ∞, and 1, respectively. The graph in question is the direct product of the line $\mathbb{Z}$ and a regular tree $\mathbb{T}$, and the actual construction is as follows. Let $\mathbb{T}$ be an infinite regular labelled tree with degree $k+1$, where $k \geq 2$. The *distance* $\delta_{\mathbb{T}}(t_1,t_2)$ between two vertices t_1 and t_2 is defined to be the number of edges in the unique path of $\mathbb{T}$ from t_1 to t_2. A nominated vertex of $\mathbb{T}$ is called the *origin* and labelled $\emptyset$ (the empty word). Vertices adjacent to $\emptyset$ are labelled $0,1,2,\ldots,k$ respectively. More generally, vertices having distance n (≥ 1) from the origin are labelled by words $\alpha_1\alpha_2\cdots\alpha_n$ where $\alpha_1 \in \{0,1,2,\ldots,k\}$ and $\alpha_i \in \{1,2,\ldots,k\}$ for $i \geq 2$; these labels are attached to the vertices in such a way that the vertex $\alpha_1\alpha_2\cdots\alpha_n$ has as neighbours the vertex $\alpha_1\alpha_2\cdots\alpha_{n-1}$ and $\alpha_1\alpha_2\cdots\alpha_n\alpha$ as α ranges over $\{1,2,\ldots,k\}$. We write $V(\mathbb{T})$ for the vertex set of $\mathbb{T}$. The second component of the graph under study is the line $\mathbb{Z}=\{z : z=\ldots,-1,0,1,\ldots\}$ with distance function $\delta_{\mathbb{Z}}(z_1,z_2)=|z_1-z_2|$. We denote by $\mathbb{L}$ the graph with vertex set $V(\mathbb{L})=\{(t,z) : t \in V(\mathbb{T}), z \in \mathbb{Z}\}$ and edge set given by the adjacency relation $(t_1,z_1) \sim (t_2,z_2)$ if and only if $\delta_{\mathbb{T}}(t_1,t_2)+\delta_{\mathbb{Z}}(z_1,z_2)=1$. We write $\mathbb{L}=\mathbb{T}\times\mathbb{Z}$ and note that two vertices of $\mathbb{L}$ are adjacent if and only if either their $\mathbb{T}$-components are equal and their $\mathbb{Z}$-components are adjacent in $\mathbb{Z}$, or vice versa. The *origin* of $\mathbb{L}$ is the vertex $(\emptyset,0)$. We call an edge of $\mathbb{L}$ a $\mathbb{T}$-edge (respectively a $\mathbb{Z}$-edge) if it joins two vertices which differ only in their $\mathbb{T}$-component (respectively $\mathbb{Z}$-component).

We shall consider bond percolation on $\mathbb{L}$, but rather than restricting ourselves to isotropic percolation with constant density, we allow a natural anisotropy as follows. Let τ and λ satisfy $0 \leq \tau, \lambda \leq 1$, and declare each $\mathbb{T}$-edge (respectively $\mathbb{Z}$-edge) to be open with probability τ (respectively λ) independently of the states of all other edges. We shall generally assume

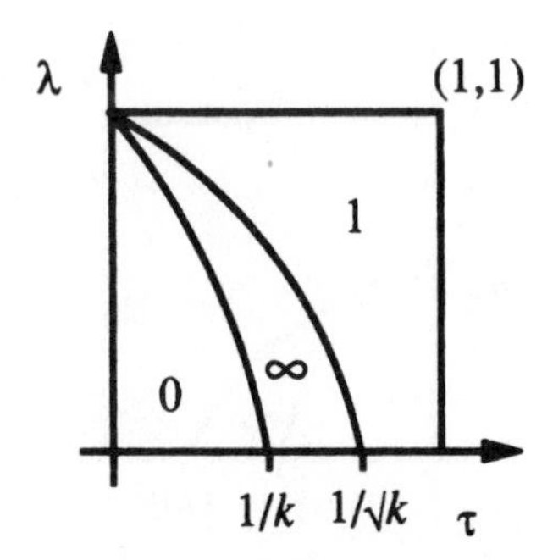

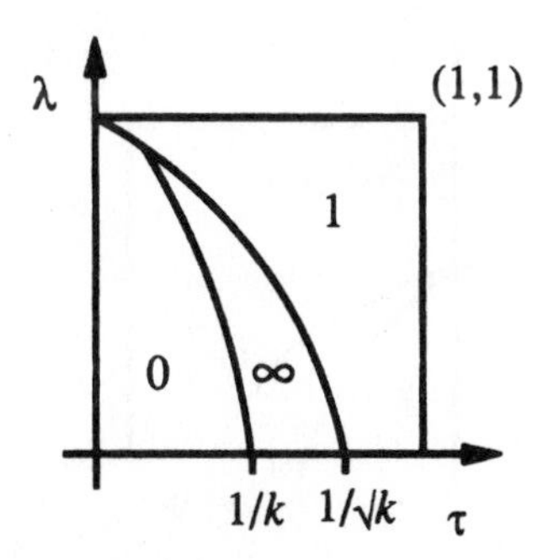

FIG. 1. The set of possible values of (τ, λ) may be partitioned into three regions corresponding to the cases $N = 0$, $N = \infty$, and $N = 1$, respectively. The figure on the left is probably correct, although we have not ruled out the possibility that the figure on the right is correct for small values of k.

that $0 < \tau$, $\lambda < 1$ unless we state otherwise. We write $P_{\tau,\lambda}$ for the ensuing probability measure. Similarly we write P_τ and P_λ for the induced measures on the edge states of subgraphs of the form $\mathbb{T} \times \{z\}$ and $\{t\} \times \mathbb{Z}$ respectively for any given $z \in \mathbb{Z}$ and $t \in V(\mathbb{T})$. More generally, the individual subscripts τ and λ are used to denote quantities associated with projections of $\mathbb{L}$ onto copies of $\mathbb{T}$ or of $\mathbb{Z}$ respectively.

We shall explore the existence and number N of infinite open clusters in $\mathbb{L}$ for various ranges of values of the parameters (τ, λ). It is easy to show in the usual way that $P_{\tau,\lambda}(N = 0) = 1$ for all sufficiently small τ and λ. Also, it is not difficult to adapt the arguments of Newman and Schulman (1981) to see that, for any given (τ, λ), one of the following holds: (i) $N = 0$ a.s., (ii) $N = 1$ a.s., (iii) $N = \infty$ a.s. It turns out that the set of values of (τ, λ) (i.e. the unit square) may be partitioned into three regions each with non-empty interior corresponding to the three cases $N = 0$, $N = 1$, and $N = \infty$. See Figure 1.

We make a number of remarks about Figure 1:

1. The $\lambda = 0$ boundary of the unit square has of course $N = \infty$ for $1/k < \tau < 1$; among the results of this paper is that here $N = \infty$ is stable (respectively unstable) relative to $N = 1$ under perturbations of λ when $1/k < \tau < 1/\sqrt{k}$ (respectively $\tau \geq 1/\sqrt{k}$). Less interesting is the stability of $N = 0$ for $\lambda = 0$, $\tau < 1/k$ or for $\tau = 0$, $\lambda < 1$ and the stability of $N = 1$ for $\tau = 1$ or for $\lambda = 1$, $\tau > 0$.
2. The upper left point $(\tau, \lambda) = (0, 1)$, which has $N = \infty$ while being the endpoint of both $N = 0$ and $N = 1$ boundary segments, is clearly special. For values of k sufficiently large ($k \geq 6$ certainly suffices), the

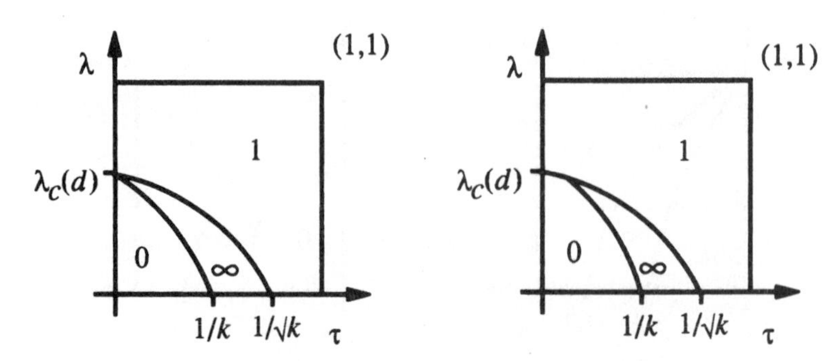

FIG. 2. As in the case of Figure 1, the figure on the left is probably correct, although we have not ruled out the possibility of the figure on the right.

correct picture is the one on the left, in that any neighbourhood of the point $(\tau, \lambda) = (0, 1)$ contains points in each of the three regions. For small values of k, we have not ruled out the possibility that the second diagram in Figure 1 is correct (with only the $N = 0$ and $N = 1$ regions reaching to $(0, 1)$). The first diagram is of course correct for some smallest value of k, and we have no evidence that it is not always the correct picture.

3. A cautionary remark is that although Figure 1 shows the boundary between $N = \infty$ and $N = 1$ as the graph of a function, we are unaware of any (say monotonicity) argument which guarantees this. If the co-existence of a positive-density (defined using ergodicity in the $\mathbb{Z}$-direction) infinite cluster with infinitely many zero-density infinite clusters (as considered in Newman and Schulman 1981) could be ruled out, then monotonicity for the existence of a positive-density infinite cluster would yield such a conclusion. Note however that such a result would imply that along the common boundary of the $N = \infty$ and $N = 1$ regimes there is a line of discontinuities of $P_{\tau,\lambda}((\emptyset, 0)$ belongs to a positive-density infinite cluster)!

A somewhat different picture emerges if the $\mathbb{Z}$-component of $\mathbb{L}$ is replaced by the d-dimensional cubic lattice $\mathbb{Z}^d$ where $d \geq 2$, since such a lattice is capable of sustaining an infinite open cluster without support from the $\mathbb{T}$-edges. Our analysis may be adapted to this situation at little extra cost, and we believe that the correct phase diagram, at least for sufficiently large values of k, is as drawn on the left side of Figure 2, although for no value of k have we ruled out the possibility that the right-hand picture is correct. In this figure, $\lambda_c(d)$ is the critical value of λ for bond percolation on $\mathbb{Z}^d$.

This paper is laid out in the following way. In Section 2 we introduce the necessary notation and we review some useful facts about percolation theory. In Section 3 we explore conditions which are (respectively) necessary and sufficient for the (a.s.) existence of an infinite open cluster in $\mathbb{L}$. This amounts to finding lower and upper bounds for the lower curve in Figure 1. We turn then to conditions which are (respectively) necessary and sufficient for the (a.s.) existence of infinitely many infinite open clusters; this amounts to establishing upper and lower bounds for the upper curve in Figure 1. We present such results in Section 4. There follows a final section devoted to the graph $\mathbb{T} \times \mathbb{Z}^d$ where $d \geq 2$. We remark that results similar to those of this paper may be derived for Ising (and Potts) models on $\mathbb{T} \times \mathbb{Z}$ and $\mathbb{T} \times \mathbb{Z}^d$ (Wu 1989; Newman and Wu 1989).

2. Percolation Notation and Background

For any graph G, we write $V(G)$ for its vertex set and $\langle u, v \rangle$ for the edge between neighbours u and v. We shall explore percolation on $\mathbb{L}$ and on the square lattice $\mathbb{Z}^2$, and have already defined the appropriate percolation model on $\mathbb{L}$. For $\mathbb{Z}^2$ we shall be interested in anisotropic percolation in which each edge $\langle (z_1, z_2), (z_1 + 1, z_2) \rangle$ is open with probability τ, and each edge $\langle (z_1, z_2), (z_1, z_2 + 1) \rangle$ is open with probability λ. We write P and E for the corresponding probability measure and expectation.

In studying percolation on $\mathbb{L}$, we shall use certain results about percolation on $\mathbb{Z}^2$. It is easy to see why such results are relevant. Let π be a doubly infinite path in $\mathbb{T}$ (paths are defined to be self-avoiding). Then π induces a subgraph of $\mathbb{L}$, viz. that with vertex set $\Pi_\pi = \{(t, z) : t \in V(\pi), z \in \mathbb{Z}\}$, and it is easily seen that this subgraph is isomorphic to $\mathbb{Z}^2$; furthermore each 'horizontal' edge of Π_π is open with probability τ, and each 'vertical' edge with probability λ.

We write $\theta(\tau, \lambda)$ for the probability that the origin of $\mathbb{Z}^2$ is in an infinite open cluster of $\mathbb{Z}^2$ and $\theta_{\mathbb{L}}(\tau, \lambda)$ for the probability that the origin of $\mathbb{L}$ is in an infinite open cluster of $\mathbb{L}$.

It is well known that $\theta(\tau, \lambda) > 0$ if and only if $\tau + \lambda > 1$, and that the infinite open cluster is (a.s.) unique under this assumption. See Kesten (1982) and Grimmett (1989) for these and related results and techniques. It is completely standard that there exists (a.s.) an infinite open cluster in any connected graph G if and only if each vertex of G has a strictly positive probability of being in such a cluster. For subsets A and B of the vertex set of G, we write $A \leftrightarrow B$ if there exists an open path in G joining some vertex in A to some vertex in B.

We shall need the idea of the (horizontal) correlation length of bond percolation on $\mathbb{Z}^2$. Define the 'strip'

$$T_m(n) = \{(z_1, z_2) \in \mathbb{Z}^2 : 0 \leq z_1 \leq n, |z_2| \leq m\} \tag{2.1}$$

of length n and height $2m$, and turn $T_m(n)$ into a graph by adding all appropriate edges of $\mathbb{Z}^2$ except those in the 'left' and 'right' sides of $T_m(n)$ (i.e. those of the form $\langle(0,y),(0,y+1)\rangle$ and $\langle(n,y),(n,y+1)\rangle$). Let

$$\phi_m(\tau,\lambda) = \lim_{n\to\infty}\left\{-\frac{1}{n}\log P\big((0,0)\leftrightarrow(n,0) \text{ in } T_m(n)\big)\right\}. \tag{2.2}$$

The limit exists by subadditivity, and furthermore

$$P\big((0,0)\leftrightarrow(n,0) \text{ in } T_m(n)\big) \le e^{-n\phi_m(\tau,\lambda)} \text{ for all } n. \tag{2.3}$$

As in the case of isotropic percolation on $\mathbb{Z}^2$ (see Aizenman, Chayes, Chayes, and Newman 1988 and Grimmett 1989) it is the case that

$$\phi_m(\tau,\lambda) \downarrow \phi(\tau,\lambda) \text{ as } m\to\infty \tag{2.4}$$

where

$$\phi(\tau,\lambda) = \lim_{n\to\infty}\left\{-\frac{1}{n}\log P\big((0,0)\leftrightarrow(n,0)\big)\right\} \tag{2.5}$$

is the reciprocal of the (horizontal) correlation length. Note that, as usual, $\phi(\tau,\lambda)$ is strictly decreasing in τ and λ when $\tau+\lambda<1$, and

$$\phi(\tau,\lambda)\downarrow 0 \text{ as } \lambda\uparrow 1-\tau \text{ or } \tau\uparrow 1-\lambda; \tag{2.6}$$

see Grimmett (1989, Ch. 5) for the corresponding results for isotropic percolation. We note that ϕ may be defined equivalently by

$$\phi(\tau,\lambda) = \lim_{n\to\infty}\left\{-\frac{1}{n}\log P\big((0,0)\leftrightarrow L_n \text{ in } H\big)\right\} \tag{2.7}$$

where L_n is the vertical line $\{(n,z) : z\in\mathbb{Z}\}$ and H is the half-plane $\{(x,y) : x\ge 0, y\in\mathbb{Z}\}$.

Inequality (2.3) provides an upper bound for

$$p_m(n) = P\big((0,0)\leftrightarrow(n,0) \text{ in } T_m(n)\big). \tag{2.8}$$

A comparable lower bound is easily obtained as follows. Let r and s be positive integers, and let A be the event that all edges of the form $\langle(r-1,y),(r-1,y+1)\rangle$, $\langle(r+1,y),(r+1,y+1)\rangle$ for $-m\le y<m$ together with the edges $\langle(r-1,0),(r,0)\rangle$ and $\langle(r,0),(r+1,0)\rangle$ are open. We have by the FKG inequality that

$$\lambda^{4m}\tau^2 p_m(r+s) \le P\big((0,0)\leftrightarrow(r+s,0) \text{ in } T_m(r+s), \text{ and } A\big).$$

The latter probability is no greater than $p_m(r)p_m(s)$, so that

$$\lambda^{4m}\tau^2 p_m(r+s) \le p_m(r)p_m(s),$$

implying by standard arguments that

$$p_m(n) \ge \lambda^{4m}\tau^2 e^{-n\phi_m(\tau,\lambda)} \quad \text{for all} \quad n. \tag{2.9}$$

Finally, here are two bounds involving the function $\phi(\tau, \lambda)$. Clearly $p_m(n) \ge \tau^n$, so that

$$e^{-\phi(\tau,\lambda)} \ge \tau. \tag{2.10}$$

It is not hard to improve this in the following standard way. It is sometimes possible to find open connections (in H) from the origin to the line L_n by observing the (possibly empty) vertical line of open edges through the origin, finding some open edge leading rightwards from this line, and so on. The probability that this construction succeeds in reaching the line L_n is $\{1 - E((1-\tau)^L)\}^n$ where L is the number of vertices in the vertical line of open edges through the origin. An easy calculation shows that

$$E\big((1-\tau)^L\big) = (1-\tau)\left\{\frac{1-\lambda}{1-\lambda(1-\tau)}\right\}^2$$

and hence

$$e^{-\phi(\tau,\lambda)} \ge 1 - \frac{(1-\tau)(1-\lambda)^2}{(1-\lambda(1-\tau))^2}. \tag{2.11}$$

3. Existence of Infinite Clusters in $\mathbb{L}$

Our first results provide (respectively) necessary and sufficient conditions for the existence in $\mathbb{L}$ of an infinite open cluster.

PROPOSITION 1. *If*

$$\tau k(1 + \lambda + \sqrt{2\lambda(1+\lambda)}) < 1 - \lambda \tag{3.1}$$

then there is a.s. no infinite open cluster in $\mathbb{L}$.

PROPOSITION 2. *If*

$$k e^{-\phi(\tau,\lambda)} > 1 \tag{3.2}$$

where ϕ is given by (2.5) or (2.7), then there is a.s. an infinite open cluster in $\mathbb{L}$.

Before giving their proofs, we make a number of remarks concerning these two propositions:

1. Proposition 1 provides a lower bound $\tau = \underline{\tau}_l(\lambda)$ for the lower curve of Figure 1 with the endpoint properties

$$\underline{\tau}_l(0) = \frac{1}{k}, \quad \underline{\tau}_l'(0) = -\infty, \quad \underline{\tau}_l(1) = 0, \quad \text{and} \quad \underline{\tau}_l'(1) = -\frac{1}{4k}. \tag{3.3}$$

2. Proposition 2 provides an upper bound $\tau = \overline{\tau}_l(\lambda)$ for the same curve with a quality depending on how accurate a bound is used for $\phi(\tau, \lambda)$. The trivial bound (2.10) yields the obvious result that $\tau > 1/k$ implies percolation. The improved bound (2.11) for $\phi(\tau, \lambda)$ gives a $\overline{\tau}_l(\lambda)$ with

$$\begin{gathered} \overline{\tau}_l(0) = \frac{1}{k}, \quad \overline{\tau}_l'(0) = -2(k-1)/k^2, \\ \overline{\tau}_l(1) = 0, \quad \text{and} \quad \overline{\tau}_l'(1) = -\left(\sqrt{\frac{k}{k-1}} - 1\right). \end{gathered} \tag{3.4}$$

3. Various small improvements in Proposition 1, (2.11), and Proposition 2 may be obtained. We do not present these here because the increased cost in their proofs seems to be out of proportion to the information gained.

PROOF OF PROPOSITION 1: We shall find an upper bound for $\mathcal{X}(\tau, \lambda)$, the mean number of vertices in the open cluster C of $\mathbb{L}$ at the origin. Our target is to show that $\mathcal{X}(\tau, \lambda) < \infty$ if (3.1) holds, since this implies that C is a.s. finite. We shall make use later of the same method of proof when finding a condition which guarantees the a.s. existence of infinitely many infinite clusters. Clearly

$$\mathcal{X}(\tau, \lambda) = \sum_{(t,z) \in V(\mathbb{L})} P_{\tau,\lambda}((\emptyset, 0) \leftrightarrow (t, z)). \tag{3.5}$$

Now $(\emptyset, 0) \leftrightarrow (t, z)$ if and only if there exists a (self-avoiding) path of $\mathbb{L}$ from $(\emptyset, 0)$ to (t, z) which is open. Any such path contains $\mathbb{T}$-edges and $\mathbb{Z}$-edges, but may be projected onto the section $\mathbb{T} \times \{0\}$ to give a route from $(\emptyset, 0)$ to the vertex $(t, 0)$ (*routes* are paths with the self-avoiding condition removed). This route is a sequence $(t_0, 0), (t_1, 0), \ldots, (t_n, 0)$ where $t_0 = \emptyset$, $t_n = t$, and t_i is adjacent to t_{i+1} in $\mathbb{T}$ for $0 \le i < n$. We denote this route by $\pi_{\mathbf{t}} = (t_0, t_1, \ldots, t_n)$, which we think of as a route in $\mathbb{T}$ from $\emptyset$ to t. The aforesaid path in $\mathbb{L}$ from $(\emptyset, 0)$ to (t, z) proceeds along $\mathbb{Z}$-edges from $(\emptyset, 0)$ to some $(\emptyset, z_0)$, thence along $\mathbb{T}$-edges to (t_1, z_0), thence along $\mathbb{Z}$-edges to some (t_1, z_1), thence to (t_2, z_1), and so on until it arrives at $(t_n, z_n) = (t, z)$. We denote this path by $\pi(\mathbf{t}, \mathbf{z})$. It follows from (3.5) that

$$\begin{aligned} \mathcal{X}(\tau, \lambda) &\le \sum_{(t,z)} \sum_{\substack{\mathbf{t}: \\ t_n = t}} \sum_{\substack{\mathbf{z}: \\ z_n = z}} P_{\tau,\lambda}(\pi(\mathbf{t}, \mathbf{z}) \text{ is open}) \\ &= \sum_{n=0}^{\infty} \sum_{\mathbf{t}} \sum_{\mathbf{z}} P_{\tau,\lambda}(\pi(\mathbf{t}, \mathbf{z}) \text{ is open}) \end{aligned} \tag{3.6}$$

where the final two summations are over all appropriate sequences $\mathbf{t} = (t_0, t_1, \ldots, t_n)$, $\mathbf{z} = (z_0, z_1, \ldots, z_n)$, where $(t_0, z_0) = (\emptyset, 0)$. We sum first over possible values for z_n. The number of choices for z_n is restricted by the fact that $\pi(\mathbf{t}, \mathbf{z})$ may already have visited the line $\{t_n\} \times \mathbb{Z}$ thereby removing certain possible vertices from consideration. In any case, the set of possibilities for z_n is no larger than the whole line $\{t_n\} \times \mathbb{Z}$, so that

$$\sum_{z_n} P_{\tau,\lambda}(\pi(\mathbf{t}, \mathbf{z}) \text{ is open}) \le \mathcal{X}_\lambda P_{\tau,\lambda}(\pi(\mathbf{t}, \mathbf{z})' \text{ is open}) \tag{3.7}$$

where

$$\mathcal{X}_\lambda = 1 + 2\sum_{i=1}^{\infty} \lambda^i = \frac{1+\lambda}{1-\lambda} \tag{3.8}$$

is the mean number of vertices on $\{t_n\} \times \mathbb{Z}$ joined by open $\mathbb{Z}$-paths to (t_n, z_{n-1}), and $\pi(\mathbf{t}, \mathbf{z})'$ is the path $\pi(\mathbf{t}, \mathbf{z})$ with the $\mathbb{Z}$-edges from (t_n, z_{n-1}) to (t_n, z_n) removed. Progressive summation over all the z_i's yields similarly

$$\mathcal{X}(\tau, \lambda) \le \sum_{n=0}^{\infty} \sum_{\mathbf{t}} \tau^n \mathcal{X}_\lambda^{n+1}, \tag{3.9}$$

the term τ^n coming from the fact that $\pi(\mathbf{t}, \mathbf{z})$ uses exactly n $\mathbb{T}$-edges of $\mathbb{L}$. This bound for $\mathcal{X}(\tau, \lambda)$ may be improved as follows. For a given route t_0, $t_1, \ldots, t_n$ with $t_0 = \emptyset$, define for $2 \le i \le n$

$$I_i = \begin{cases} 1 & \text{if } t_i = t_{i-2}, \\ 0 & \text{otherwise,} \end{cases}$$

and

$$S(\mathbf{t}) = \sum_{i=2}^{n} I_i.$$

If $I_n = 1$ then the projected walk $\pi_{\mathbf{t}}$ moves from t_{n-2} to t_{n-1} and back to t_{n-2} $(= t_n)$. In this circumstance, the sum over possible choices for z_{n-1} in (3.6) contributes no more than $\mathcal{X}_\lambda - 1$, since the path $\pi(\mathbf{t}, \mathbf{z})$ is self-avoiding and thus $z_{n-1} \ne z_{n-2}$. It follows similarly that

$$\begin{aligned} \mathcal{X}(\tau, \lambda) &\le \sum_{n=0}^{\infty} \sum_{\mathbf{t}} \tau^n (\mathcal{X}_\lambda - 1)^{S(\mathbf{t})} \mathcal{X}_\lambda^{n+1-S(\mathbf{t})} \\ &= \mathcal{X}_\lambda \sum_{n=0}^{\infty} (\tau \mathcal{X}_\lambda)^n \sum_{\mathbf{t}} (1 - \mathcal{X}_\lambda^{-1})^{S(\mathbf{t})} \end{aligned} \tag{3.10}$$

in place of (3.9).

Let π be a route in $\mathbb{T}$ beginning at the origin, thought of as a sequence of *directed* steps. We classify each step of π as either an 'outstep' or an 'instep' according to the following rule. A step from t_1 to t_2 where $t_1 \neq \emptyset$ is an *outstep* if and only if $\delta_{\mathbb{T}}(\emptyset, t_2) > \delta_{\mathbb{T}}(\emptyset, t_1)$. A step from $\emptyset$ to t is an *outstep* if t is not labelled 0 and an *instep* otherwise. From each $t \in \mathbb{T}$ there are exactly k possible outsteps.

Returning to the route $\pi_{\mathbf{t}}$ above, we define $J_i = 1$ if the step of $\pi_{\mathbf{t}}$ from t_i to t_{i+1} is an instep and $J_i = 2$ otherwise. We set

$$T(\mathbf{J}) = |\{i : J_{i-1} = 2, J_i = 1\}|,$$

the number of times an outstep is followed by an instep. Note that $T(\mathbf{J}) \leq S(\mathbf{t})$ so that

$$\begin{aligned} \mathcal{X}(\tau,\lambda) &\leq \mathcal{X}_\lambda \sum_{n=0}^{\infty} (\tau \mathcal{X}_\lambda)^n \sum_{\mathbf{t}} (1 - \mathcal{X}_\lambda^{-1})^{T(\mathbf{J})} \\ &\leq \mathcal{X}_\lambda \sum_{n=0}^{\infty} (k\tau \mathcal{X}_\lambda)^n \sum_{\mathbf{J}} (1 - \mathcal{X}_\lambda^{-1})^{T(\mathbf{J})} \end{aligned} \tag{3.11}$$

where the final summation is over all sequences $\mathbf{J} = (J_0, J_1, \cdots, J_{n-1})$ of 1's and 2's; the second inequality holds since each sequence $\mathbf{J}$ corresponds to at most k^n sequences $\mathbf{t}$. However,

$$\sum_{\mathbf{J}} (1 - \mathcal{X}_\lambda^{-1})^{T(\mathbf{J})} = \begin{pmatrix} 1 & 1 \end{pmatrix} \begin{pmatrix} 1 & 1 \\ 1 - \mathcal{X}_\lambda^{-1} & 1 \end{pmatrix}^{n-1} \begin{pmatrix} 1 \\ 1 \end{pmatrix},$$

which behaves for large n in the manner of η^{n-1} where $\eta = 1 + \sqrt{1 - \mathcal{X}_\lambda^{-1}}$ is the larger eigenvalue of the matrix

$$\begin{pmatrix} 1 & 1 \\ 1 - \mathcal{X}_\lambda^{-1} & 1 \end{pmatrix}.$$

Hence $\mathcal{X}(\tau, \lambda) < \infty$ if

$$k\tau \mathcal{X}_\lambda (1 + \sqrt{1 - \mathcal{X}_\lambda^{-1}}) < 1. \tag{3.12}$$

Substituting from (3.8) for $\mathcal{X}_\lambda$, we obtain the assertion of the proposition. ∎

PROOF OF PROPOSITION 2: We shall show that $\theta_{\mathbb{L}}(\tau, \lambda) > 0$ if $ke^{-\phi(\tau,\lambda)} > 1$. Consider the subtree $\mathbb{T}^+$ of $\mathbb{T}$ being the component containing $\emptyset$ of the graph obtained from $\mathbb{T}$ by deleting the edge $\langle \emptyset, 0 \rangle$, and fix a positive integer L; later we shall take the limit as $L \to \infty$. We construct a branching process on $\mathbb{T}^+$ as follows. If $t \in V(\mathbb{T}^+)$ is such that $\delta_{\mathbb{T}}(\emptyset, t) = L$, we declare t to be *green* if there exists an open path of $\mathbb{T}^+ \times \mathbb{Z}$ joining $(\emptyset, 0)$ to $(t, 0)$ and (apart from its endvertices) using only vertices of $\mathbb{L}$ in $\{(u, z) \in V(\mathbb{L}) : 0 < \delta_{\mathbb{T}}(\emptyset, u) < L, \quad z \in \mathbb{Z}\}$. Proceeding inductively, suppose that $t \in V(\mathbb{T}^+)$ is such that $\delta_{\mathbb{T}}(\emptyset, t) = aL$ for some positive integer a. There is a unique $v \in V(\mathbb{T}^+)$ such that $\delta_{\mathbb{T}}(\emptyset, v) = (a-1)L$, $\delta_{\mathbb{T}}(v, t) = L$. We declare t to be *green* if and only if (i) v is green, and (ii) there is an open path from $(v, 0)$ to $(t, 0)$ using (apart from its endvertices) only vertices of $\mathbb{L}$ in $\{(u, z) : (a-1)L < \delta_{\mathbb{T}}(\emptyset, u) < a\mathbb{L}, \quad z \in \mathbb{Z}\}$. It should be clear that the set of green vertices constitutes a branching process on $\mathbb{T}^+$ with mean family-size at least

$$\mu_L = k^L P\big((0,0) \leftrightarrow (L,0) \text{ in } T_\infty(L)\big)$$

where P is the probability measure of anisotropic percolation on $\mathbb{Z}^2$ and

$$T_\infty(L) = \lim_{m \to \infty} T_m(L),$$

$T_m(L)$ being the strip given in (2.1). If $\mu_L > 1$ for some $L \geq 1$ then this branching process is supercritical and has therefore strictly positive probability of being infinite. This implies that $\theta_{\mathbb{L}}(\tau, \lambda) > 0$. Now

$$\begin{aligned} -\frac{1}{L} \log \mu_L &= -\log k - \frac{1}{L} \log P\big((0,0) \leftrightarrow (L,0) \text{ in } T_\infty(L)\big) \\ &\leq -\log k - \frac{1}{L} \log P\big((0,0) \leftrightarrow (L,0) \text{ in } T_m(L)\big) \\ &\to -\log k + \phi_m(\tau, \lambda) \qquad \text{as } L \to \infty \\ &\to -\log(ke^{-\phi(\tau,\lambda)}) \qquad \text{as } m \to \infty \end{aligned}$$

by (2.2) and (2.4). Therefore, if $ke^{-\phi(\tau,\lambda)} > 1$ then $\mu_L > 1$ for all large L, and the result is proved. ■

4. Existence of Infinitely Many Infinite Clusters

We establish in Propositions 4 and 5 below conditions which are (respectively) sufficient and necessary for the existence of infinitely many infinite open clusters in $\mathbb{L}$. First we state a lemma relating this phenomenon to the decay of the connectivity function; its proof is given after the statement of Proposition 5.

LEMMA 3. *Let C be the open cluster of $\mathbb{L}$ at the origin, and for $t \in V(\mathbb{T})$ let*

$$D_t = \{(t,z) \in \mathbb{L} : (\emptyset,0) \leftrightarrow (t,z)\}$$

denote the intersection of C with $\{t\} \times \mathbb{Z}$. If, for some τ, λ, it is the case that $|D_\emptyset| < \infty$ a.s., then
(i) $|D_t| < \infty$ a.s., for all $t \in V(\mathbb{T})$,
(ii) $P_{\tau,\lambda}((\emptyset,0) \leftrightarrow (t,z)) \to 0$ as $\delta_{\mathbb{T}}(\emptyset,t) + |z| \to \infty$.
If further $\theta_{\mathbb{L}}(\tau,\lambda) > 0$, then there are a.s. infinitely many infinite open clusters in $\mathbb{L}$, each of which intersects each $\{t\} \times \mathbb{Z}$ in only finitely many vertices.

PROPOSITION 4. *If*

$$\tau\sqrt{k}\left(1+\lambda+\sqrt{2\lambda(1+\lambda)}\right) < 1-\lambda, \tag{4.1}$$

then $|D_\emptyset| < \infty$ a.s. Thus if in addition $\theta_{\mathbb{L}}(\tau,\lambda) > 0$, then there exist a.s. infinitely many infinite open clusters in $\mathbb{L}$.

Note that (4.1) differs from (3.1) only in the replacement of k by $\sqrt{k}$. As was the case with (3.1), we may improve condition (4.1) to obtain a weaker condition sufficient for the conclusion. Such improvements incur extra costs without the benefit of substantial improvement towards optimality. Proposition 4 provides a lower bound $\tau = \underline{\tau}_u(\lambda)$ for the upper curve of Figure 1 satisfying

$$\begin{gathered}\underline{\tau}_u(0) = \frac{1}{\sqrt{k}}, \quad \underline{\tau}_u'(0) = -\infty, \\ \underline{\tau}_u(1) = 0, \quad \text{and} \quad \underline{\tau}_u'(1) = -\frac{1}{4\sqrt{k}}.\end{gathered} \tag{4.2}$$

Proposition 4 combined with Proposition 2 implies the existence for all $k \geq 2$ of a region of values of (τ,λ) for which there exist infinitely many infinite open clusters; to see this, simply note from (3.4) and (4.2) that $\overline{\tau}_l(0) < \underline{\tau}_u(0)$. For sufficiently large k (i.e. $k \geq 6$) our estimates imply that this region extends all the way to the point $(\tau,\lambda) = (0,1)$ since $|\overline{\tau}_l'(1)| < |\underline{\tau}_u'(1)|$ for $k \geq 6$. It can also be checked that, for large k, $\overline{\tau}_l(\lambda) < \underline{\tau}_u(\lambda)$ for all $0 < \lambda < 1$.

We turn next to conditions which are sufficient for the a.s. uniqueness of the infinite cluster. It is not too difficult to show that there is a.s. a unique infinite open cluster when $\tau+\lambda > 1$, making use of the fact that each infinite path π in $\mathbb{T}$ gives rise to a subgraph $\Pi_\pi = \{(t,z) : t \in V(\pi), z \in \mathbb{Z}\}$ of $\mathbb{L}$ which is isomorphic to $\mathbb{Z}^2$ and therefore contains a.s. a unique infinite open cluster. We weaken the condition $\tau + \lambda > 1$ in the next proposition.

PROPOSITION 5. *If*

$$ke^{-2\phi(\tau,\lambda)} > 1 \tag{4.3}$$

then there exists a.s. a unique infinite open cluster in $\mathbb{L}$.

As with Proposition 4, the condition (4.3) of Proposition 5 differs from (3.2), the corresponding condition for actual percolation, in the replacement by $\sqrt{k}$ of k. Proposition 5 combined with (2.11) provides an upper bound $\overline{\tau}_u(\lambda)$ for the upper curve of Figure 1 satisfying

$$\begin{gathered}\overline{\tau}_u(0) = \frac{1}{\sqrt{k}}, \quad \overline{\tau}'_u(0) = -2(\sqrt{k}-1)/k, \\ \overline{\tau}_u(1) = 0, \quad \text{and} \quad \overline{\tau}'_u(1) = -\left(\left(\frac{\sqrt{k}}{\sqrt{k}-1}\right)^{1/2} - 1\right).\end{gathered} \tag{4.4}$$

We remark that it is natural to conjecture that when $\theta_{\mathbb{L}}(\tau,\lambda) > 0$, either the infinite cluster is unique or else the situation of Lemma 3 is valid; however this has not been proved.

PROOF OF LEMMA 3: We first prove (i). Let us suppose that

$$P_{\tau,\lambda}(|D_\emptyset| < \infty) = 1$$

but that there exist $t \in \mathbb{T}$ such that

$$P_{\tau,\lambda}(|D_t| = \infty) > 0.$$

We may choose such a t such that $\delta_{\mathbb{T}}(\emptyset, t) = m$ is a minimum, and we write s for the unique vertex of $\mathbb{T}$ satisfying $\delta_{\mathbb{T}}(\emptyset, s) = m-1$, $\delta_{\mathbb{T}}(s,t) = 1$. Then D_s is a.s. finite but $P_{\tau,\lambda}(|D_t| = \infty) = \eta > 0$; we shall show the event $\{|D_s| < \infty\} \cap \{|D_t| = \infty\}$ has probability zero, thus contradicting the minimality of $\delta_{\mathbb{T}}(\emptyset, t)$. Pick ϵ satisfying $0 < \epsilon < \eta$ and find a positive integer M such that

$$P_{\tau,\lambda}\big((\emptyset,0) \leftrightarrow (s,z) \text{ for some } |z| > M\big) < \epsilon. \tag{4.5}$$

On the other hand

$$P_{\tau,\lambda}\big((\emptyset,0) \leftrightarrow (t,z) \text{ for infinitely many } |z| > M\big) = \eta.$$

There is probability at least $\eta - \epsilon$ that there exists an infinite set Z of integers z with $|z| > M$ such that $(\emptyset, 0) \leftrightarrow (t,z)$ for all $z \in Z$ in the graph obtained from $\mathbb{L}$ by deleting (i.e. without examining the states of) the edges in the set $E = \{\langle (s,z),(t,z)\rangle : |z| > M\}$. Almost surely infinitely many edges in E having an endvertex of the form (t, z) for $z \in Z$ are open. Hence, $|D_s| = \infty$ occurs with probability $\eta - \epsilon > 0$, a contradiction.

Before discussing (ii) we prove the final statement of the lemma. If there were a.s. a unique infinite open cluster, then by ergodicity in the $\mathbb{Z}$-direction the set of $(\emptyset, z)$ belonging to this cluster would have positive density θ_{L} and hence would be infinite, so that $D_\emptyset$ would be infinite with probability $\theta_{\mathsf{L}} > 0$, a contradiction. However, by the arguments of Newman and Schulman (1981), the only alternative is that there exist a.s. infinitely many infinite clusters; each such cluster must have a.s. finite intersection with $\{t\} \times \mathbb{Z}$ by (i).

It remains to prove (ii). We first note that by (i) the probability in (ii) tends to zero as $|z| \to \infty$ for fixed t. Thus we assume that for some sequence (t_i, z_i) with $\delta_{\mathsf{T}}(\emptyset, t_i) \to \infty$, it is the case that

$$P_{\tau,\lambda}\big((\emptyset, 0) \leftrightarrow (t_i, z_i)\big) \geq \eta > 0 \quad \text{for all} \quad i,$$

and we search for a contradiction. For each i, we may choose R_i so that for any z, the event A_i^z, that $(\emptyset, z) \leftrightarrow (t_i, z_i + z)$ in the region $\mathcal{R}_i = \{(t, z') : \delta_{\mathsf{T}}(\emptyset, t) \leq R_i, z' \in \mathbb{Z}\}$, is such that $P_{\tau,\lambda}(A_i^z) \geq \eta/2$. By choosing a subsequence if necessary, we can and will assume that $R_i < \delta_{\mathsf{T}}(\emptyset, t_{i+1})$ for each i. For $z \geq 0$, let B_i^z be the event that both $(\emptyset, 0)$ and $(\emptyset, z)$ are connected in the region $\mathcal{R}_i$ to vertices in $\{(t_i, z') : z_i \leq z' \leq z_i + z\}$. Then by the Harris-FKG inequality,

$$P_{\tau,\lambda}(B_i^z) \geq P_{\tau,\lambda}(A_i^0 \cap A_i^z) \geq (\eta/2)^2 \quad \text{for all} \quad z \geq 0$$

so that B_i^z occurs for infinitely many i's with probability at least $(\eta/2)^2$. We show next that

$$\begin{aligned} P_{\tau,\lambda}\big((\emptyset, 0) \leftrightarrow (\emptyset, z)\big) &\geq P_{\tau,\lambda}(B_i^z \text{ occurs for infinitely many } i\text{'s}) \\ &\geq (\eta/2)^2; \end{aligned} \tag{4.6}$$

the contradiction follows since we have already concluded from (i) that $P_{\tau,\lambda}((\emptyset, 0) \leftrightarrow (\emptyset, z)) \to 0$ as $|z| \to \infty$.

To obtain (4.6) let us for a given z define $b_1, b_2, \ldots$ to be the sequence of i's for which B_i^z occurs ($b_k = \infty$ if B_i^z occurs fewer then k times). Then (4.6) is an easy consequence of the limit as $j \to \infty$ of the inequalities

$$\begin{aligned} &P_{\tau,\lambda}\big(b_j < \infty \text{ and } (\emptyset, 0) \not\leftrightarrow (\emptyset, z) \text{ in } \mathcal{R}_{b_j}\big) \\ &\qquad \leq (1 - \lambda^z) P_{\tau,\lambda}\big(b_{j-1} < \infty \text{ and } (\emptyset, 0) \not\leftrightarrow (\emptyset, z) \text{ in } \mathcal{R}_{b_{j-1}}\big) \\ &\qquad \leq (1 - \lambda^z)^j. \end{aligned}$$

To obtain this estimate, condition on b_j and on the states of all edges in $\mathcal{R}_{b_j}$ except the z $\mathbb{Z}$-edges connecting the vertices in $\{(t_{b_j}, z') : z_i \leq z' \leq$

$z_i + z\}$. The states of these z edges are independent of the value of b_j, and $(\emptyset, 0) \leftrightarrow (\emptyset, z)$ in $\mathcal{R}_{b_j}$ if all z edges are open, which occurs with probability λ^z. The desired estimate follows. ■

PROOF OF PROPOSITION 4: The proof resembles very closely that of Proposition 1. Arguing as in that proof, we find that

$$E_{\tau,\lambda}|D_\emptyset| \leq \mathcal{X}_\lambda \sum_{n=0}^{\infty} (\tau \mathcal{X}_\lambda)^n \sum_{\mathbf{t}} (1 - \mathcal{X}_\lambda^{-1})^{T(\mathbf{J})}$$

where, unlike (3.11), the second summation is over all routes $\mathbf{t} = (t_0, t_1, \ldots, t_n)$ in $\mathbb{T}$ satisfying $t_0 = \emptyset$ and $t_n = \emptyset$. In such a case, n is even. We claim that any such path contains no more than $\frac{1}{2}n$ outsteps. To see this, note that in excursions of $\mathbf{t}$ from $\emptyset$ beginning with an outstep, the number of outsteps equals the number of insteps, whereas in excursions beginning with an instep, the insteps outnumber the outsteps by 2. Hence each sequence $\mathbf{J}$ corresponds to at most $k^{n/2}$ sequences $\mathbf{t}$, giving that

$$E_{\tau,\lambda}|D_\emptyset| \leq \mathcal{X}_\lambda \sum_{\substack{n=0 \\ n \text{ even}}}^{\infty} (\tau \mathcal{X}_\lambda \sqrt{k})^n \sum_{\mathbf{J}} (1 - \mathcal{X}_\lambda^{-1})^{T(\mathbf{J})}$$

which, by the previous argument, converges if

$$\tau \mathcal{X}_\lambda \sqrt{k} \left(1 + \sqrt{1 - \mathcal{X}_\lambda^{-1}}\right) < 1.$$

Substituting $\mathcal{X}_\lambda = (1+\lambda)/(1-\lambda)$, we conclude that, under (4.1), $E_{\tau,\lambda}|D_\emptyset| < \infty$ and therefore $P_{\tau,\lambda}(|D_\emptyset| = \infty) = 0$ as required. ■

PROOF OF PROPOSITION 5: We suppose that

$$ke^{-2\phi} > 1. \tag{4.7}$$

By (2.4), we may pick a positive integer m such that

$$ke^{-2\phi_m} > 1$$

where

$$\phi_m(\tau, \lambda) = \lim_{n \to \infty} \left\{ -\frac{1}{n} \log P\big((0,0) \leftrightarrow (n,0) \text{ in } T_m(n)\big) \right\}$$

as in (2.2). We have from (2.9) that

$$p_m(n) = P\big((0,0) \leftrightarrow (n,0) \text{ in } T_m(n)\big)$$

satisfies

$$p_m(n) \geq \lambda^{4m}\tau^2 e^{-n\phi_m} \text{ for all } n \geq 0,$$

and therefore we may pick a positive integer n such that

$$k^n p_m(n)^2 \geq \lambda^{8m}\tau^4(ke^{-2\phi_m})^n > 1. \tag{4.8}$$

Next we recall and introduce some notation. For any vertex (t,z) of $\mathbb{L}$ with $t \neq \emptyset$, there is a unique vertex s of $\mathbb{T}$ such that $\delta_{\mathbb{T}}(s,\emptyset) = \delta_{\mathbb{T}}(t,\emptyset) - 1$ and $\delta_{\mathbb{T}}(s,t) = 1$. Let $\mathbb{T}^+(t)$ denote the subtree of $\mathbb{T}$ being the component containing t of the graph obtained by deleting from $\mathbb{T}$ the edge $\langle s,t\rangle$. We write $\mathbb{T}^+(t,z)$ for the subgraph of $\mathbb{L}$ induced by the vertex set $\{(u,z) : u \in V(\mathbb{T}^+(t))\}$, and we denote by $\mathbb{T}^+_m(t,z)$ the subgraph of $\mathbb{L}$ induced by the vertex set $\{(u,y) : u \in V(\mathbb{T}^+(t)), |z-y| \leq m\}$. We introduce similar notation for a vertex of the form $(\emptyset, z)$ in terms of the tree $\mathbb{T}^+(\emptyset)$ obtained as the component containing $\emptyset$ of the graph obtained from $\mathbb{T}$ by deleting the edge $\langle \emptyset, 0\rangle$.

Let $(t,z) \in V(\mathbb{L})$. We construct a (random) set of vertices of $\mathbb{T}^+_m(t,z)$ in the following way. We begin by colouring (t,z) *red*. Next we examine vertices of $\mathbb{T}^+_m(t,z)$ of the form (u,z) where $\delta_{\mathbb{T}}(t,u) = n$. There is a unique path $\pi_z(t,u)$ of $\mathbb{T} \times \{z\}$ joining (t,z) to (u,z); with this path we associate a 'strip'

$$S_z(t,u) = \{(s,y) \in V(\mathbb{L}) : (s,z) \in \pi_z(t,u), |y-z| \leq m\}$$

together with all associated edges of $\mathbb{L}$ with at least one endvertex of the form (s,y) with $s \neq t,u$ and $|y-z| \leq m$. We colour the vertex (u,z) *red* if and only if $(t,z) \leftrightarrow (u,z)$ in $S_z(t,u)$. Having coloured the vertices (u,z) with $\delta_{\mathbb{T}}(t,u) = n$, we turn to those vertices (w,z) of $\mathbb{T}^+_m(t,z)$ with $\delta_{\mathbb{T}}(t,w) = 2n$. Let (w,z) be such a vertex. There exists a unique vertex u of $\mathbb{T}$ with $\delta_{\mathbb{T}}(t,u) = \delta_{\mathbb{T}}(u,w) = n$. We colour (w,z) *red* if and only if (a) (u,z) is red, and (b) $(u,z) \leftrightarrow (w,z)$ in the strip $S_z(u,w)$. We proceed inductively to obtain a set of red vertices. Clearly the set of red vertices is the set of members of a branching process with mean family-size $k^n p_m(n)$, and we denote this set by $T(t,z)$.

We say that $T(t,z)$ and $T(t,y)$ *overlap infinitely often* (i.o.) if there exist infinitely many vertices $w \in \mathbb{T}^+(t)$ such that (w,z) is red and (w,y) is red. Let $0 \leq \epsilon < 1$. We call the (random) set $T(t,z)$ *ϵ-robust* if (conditional on $T(t,z)$) the probability that $T_1 = T(t,z)$ and $T_2 = T(t, z+2m+1)$ overlap i.o. is strictly larger than ϵ. We note that the (unconditional) probability that T_1 and T_2 overlap i.o. is exactly the probability that the set of vertices w of $\mathbb{T}^+(t)$ such that both (w, z) and $(w, z + 2m + 1)$ are red is infinite. Thus this probability equals the probability that a branching process with mean family-size $k^n p_m(n)^2$ is infinite; such a process is supercritical by (4.8), and therefore the probability in question is strictly positive. Thus

$$\begin{aligned} 0 &< P_{\tau,\lambda}(T_1 \text{ and } T_2 \text{ overlap i.o.}) \\ &= E_{\tau,\lambda}\big(P_{\tau,\lambda}(T_2 \text{ overlaps } T_1 \text{ i.o.} | T_1)\big) \\ &\leq P_{\tau,\lambda}(T_1 \text{ is } \epsilon\text{-robust}) + \epsilon P_{\tau,\lambda}(T_1 \text{ is 0-robust but not } \epsilon\text{-robust}) \end{aligned}$$

implying that

$$P_{\tau,\lambda}(T_1 \text{ is } \epsilon\text{-robust}) \geq \frac{P_{\tau,\lambda}(T_1 \text{ and } T_2 \text{ overlap i.o.}) \ - \epsilon}{1 - \epsilon},$$

where the final quantity is strictly positive for all sufficiently small non-negative ϵ. Clearly $P_{\tau,\lambda}(T_1 \text{ is } \epsilon\text{-robust})$ is a decreasing function of ϵ, and we claim that

$$P_{\tau,\lambda}(T_1 \text{ is 0-robust}) = P_{\tau,\lambda}(|T_1| = \infty). \tag{4.9}$$

Certainly the left-hand side is no larger than the right-hand side; to prove equality it suffices to show that

$$P_{\tau,\lambda}(T_1 \text{ is 0-robust}) \geq P_{\tau,\lambda}(|T_1| = \infty). \tag{4.10}$$

To see this we argue as follows. Let ϵ be small and positive. We grow T_1 generation by generation. Each time we reach a new red vertex, there is a strictly positive probability that this vertex is the root of an ϵ-robust tree in future generations. If T_1 is infinite than a.s. we encounter such a red vertex at some stage. If N is the generation number of the first such vertex reached, then there is probability at least $\tau^{nN}\epsilon$ (> 0), that T_2 overlaps T_1 i.o. For any given T_1, we write $\rho(T_1)$ for the supremum of the values of ϵ for which T_1 is ϵ-robust, with the convention that $\rho(T_1) = -1$ if T_1 is either finite or not 0-robust. We have proved that

$$P_{\tau,\lambda}(\rho(T_1) > 0) = P_{\tau,\lambda}(T_1 \text{ is infinite}). \tag{4.11}$$

Calculations related to these may be found in Lyons (1988).

Having set the scene, we move on to the proof proper. Let (t, z) be a vertex of $\mathbb{L}$ which is in an infinite open cluster, say $C(t, z)$, of $\mathbb{L}$. We claim that $C(t, z)$ contains a.s. some vertex of $\mathbb{T} \times \{y\}$ for infinitely many values of y $(\in \mathbb{Z})$. Suppose to the contrary that $C(t, z)$ is confined to some layer $\mathbb{T} \times \{M, \ldots, N\}$. Then there exists some positive integer I which is maximal with the property that $C(t, z)$ contains infinitely many vertices of $\mathbb{T} \times \{I\}$. By an argument similar to that in the proof of part (i) of Lemma 3, this event has probability 0 for any given value I, so that a.s. no such I exists. Thus $C(t, z)$ contains a.s. vertices of $\mathbb{T} \times \{y\}$ for every value of y. Growing $C(t, z)$ in the usual algorithmic way (see for example Aizenman, Kesten, and Newman 1987 or Grimmett 1989), we find that $C(t, z)$ contains a.s. some vertex (u, y) which is the root of a robust (i.e. 0-robust) $T(u, y)$ in $\mathbb{T}_m^+(u, y)$ (in fact this will hold for infinitely many values of y); this holds since, each time the growth process reaches a new plane $\mathbb{T} \times \{y\}$, arriving from $\mathbb{T} \times \{y - 1\}$ say, there is a strictly positive probability that the hitting point (v, y) is joined to $(v, y + m)$ by a direct path of open edges and in addition $(v, y + m)$ is the root of a robust $T(v, y + m)$ in $\mathbb{T}_m^+(v, y + m)$. By a similar argument we may (and will) assume that $\delta_{\mathbb{T}}(u, \emptyset)$ is a multiple of n.

Suppose now that (s, y) and (t, z) are distinct vertices of $\mathbb{L}$ which are in infinite open clusters. We wish to show that $(s, y) \leftrightarrow (t, z)$ a.s. on this event. Almost surely, $C(s, y)$ and $C(t, z)$ contain vertices (a, i) and (b, j) (respectively) which are the roots of robust sets $T(a, i)$ and $T(b, j)$ in $\mathbb{T}_m^+(a, i)$ and $\mathbb{T}_m^+(b, j)$ respectively (and with $\delta_{\mathbb{T}}(a, \emptyset)$ and $\delta_{\mathbb{T}}(b, \emptyset)$ multiples of n, and, if desired, with $|i - j| > 2m$). It suffices therefore to show that there is probability 0 that there exist two such distinct vertices (a, i) and (b, j) which are the roots of such robust sets but which are not connected by an open path of $\mathbb{L}$. Let (a, i) and (b, j) be distinct vertices of $\mathbb{L}$. We say that (a, i) is *related* to (b, j) if $V(\mathbb{T}^+(a)) \cap V(\mathbb{T}^+(b)) \neq \emptyset$, and *unrelated* otherwise. Suppose first that (a, i) and (b, j) are unrelated, and let $T_1 = T(a, i)$ and $T_2 = T(b, j)$. There exists a shortest path π of $\mathbb{L}$ from (a, i) to (b, j) using no edges of $\mathbb{T}^+(a, i)$ or $\mathbb{T}^+(b, j)$ and which is open with probability $\lambda^{|i-j|}\tau^{\delta_{\mathbb{T}}(a,b)} = \sigma$, say. If T_1 is infinite then it is a.s. ϵ-robust for some (random) $\epsilon > 0$ (any ϵ in $(0, \rho(T_1))$ will do) by (4.11). Consider the graphs $\mathbb{T}_m^+(a, i + k(2m + 1))$ as k ranges over the positive integers. Conditional on T_1, we have that if $|T_1| = \infty$ then each vertex $(a, i + k(2m + 1))$ has a strictly positive probability (depending on T_1) of being the root of a 'red' branching process in $\mathbb{T}_m^+(a, i + k(2m+1))$ which overlaps T_1 i.o., and furthermore the corresponding events are independent for different values of k. On the event that $(a, i + k(2m+1))$ is such a vertex, it is the case that $(a, i + k(2m+1)) \leftrightarrow (a, i)$ a.s., since there is (conditional) probability 1 that, given two 'red' processes which overlap i.o., we may find two points, one from each process, which lie in the same copy of $\mathbb{Z}$ and with the property that the path of $\mathbb{Z}$-edges joining them is open. (We are using here the fact that the strips $S_z(t, u)$ did not include $\mathbb{Z}$-edges along $\{t\} \times \mathbb{Z}$ or $\{u\} \times \mathbb{Z}$.) Let A_k be the event that (i) $(a, i + k(2m+1))$ is the root of a red tree which overlaps T_1 i.o., (ii) $(b, j + k(2m+1))$ is the root of a red tree which overlaps

T_2 i.o., and (iii) the path π_k with vertex set $\{(t, z+k(2m+1) : (t,z) \in \pi\}$ is open. Conditional on T_1 and T_2, the events $\{A_k : k \geq 1\}$ are independent and each has probability at least $\frac{1}{2}\rho(T_1)\rho(T_2)\sigma$ (> 0); hence, a.s. some A_k occurs, so that $(a, i) \leftrightarrow (b, j)$ a.s. by the remarks above. Therefore

$$\begin{aligned} &P_{\tau,\lambda}\big((a,i) \leftrightarrow (b,j);\ |T_1| = |T_2| = \infty\big) \\ &= \iint\limits_{T_1,T_2:|T_1|=|T_2|=\infty} dP_{\tau,\lambda}(T_1)dP_{\tau,\lambda}(T_2)P_{\tau,\lambda}\big((a,i) \leftrightarrow (b,j)|T_1,T_2\big) \\ &= \iint\limits_{T_1,T_2:|T_1|=|T_2|=\infty} dP_{\tau,\lambda}(T_1)dP_{\tau,\lambda}(T_2) \\ &= P_{\tau,\lambda}(|T_1| = \infty, |T_2| = \infty) \end{aligned}$$

as required.

Suppose finally that (a, i) and (b, j) are related and are the roots of infinite red processes $T(a, i)$ and $T(b, j)$, respectively. Suppose also that $\delta_{\mathbb{T}}(a, \emptyset)$ and $\delta_{\mathbb{T}}(b, \emptyset)$ are multiples of n, and that $|i - j| > 2m$. If $T(a, i)$ and $T(b, j)$ overlap i.o., then $(a, i) \leftrightarrow (b, j)$ a.s. by an earlier argument, and so it suffices to assume that $T(a, i)$ and $T(b, j)$ do not overlap i.o. In this case there exists a (random) positive integer R such that $T(a, i)$ and $T(b, j)$ contain *no* overlaps in the set $S \times \mathbb{Z}$ where $S = V(\mathbb{T}^+(a)) \cap V(\mathbb{T}^+(b)) \cap \{t : \delta_{\mathbb{T}}(a, t) \geq R\}$. We may pick c, $d \in S$ such that (c, i) and (d, j) are the roots of infinite red processes in $\mathbb{T}^+_m(c, i)$ and $\mathbb{T}^+_m(d, j)$ and such that (c, i) and (d, j) are unrelated. The chance that such (c, i) and (d, j) are in different infinite open clusters of $\mathbb{L}$ is 0, by the preceding argument, and the proof is complete. ∎

5. Percolation in $\infty + d$ Dimensions

In this section we consider the lattice $\mathbb{L}_d = \mathbb{T} \times \mathbb{Z}^d$ for $d > 1$ and discuss briefly how the arguments and results differ from the case $d = 1$. We continue to denote vertices in $\mathbb{Z}^d$ by z, z_1, and so on.

To modify the analysis which led to Propositions 1 and 4, we note that $(\emptyset, 0) \leftrightarrow (t, z)$ in $\mathbb{L}_d$ if and only if for some n there is a route $\mathbf{t} = (t_0, t_1, \ldots, t_n)$ in $\mathbb{T}$ from $\emptyset$ to t and a sequence $\mathbf{z} = (z_{-1} = 0, z_0, z_1, \ldots, z_n = z)$ such that:

(a) $\langle (t_{i-1}, z_{i-1}), (t_i, z_{i-1}) \rangle$ is open, for $i = 1, \ldots, n$,
(b) $(t_i, z_{i-1}) \leftrightarrow (t_i, z_i)$ in $\{t_i\} \times \mathbb{Z}^d$, for $i = 0, \ldots, n$,
(c) if $t_i = t_j$, then $(t_i, z_i) \not\leftrightarrow (t_j, z_j)$ in $\{t_i\} \times \mathbb{Z}^d$, for $0 \leq i < j \leq n$.

Condition (c) is a 'cluster self-avoiding' property; it implies among other things that (as in the $d = 1$ case) $z_{i-1} \neq z_{i-2}$ when $t_i = t_{i-2}$. By successively conditioning on the $\{t_i\} \times \mathbb{Z}^d$ clusters of (t_i, z_{i-1}), one sees that inequality (3.10) remains valid, but with $\mathcal{X}_\lambda$ replaced by $\mathcal{X}_{\lambda,d}$, the expected cluster size for standard bond percolation on $\mathbb{Z}^d$ with isotropic bond density λ. The remainder of the analysis remains valid and leads to the following extension of Propositions 1 and 4.

PROPOSITION 6. *If*

$$\tau k \mathcal{X}_{\lambda,d}\left(1+\sqrt{1-\mathcal{X}_{\lambda,d}^{-1}}\right) < 1 \tag{5.1}$$

then there is a.s. no infinite open cluster in $\mathbb{L}_d$. *If* $\theta_{\mathbb{L}_d}(\tau,\lambda) > 0$ *but*

$$\tau\sqrt{k}\mathcal{X}_{\lambda,d}\left(1+\sqrt{1-\mathcal{X}_{\lambda,d}^{-1}}\right) < 1, \tag{5.2}$$

then there exists a.s. infinitely many infinite open clusters in $\mathbb{L}_d$.

Proposition 6 provides lower bounds $\tau = \underline{\tau}_{l,d}(\lambda)$ and $\tau = \underline{\tau}_{u,d}(\lambda)$ for the lower and upper curves of Figure 2. These curves are only implicitly defined since they are expressed in terms of $\mathcal{X}_{\lambda,d}$. As $\lambda \to 0$, $\mathcal{X}_{\lambda,d} = 1 + 2d\lambda + o(\lambda)$ so that, just as when $d = 1$,

$$\underline{\tau}_{l,d}(0) = \frac{1}{k}, \quad \underline{\tau}'_{l,d}(0) = -\infty; \quad \underline{\tau}_{u,d}(0) = \frac{1}{\sqrt{k}}, \quad \underline{\tau}'_{u,d}(0) = -\infty. \tag{5.3}$$

On the other hand, for $d > 1$, the critical probability $\lambda_c(d)$ for percolation satisfies $\lambda_c(d) < 1$ and $\mathcal{X}_{\lambda,d}$ diverges as $\lambda \uparrow \lambda_c$ (Menshikov 1986; Menshikov, Molchanov, and Sidorenko 1986; Aizenman and Barsky 1987). It follows that

$$\underline{\tau}_{l,d}(\lambda) \sim \frac{1}{2k}\mathcal{X}_{\lambda,d}^{-1}, \quad \underline{\tau}_{u,d}(\lambda) \sim \frac{1}{2\sqrt{k}}\mathcal{X}_{\lambda,d}^{-1} \quad \text{as } \lambda \uparrow \lambda_c(d), \tag{5.4}$$

so that $\underline{\tau}_{l,d}(\lambda_c(d)) = 0 = \underline{\tau}_{u,d}(\lambda_c(d))$. Since $\mathcal{X}_{\lambda,d}^{-1} = O(\lambda_c(d) - \lambda)$ as $\lambda \uparrow \lambda_c(d)$ (Aizenman and Newman 1984), we see that the derivatives (with respect to λ) of the two lower curves are finite for any d; however since for $d < 6$ it is expected that $\mathcal{X}_{\lambda,d}^{-1}$ behaves as $(\lambda_c(d) - \lambda)^\gamma$ with critical exponent $\gamma > 1$, these derivatives should be zero at $\lambda_c(d)$. These derivatives have recently been proved to be non-zero in sufficiently high dimensions (see Hara and Slade 1989a,b).

The analysis which led to Propositions 2 and 5 extends almost unchanged to the cases of two and more dimensions. We consider anisotropic bond percolation on $\mathbb{Z} \times \mathbb{Z}^d$ with edge density τ for edges in the first ($\mathbb{Z}$) component and λ for $\mathbb{Z}^d$-edges, and we define $\phi_m^d(\tau,\lambda)$ exactly as in (2.2) with

$$T_m(n) = \{(z_1, z_2) \in \mathbb{Z} \times \mathbb{Z}^d : 0 \le z_1 \le n, z_2 \in [-m,m]^d\}$$

(once again without the edges in its 'left' and 'right' boundary faces). Then the (horizontal) correlation length ϕ^d is given by

$$\begin{aligned}\phi^d(\tau,\lambda) &= \lim_{m\to\infty} \phi^d_m(\tau,\lambda)\\ &= \lim_{n\to\infty}\left\{-\frac{1}{n}\log P\big((0,0)\leftrightarrow(n,0)\text{ in } \mathbb{Z}\times\mathbb{Z}^d\big)\right\}\\ &= \lim_{n\to\infty}\left\{-\frac{1}{n}\log P\big((0,0)\leftrightarrow\{n\}\times\mathbb{Z}^d\text{ in }[0,\infty)\times\mathbb{Z}^d\big)\right\}.\end{aligned}$$

PROPOSITION 7. *If*

$$ke^{-\phi^d(\tau,\lambda)} > 1, \tag{5.5}$$

then there is a.s. an infinite open cluster in $\mathbb{L}_d$. If

$$ke^{-2\phi^d(\tau,\lambda)} > 1, \tag{5.6}$$

then there is a.s. a unique infinite open cluster in $\mathbb{L}_d$.

Proposition 7 provides implicitly defined upper bounds $\tau = \overline{\tau}_{l,d}(\lambda)$ and $\tau = \overline{\tau}_{u,d}(\lambda)$ for the lower and upper curves of Figure 2. Since $e^{-\phi^d} \geq \tau$ (recall (2.10)), one knows that

$$\overline{\tau}_{l,d}(0) = 1/k, \quad \overline{\tau}_{u,d}(0) = 1/\sqrt{k}. \tag{5.7}$$

The generalization of (2.11) is

$$e^{-\phi^d(\tau,\lambda)} \geq 1 - E\big((1-\tau)^{|C_\lambda|}\big) \tag{5.8}$$

where $|C_\lambda|$ is the cluster size at the origin of isotropic bond percolation on $\mathbb{Z}^d$ with edge-density λ. Inequality (5.8) can be used to estimate the slopes of $\overline{\tau}_{l,d}$ and $\overline{\tau}_{u,d}$ at $\lambda = 0$, but it does not provide the correct values of $\overline{\tau}_{l,d}(\lambda)$ and $\overline{\tau}_{u,d}(\lambda)$ at $\lambda = \lambda_c(d)$, since if $|C_{\lambda_c(d)}| < \infty$ a.s. (as is known for $d = 2$ and presumed for all $d > 2$; see Barsky, Grimmett, and Newman 1989 for the corresponding result for half-spaces, and Hara and Slade 1989a,b for the full-space result in high dimensions) the right hand side of (5.8) cannot be made larger than $1/k$ (or $1/\sqrt{k}$) as $\lambda \uparrow \lambda_c(d)$ unless τ is bounded away from zero. To see that

$$\overline{\tau}_{l,d}(\lambda_c(d)) = 0, \quad \overline{\tau}_{u,d}(\lambda_c(d)) = 0, \tag{5.9}$$

simply note that $\phi^d(\tau,\lambda) = 0$ if (τ,λ) is such that there is percolation in $\mathbb{Z}\times\mathbb{Z}^d$, and this is easily seen to occur for any small τ if λ is sufficiently close to $\lambda_c(d)$. Careful versions of such arguments show that λ need be no closer than some multiple of τ as $\tau \downarrow 0$ which implies finite bounds for the derivatives (with respect to λ) of $\overline{\tau}_{l,d}$ and $\overline{\tau}_{u,d}$ at $\lambda = \lambda_c(d)$.

Unfortunately, because of the behaviour of $\mathcal{X}_{\lambda,d}$ discussed previously, we cannot combine our present knowledge about $\overline{\tau}_{l,d}$ and $\underline{\tau}_{u,d}$ near $\lambda = \lambda_c(d)$ to conclude that the region of infinitely many infinite open clusters extends all the way to $(\tau, \lambda) = (0, \lambda_c(d))$. The best we can say for $d \geq 2$ is that for any $k \geq 2$ there *is* such a region (since $\overline{\tau}_{l,d} < \underline{\tau}_{u,d}$ near $\lambda = 0$) and that this region certainly approaches $(\tau, \lambda) = (0, \lambda_c(d))$ as $k \to \infty$. This last fact follows by taking a fixed value of λ near to $\lambda_c(d)$, and combining the inequalities

$$\underline{\tau}_{u,d}(\lambda) \geq \frac{1}{2\mathcal{X}_{\lambda,d}\sqrt{k}}$$

and

$$\overline{\tau}_{l,d}(\lambda) \leq \tau_0 = \frac{1}{k\mathcal{X}_{\lambda,d}} + O(k^{-2}) \quad \text{as} \quad k \to \infty$$

where τ_0 is the root of the equation

$$1 - E\big((1-\tau)^{|C_\lambda|}\big) = \frac{1}{k}.$$

Acknowledgment

G.R.G. acknowledges support from the Department of Mathematics and the Center for the Study of Complex Systems at the University of Arizona, where this work was carried out. Both authors thank the Arizona Center for Mathematical Sciences (ACMS) for support; the ACMS is sponsored by AFOSR contract F49620-86-C0130 under the University Research Initiative Program. C.M.N. acknowledges support under NSF Grant DMS-8514834. The authors thank Dan Stein for suggesting most of the title, and Chuntao Wu for a careful reading of the manuscript.

REFERENCES

Aizenman, M. and Barsky, D. J. (1987). Sharpness of the phase transition in percolation models. *Communications in Mathematical Physics* 108, 489–526.

Aizenman, M., Chayes, J. T., Chayes, L., and Newman, C. M. (1988). Discontinuity of the magnetization in one-dimensional $1/|x-y|^2$ Ising and Potts models. *Journal of Statistical Physics* 50, 1–40.

Aizenman, M., Kesten, H., and Newman, C. M. (1987). Uniqueness of the infinite cluster and continuity of connectivity functions for short and long range percolation. *Communications in Mathematical Physics* 111, 505–532.

Aizenman, M. and Newman, C. M. (1984). Tree graph inequalities and critical behavior in percolation models. *Journal of Statistical Physics* 36, 107–143.

Barsky, D. J., Grimmett, G. R., and Newman, C. M. (1989). Percolation in half-spaces: equality of critical probabilities and continuity of the percolation probability. In preparation.

Broadbent, S. R. and Hammersley, J. M. (1957). Percolation processes I. Crystals and mazes. *Proceedings of the Cambridge Philosophical Society* 53, 629–641.

Burton, R. M. and Keane, M. (1989). Density and uniqueness in percolation. *Communications in Mathematical Physics*, to appear.

Gandolfi, A., Grimmett, G. R., and Russo, L. (1988). On the uniqueness of the infinite open cluster in the percolation model. *Communications in Mathematical Physics* 114, 549–552.

Grigorchuk, R. I. (1983). On Milnor's problem of group growth. *Soviet Mathematics Doklady* 28, 23–26.

Grimmett, G. R. (1989). *Percolation.* Springer-Verlag, New York.

Hara, T. and Slade, G. (1989a). The mean field critical behavior of percolation in high dimensions. *Proceedings of the 1988 International Conference on Mathematical Physics*, Swansea, Adam Hilger, to appear.

——— (1989b). Mean field critical phenomena for percolation in high dimensions. Preprint.

Kesten, H. (1982). *Percolation Theory for Mathematicians.* Birkhäuser, Boston.

Lyons, R. (1988). Random walks and percolation on trees. Preprint.

Menshikov, M. V. (1986). Coincidence of critical points in percolation problems. *Soviet Mathematics Doklady* 33, 856–859.

Menshikov, M. V., Molchanov, S. A., and Sidorenko, A. F. (1986). Percolation theory and some applications. *Itogi Nauki i Techniki*, Series of Probability Theory, Mathematical Statistics, Theoretical Cybernetics, 24, 53–110.

Newman, C. M. and Schulman, L. S. (1981). Infinite clusters in percolation models. *Journal of Statistical Physics* 26, 613–628.

Newman, C.M. and Wu, C. (1989). In preparation.

Wu, C. (1989). Ph. D. Thesis, University of Arizona.

School of Mathematics
University of Bristol
University Walk
Bristol BS8 1TW.

Department of Mathematics
University of Arizona
Tucson
Arizona 85721.

Monte Carlo Methods Applied to Quantum-Mechanical Order-Disorder Phenomena in Crystals

D.C. Handscomb

1. The Metropolis Method in Classical Statistical Mechanics

Many years ago, John Hammersley induced and encouraged me to work on this application of Monte Carlo as my doctoral research topic; as it seems that some physicists have recently become interested in the idea after its spending a long period in oblivion, it may be worth dusting off and explaining again.

The idea stems from a technique originally developed in 1953 (Metropolis et al. 1953; see also Hammersley and Handscomb 1964, Chapter 9) by Metropolis et al. (here referred to as $\mathrm{MR^2T^2}$) for the study of the equation of state of a hard-sphere gas model. Suppose that one has a physical system that can occupy any of a [large] number of possible configurations C, each having an energy given by the Hamiltonian function $H(C)$. According to the laws of classical statistical mechanics, when at temperature T the system occupies each possible configuration C with a probability proportional to

$$P(C) := \mathrm{e}^{-\beta H(C)}, \tag{1.1}$$

where $\beta := (kT)^{-1}$, k being Boltzmann's constant; the expected value of any observable $\phi(C)$ (including $H(C)$ itself) is thus given by the expression

$$\mathrm{E}\phi(C) = \frac{\sum_C \phi(C)P(C)}{\sum_C P(C)}. \tag{1.2}$$

To evaluate (1.2) directly is prohibitively expensive, and needlessly extravagant since all but a few terms in each summation are negligible unless T is very large. The form of (1.2) suggests, however, that one should be able to evaluate it approximately by a sampling ('Monte Carlo') method — if one could somehow easily generate a sample (with replacement) of configurations from a probability distribution in which configuration C occurred with probability proportional to $P(C)$, the sample average of $\phi(C)$ would then be an unbiassed estimator of (1.2). The ingenious technique put forward in MR^2T^2 generates such a sample by performing a 'random walk' through configuration space: starting from an arbitrary configuration C_0, one generates a sequence of configurations C_1, C_2, ..., each one move away from its predecessor, according to the following rules:

Let $A(C)$ denote the set of all configurations accessible from C in one move; assume that a 'move' is defined so that:

- if $C' \in A(C)$ then $C \in A(C')$;
- the size of $A(C)$ is independent of C;
- there is no proper subset Σ of configurations such that

$$\bigcup_{C \in \Sigma} A(C) \subseteq \bigcup_{C \in \Sigma} C.$$

Having arrived after j moves at C_j, select a new configuration C'_j at random from a uniform distribution on $A(C_j)$.

If $P(C'_j) \geq P(C_j)$, then take $C_{j+1} = C'_j$.

If $P(C'_j) < P(C_j)$, so that the chosen move would take one to a higher-energy configuration, then draw a random number ξ_j from a uniform distribution on $[0, 1]$; if $\xi_j < P(C'_j)/P(C_j)$, then again take $C_{j+1} = C'_j$; otherwise take $C_{j+1} = C_j$, so that the previous configuration is repeated.

These rules can easily be shown to give rise to an irreducible acyclic Markov chain in which the one-step transition probabilities $\Pr(C \to C')$ satisfy the equation

$$P(C)\Pr(C \to C') = P(C')\Pr(C' \to C) \quad \forall C, C' : \; C' \in A(C), \tag{1.3}$$

so that

$$P(C) = \sum_{C'} P(C')\Pr(C' \to C) \quad \forall C, \tag{1.4}$$

and the chain has an equilibrium distribution proportional to $P(C)$.

Notice that in carrying out this procedure it is never necessary to evaluate $P(C)$ or the Hamiltonian function completely, since

$$\frac{P(C')}{P(C)} = \mathrm{e}^{-\beta(H(C')-H(C))}, \tag{1.5}$$

so that all that one needs to compute is the change in energy $H(C'_j) - H(C_j)$ produced by each proposed move. In the original hard-sphere gas model of MR^2T^2, for instance, each move consists of shifting just one sphere to a new position — the change in energy and the probability of accepting the move depend only on this sphere and those with which it comes in contact.

This procedure is thus very easy to implement. It is very effective in generating a suitable sample of configurations. There are only two real difficulties:

- starting from a low-probability configuration, it may take many moves to reach a high-probability one, so that there is some problem in deciding how many initial steps of the chain should be discarded as transient and unrepresentative before one begins sampling;
- since the sample is by its very nature highly correlated, it is not easy to make a statistical assessment of the error in using the sample average of $\phi(C_j)$ as an estimator of $\mathrm{E}\phi(C)$.

The principle of $\mathrm{MR^2T^2}$ has been applied to the study of order-disorder phenomena in binary alloys (cf. Fosdick 1959) and in the Ising model of a ferromagnet or antiferromagnet (cf. Ehrman et al. 1960).

The Ising model, for instance, models the ferromagnet by a fairly large array of n sites on a crystal lattice, the 2^n configurations C then being all possible assignments of a positive or negative spin to each site. On the assumption that each spin interacts only with its nearest neighbours and with an external (uniform) magnetic field, the Hamiltonian takes the form[1]

$$H(C) = -J(n_{++} + n_{--} - n_{+-}) - \mu\mathcal{H}(n_+ - n_-), \tag{1.6}$$

(omitting an arbitrary constant term) where n_+ and n_- are respectively the numbers of positive and negative spins in the configuration C, and n_{++}, n_{--} and n_{+-} the numbers of nearest-neighbour pairs whose spins are respectively both positive, both negative, and one of each; equivalently

$$H(C) = -J\sum_{\mathrm{nn}} S_i S_j - \mu\mathcal{H}\sum S_i, \tag{1.7}$$

where $\sum_{\mathrm{nn}}$ denotes summation over all pairs of nearest neighbours and $S_i = \pm 1$, depending on the sign of the spin at the ith site. A possible move of the $\mathrm{MR^2T^2}$ procedure consists of reversing the sign of any one spin; the set $A(C)$ thus consists of those configurations C' differing from C in the sign of a single spin, and $H(C') - H(C)$ is found simply by looking at that spin and its immediate neighbours.

A possible measure of long-range order is the value of

$$\phi_L(C) := \left(\sum S_i\right)^2 \Big/ n^2. \tag{1.8}$$

In a completely-ordered configuration, such as occurs when $T = 0$, $\phi_L(C) = 1$, while in a random configuration ($T = \infty$), $\mathrm{E}\phi_L(C) = 1/n$. As the dimensions of the lattice tend to infinity, the latter expectation tends to zero, and there is a critical temperature T_c such that $\mathrm{E}\phi_L(C) \to 0$ whenever

[1] The coefficient J represents the internal interaction, $J > 0$ for a ferromagnet or $J < 0$ for an antiferromagnet; μ is the Bohr magneton; $\mathcal{H}$ represents the external field.

$T > T_c$. On finite lattices the transition is more blurred, nevertheless, a transition between ordered and disordered behaviour is discernable even on fairly small lattices (see Figs. 1, 2), and can be picked up by the Monte Carlo method (Ehrman et al. 1960).

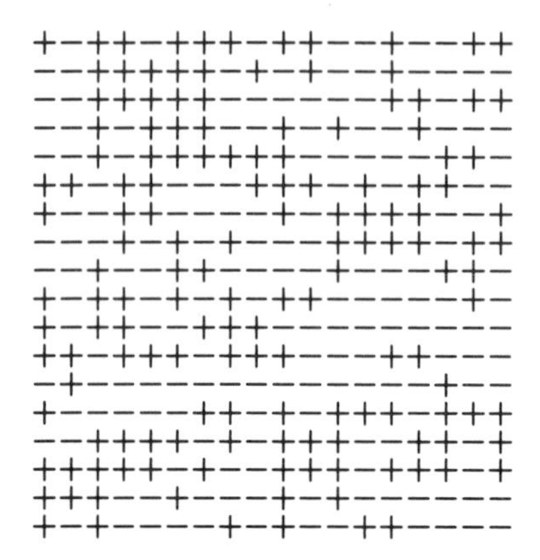

FIG. 1. A high-temperature Ising configuration.

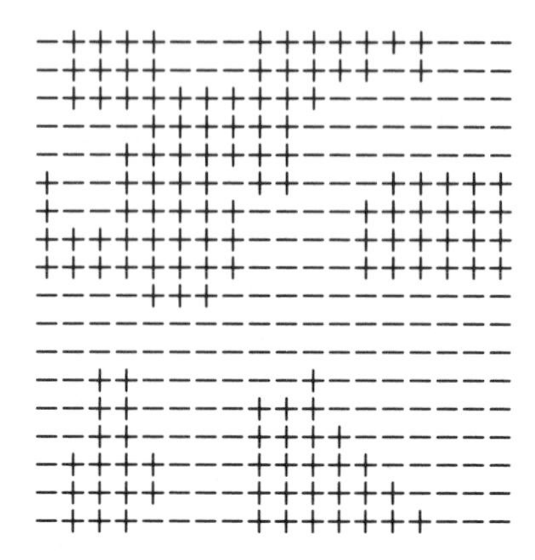

FIG. 2. A low-temperature Ising configuration.

2. Extension to a Quantum-Mechanical System

The Ising model is based on nearly-classical mechanics and does not properly represent the quantum mechanics of magnetic interactions; better models to adopt are the XY model and the Heisenberg model, in which quantum theory is applied without compromise.

In quantum statistics, the Hamiltonian H represents not a function, but a linear operator on distributions over configuration space (for classical systems H would be a diagonal operator); suppose it to take the form

$$H = H_0 + \sum_{i=1}^{N} H_i, \tag{2.1}$$

where H_0 commutes with each H_i but otherwise H_i may or may not commute with H_j. The expectation of any linear operator Φ is then given by the expression

$$\begin{aligned}\langle \Phi \rangle &= \frac{\mathrm{trace}\{\Phi \exp(-\beta H)\}}{\mathrm{trace}\{\exp(-\beta H)\}} \\ &= \frac{\sum_{r=0}^{\infty}(-\beta)^r/r! \sum_{Z_r} \mathrm{trace}\{\Phi H_{i_1} \ldots H_{i_r} \exp(-\beta H_0)\}}{\sum_{r=0}^{\infty}(-\beta)^r/r! \sum_{Z_r} \mathrm{trace}\{H_{i_1} \ldots H_{i_r} \exp(-\beta H_0)\}},\end{aligned} \tag{2.2}$$

where Z_r denotes any sequence $\{i_1, \ldots, i_r\}$ of r indices in the range $1 \leq i \leq N$. [Z_0 is the empty sequence $\{\ \}$.]

The scheme I proposed in the early 1960s (Handscomb 1962, 1964) exploits the resemblance of (2.2) to (1.2). The key idea is no longer to require a 'configuration' to correspond directly to a physical state of the system, provided only that it is possible to assign to each such 'configuration' a weight $P(C)$ and a parameter $\phi(C)$ such that (1.2) has a meaningful value.

Take as 'configurations' C all the (infinitely many) possible sequences Z_r, $0 \leq r < \infty$. If then we define

$$\phi(Z_r) := \frac{\mathrm{trace}\{\Phi H_{i_1} \ldots H_{i_r} \exp(-\beta H_0)\}}{\mathrm{trace}\{H_{i_1} \ldots H_{i_r} \exp(-\beta H_0)\}}, \tag{2.3}$$

and

$$P(Z_r) := \frac{(-\beta)^r}{r!} \mathrm{trace}\{H_{i_1} \ldots H_{i_r} \exp(-\beta H_0)\}, \tag{2.4}$$

we have

$$\langle \Phi \rangle = \mathrm{E}\phi(Z_r) = \frac{\sum_{r=0}^{\infty} \sum_{Z_r} \phi(Z_r) P(Z_r)}{\sum_{r=0}^{\infty} \sum_{Z_r} P(Z_r)}, \tag{2.5}$$

so that if one can use something like the $\mathrm{MR^2T^2}$ procedure to generate a sample of sequences with probabilities proportional to $P(Z_r)$ then the sample average of $\phi(Z_r)$ will be an unbiassed estimator of $\langle \Phi \rangle$.

In particular, it is easy to see that

$$\langle H \rangle = \langle H_0 \rangle - \langle r/\beta \rangle. \tag{2.6}$$

For this to work, one of course needs $P(Z_r)$ to yield a proper probability distribution, so that we must have

$$P(Z_r) \geq 0 \quad \forall Z_r, \tag{2.7}$$

$$\sum_{r=0}^{\infty} \sum_{Z_r} P(Z_r) < \infty. \tag{2.8}$$

We must also have

$$\phi(Z_r) \leq \infty \quad \forall Z_r \tag{2.9}$$

(which is liable to be violated when $P(Z_r) = 0$). Conditions (2.7), (2.8) and (2.9) impose restrictions on the form of H. In order to be able to apply the $\mathrm{MR^2T^2}$ process, we must further define our possible moves so that (at the very least)

$$\exists Z' \in A(Z_r) : P(Z') > 0 \quad \forall Z_r : P(Z_r) > 0; \tag{2.10}$$

otherwise there is no escape from Z_r.

If all $P(Z_r) \neq 0$, then obvious possible moves are to add a single index to the beginning, to the end, or somewhere in the middle of the current sequence or else to delete the first, the last, or another index from the current sequence. The process is less likely to become 'bogged down' in a set of a few sequences if additions and deletions are at random points of the sequence. (See Lyklema 1982.) The rule for deciding whether to accept a move has to be a little more complicated than the original $\mathrm{MR^2T^2}$ rule, since there are obviously more ways of extending a sequence than there are of shortening it, but it not difficult to formulate rules so that (1.3) is satisfied (Handscomb 1962).

In the case of the Heisenberg ferromagnet, with a single particle of spin $\frac{1}{2}$ on each site, the Ising Hamiltonian function (1.7) is replaced by the Hamiltonian operator

$$H = -\tfrac{1}{2}J\sum_{\mathrm{nn}}(\sigma_i, \sigma_j) - \mu\mathcal{H}\sum \sigma_i^z, \tag{2.11}$$

where σ_i is the Pauli spin-operator on the ith site and σ_i^z its component in the direction of the external field. Now if E_{ij} denotes the operator that interchanges the spins on the ith and jth sites, we may make the substitution

$$E_{ij} = \tfrac{1}{2}\{1 + (\sigma_i, \sigma_j)\} \tag{2.12}$$

to rewrite (2.11) as

$$H = -J\sum_{\mathrm{nn}} E_{ij} - \mu\mathcal{H}\sum \sigma_i^z. \tag{2.13}$$

Take $H_0 := -\mu\mathcal{H}\sum \sigma_i^z$ and $H_{ij} := -JE_{ij}$; then $H = H_0 + \sum_{\mathrm{nn}} H_{ij}$, where each H_{ij} commutes with H_0.

If we define Z_r to consist of a sequence $\{(i_1 j_1), \ldots, (i_r j_r)\}$ of nearest-neighbour pairs, and the corresponding interchanges together result in a permutation of the lattice that decomposes into the product of $K = K(Z_r)$ cycles, of lengths $a_1, \ldots, a_K$ ($a_k \geq 1$, $\sum a_k = n$), then a state of the lattice is invariant under this sequence of interchanges if and only if all lattice-sites in the same cycle of this permutation have the same spin, so that

$$\begin{aligned} P(Z_r) &= \frac{(-\beta)^r}{r!}\,\mathrm{trace}\{H_{i_1 j_1} \ldots H_{i_r j_r} \exp(-\beta H_0)\} \\ &= \frac{(\beta J)^r}{r!} \prod_{k=1}^{K(Z_r)} \{2\cosh(a_k L)\} \neq 0, \end{aligned} \tag{2.14}$$

where $L = \beta\mu\mathcal{H}$. The effect of adding or subtracting an interchange at the end of the sequence is clearly either to merge two cycles or to split one cycle into two; inserting or deleting an interchange elsewhere in the sequence can be shown to have a like effect. Consequently the acceptance probabilities can be made to depend only on the lengths of the cycles actually involved.

A measure of long-range order is given by $\langle\Phi_L\rangle$, where

$$\begin{aligned}\Phi_L &:= 2\left(\sum \sigma_i, \sum \sigma_i\right) \Big/ 3n(n+1) \\ &= 2\left(4\sum_{i>j} E_{ij} - n(n-4)\right) \Big/ 3n(n+1). \end{aligned} \tag{2.15}$$

This transforms into $\mathrm{E}\phi_L(Z_r)$, where

$$\phi_L(Z_r) = \frac{2}{3} \frac{\sum a_k^2 + 2\coth L \sum a_k \tanh a_k L + \{\sum a_k \tanh a_k L\}^2 - \sum a_k^2 \tanh^2 a_k L}{n(n+1)} \tag{2.16}$$

or, if $\mathcal{H} = 0$,

$$\phi_L(Z_r) = 2\sum a_k^2 \Big/ n(n+1). \tag{2.17}$$

When $T = \infty$, $\beta = 0$, we have $\langle\Phi\rangle = \operatorname{trace}\{\Phi\}/\operatorname{trace}\{\}$, so that $\langle\Phi_L\rangle = \frac{2}{n+1}$. When $T = 0$ and $\mathcal{H} = 0$, then the product of the interchanges tends towards a random permutation, so that $\langle\Phi_L\rangle = 1$. Again we expect to get a critical temperature as $n \to \infty$.

3. More Recent Work

A drawback of the scheme just presented is that it does not work for anti-ferromagnets, where $J < 0$ so that $P(Z_r)$ is not always positive, but has the sign of $(-)^r$. Also it is restricted to the Heisenberg model, and does not apply to the XY model, in which (σ_i, σ_j) is replaced by $\sigma_i^x\sigma_j^x + \sigma_i^y\sigma_j^y$.

These models may be treated by defining the new operator

$$h_{ij} = \sigma_i^+\sigma_j^- + \sigma_i^-\sigma_j^+, \tag{3.1}$$

when we can show that

$$I - E_{ij} = h_{ij}^2 - h_{ij} \tag{3.2}$$

and

$$\sigma_i^x\sigma_j^x + \sigma_i^y\sigma_j^y = (\sigma_i, \sigma_j) - \sigma_i^z\sigma_j^z = 2h_{ij}. \tag{3.3}$$

Thus, instead of looking at traces of operators whose main components are of the form $\prod E_{ij}$ we can look at those of operators with components of the form $\prod h_{ij}$. Now the effect of h_{ij} on a spin state is to give zero if the spins on sites i and j are the same, and to interchange them if they differ. Consequently (provided that the lattice is such that every closed loop of nearest-neighbour bonds is of even length) trace $\prod h_{ij} \neq 0$ only if the product has an even number of interchanges with each suffix appearing an even number of times, and not always even then. It follows that in the case of the XY model we have $P(Z_r) \neq 0$ only when r is even, when $P(Z_r) \geq 0$ whatever the sign of J. Although $\prod h_{ij}$ is not now a simple permutation operator, it is possible to work out its trace mechanically, and thus implement a form of the MR^{2}T^2 algorithm — a move in this case adding or deleting *two* interchanges, which must not always be adjacent in the sequence Z_r if the space of possible sequences is to be properly sampled (Chakravarty and Stein 1982).

The Heisenberg antiferromagnet is nearly as easy (Lee et al. 1984). Shifting the energy baseline, we use (3.2) to write

$$H = J \sum_{\mathrm{nn}} (I - E_{ij}) + H_0 = J \sum_{\mathrm{nn}} (h_{ij}^2 - h_{ij}) + H_0, \tag{3.4}$$

where now, we remember, $J < 0$. Therefore

$$P(Z_r) = \frac{(\beta |J|)^r}{r!} \operatorname{trace}\left\{ \prod (h_{i_r j_r}^2 - h_{i_r j_r}) \exp(-\beta h_0) \right\}. \tag{3.5}$$

Once again, under the same proviso, every non-zero $P(Z_r)$ will be positive.

If the lattice has closed loops of an odd number of bonds, there will be some negative weights in each case, although positive weights will normally predominate. In such a case one has to rewrite (1.2) in the form

$$\mathrm{E}\phi(C) = \frac{\sum_C \phi(C) P(C)}{\sum_C |P(C)|} \Big/ \frac{\sum_C P(C)}{\sum_C |P(C)|} \tag{3.6}$$

and estimate numerator and denominator separately — that is to say, one constructs a sample with probabilities proportional to $|P(C)|$ and uses the ratio of the sample averages of $\phi(C) \operatorname{sgn} P(C)$ and $\operatorname{sgn} P(C)$.

A completely different approach to the general problem (see Suzuki 1976, etc) is based on approximating the Trotter formula (Trotter 1959)

$$\exp \sum_j A_j = \lim_{n \to \infty} \left(\prod_j \exp \frac{1}{n} A_j \right)^n \tag{3.7}$$

by truncating the limiting process at some large n. The unanswered question here, of course, is how large n ought to be in relation to accuracy required and (possibly) the size of the lattice.

More satisfactory is the 'decoupled cell method' (Homma et al. 1986, 1987; Matsuda et al. 1988). This is closer to the original MR^2T^2 method in that one generates a sample of spin configurations form a Markov chain, the basic move being the reversal of the spin at a lattice site. The difference is that in quantum mechanics the expression (1.1), to which the probability of configuration C is proportional, is now a diagonal element of the exponential of the operator $-\beta H$, and no longer easily determined by inspection of C. In the DCM, when one has selected the site on which the spin S_i is possibly to be reversed, one then 'decouples' the lattice into the set L_i^ν of spins which are at most ν nearest-neighbour steps from the selected site and the complementary set $\bar{L}_i^\nu$, and ignores all interactions between sites in L_i^ν and $\bar{L}_i^\nu$. One can then calculate the ratio of the probabilities that $S_i = \pm 1$, given the state of $L_i^\nu \setminus S_i$, and hence the acceptance probability for reversing S_i. In Matsuda et al. (1988) it is shown that the errors in the transition probabilities due to this approximation are $O(\beta^{\nu+1})$ for small β (high temperature T).

References

Binder, K. (1975). Monte Carlo investigations of phase transitions and critical phenomena. In *Phase Transitions and Critical Phenomena*, volume 5b, ed. C. Domb and M.S. Green, 2–107, Academic Press, London.

——— (1986). Recent trends in the development and application of the Monte Carlo method. In *Monte Carlo Methods in Statistical Physics*, 2nd edition, ed. K. Binder, Chapter 10, 357–392, Springer-Verlag, New York.

Chakravarty, S. and Stein, D.B. (1982). Monte Carlo simulation of quantum spin systems. *Physical Review Letters* 49, 582–585.

Ehrman, J.R., Fosdick, L.D., and Handscomb, D.C. (1960). Computation of order parameters in an Ising lattice by the Monte Carlo method. *Journal of Mathematical Physics* 1, 547–558.

Fosdick, L.D. (1959). Calculation of order parameters in a binary alloy by the Monte Carlo method. *Physical Review* 116, 565–573.

Hammersley, J.M. and Handscomb, D.C. (1964). *Monte Carlo Methods*, Methuen, London.

Handscomb, D.C. (1962). The Monte Carlo method in quantum statistical mechanics. *Proceedings of the Cambridge Philosophical Society* 58, 594–598.

——— (1964). A Monte Carlo method applied to the Heisenberg ferromagnet. *Proceedings of the Cambridge Philosophical Society* 60, 115–122.

Harris E.A. (196?). Antiferromagnetic ordering in an Ising lattice by a Monte Carlo method. Private communication.

Homma, S., Matsuda, H., and Ogita, N. (1986). Decoupled cell method Monte Carlo simulation for quantum spin systems. *Progress in Theoretical Physics* 75, 1058–1065.

——— (1989). Monte Carlo studies of the quantum XY model on the square lattice. To appear.

Homma, S., Matsuda, H., Horiki, T., and Ogita, N. (1989). Off-diagonal induction of short range order — spin pair correlation of the quantum XXZ-model. To appear.

Homma, S., Sano, K., Matsuda, H., and Ogita, N. (1986). Decoupled cell Monte Carlo method for quantum spin systems on linear chain and on triangular lattice. *Progress in Theoretical Physics Supplement* 87, 127–138.

——— (1987). Decoupled cell Monte Carlo method for quantum spin systems. In *Ninth Taniguchi International Symposium on Quantum Monte Carlo Methods in Equilibrium and Non-Equilibrium*, ed. M. Suzuki, 153–162, Springer Series in Solid-State Science 74.

Kadowaki, S. and Ueda, A. (1986). A direct evaluation method of the partition function of quantum spin systems. *Progress in Theoretical Physics* 75, 451–454.

——— (1987). A new Monte Carlo approach to quantum spin systems — direct evaluation of the partition function. *Progress in Theoretical Physics* 78, 224–236.

Lee, D.H., Joannopoulos, J.D., and Negele, J.W. (1984). Monte-Carlo solution of antiferromagnetic quantum Heisenberg spin systems. *Physical Review B* 30, 1599–1602.

Lyklema, J.W. (1982). Quantum-statistical Monte Carlo method for Heisenberg spins. *Physical Review Letters* 49, 88–90.

——— (1983). Monte Carlo study of the one-dimensional quantum Heisenberg ferromagnet near $T = 0$. *Physical Review B* 27, 3108–3110.

——— (1984). A quantum Monte Carlo method for the Heisenberg spin system. In *Monte Carlo Methods in Quantum Problems*, ed. M.H. Kalos, Reidel, Dordrecht.

Matsuda, H., Ishii, K., Homma, S., and Ogita, N. (1988). Theoretical basis of the decoupled cell Monte Carlo simulation for quantum systems. *Progress in Theoretical Physics* 80, 583–587.

Metropolis, N.M., Rosenbluth, A.W. and M.N., Teller, A.H. and E. (1953). Equations of state calculations by fast computing machines. *Journal of Chemical Physics* 21, 1087–1092. ['$\mathrm{MR^2T^2}$']

Miyashita, S., Takasu, M., and Suzuki, M. (1987). Magnetic properties of Ising-like Heisenberg $S = \frac{1}{2}$ antiferromagnets on the triangular lattice. In *Ninth Taniguchi International Symposium on Quantum Monte Carlo Methods in Equilibrium and Non-Equilibrium*, ed. M. Suzuki, Springer Series in Solid-State Science 74.

Onogi, T., Miyashita, S., and Suzuki, M. (1987). Critical properties of the two-dimensional spin-$\frac{1}{2}$ XY ferromagnet. In *Ninth Taniguchi International Symposium on Quantum Monte Carlo Methods in Equilibrium and Non-Equilibrium*, ed. M. Suzuki, Springer Series in Solid-State Science 74.

Suzuki, M. (1976). Relationship between d-dimensional quantal spin systems and $(d+1)$-dimensional Ising systems. *Progress in Theoretical Physics* 56, 1454–1469.

——— (1976). Generalized Trotter's formula and systematic approximants of exponential operators and inner derivations with applications to many-body problems. *Communications in Mathematical Physics* 51, 183–190.

——— (1985a). Thermo field dynamics of quantum spin systems. *Journal of Statistical Physics* 42, 1047–1070.

——— (1985b). General correction theorems on decomposition formulae of exponential operators and extrapolation methods for quantum Monte Carlo systems. *Physics Letters* 113A, 299–300.

——— (1986a). Quantum statistical Monte Carlo methods and applications to spin systems. *Journal of Statistical Physics* 43, 883–909.

——— (1986b). New methods to study critical phenomena — systematic cluster mean-field approach, and thermo field dynamics of interacting quantum systems. In *Quantum Field Theory*, ed. F. Mancini, 505–531, Elsevier, Amsterdam.

——— (1987). General review of quantum statistical Monte Carlo methods. In *Ninth Taniguchi International Symposium on Quantum Monte Carlo Methods in Equilibrium and Non-Equilibrium*, ed. M. Suzuki, Springer Series in Solid-State Science 74.

Takasu, M., Miyashita, S., and Suzuki, M. (1987). Thermodynamic properties of the spin-$\frac{1}{2}$ Heisenberg antiferromagnet on the triangular lattice. In *Ninth Taniguchi International Symposium on Quantum Monte Carlo Methods in Equilibrium and Non-Equilibrium*, ed. M. Suzuki, Springer Series in Solid-State Science 74.

Talbot, E.F. (1986). Thermodynamic properties of the two-dimensional quantum Heisenberg ferromagnet and the effects of bond dilution. *Physical Review B* 33, 4906–4918.

Trotter, H.F. (1959). On the product of semigroups of operators. *Proceedings of the American Mathematical Society* 10, 545–551.

Oxford University Computing Laboratory
8–11 Keble Road
Oxford OX1 3QD.

The Diffusion of Euclidean Shape

Wilfrid S. Kendall

1. Introduction

This paper is a preliminary report on the results of an investigation into the diffusion of Euclidean shape, using computer algebra to reduce complicated intermediate calculations to an informative final form. The computer algebra takes the form of an extension to the *symbolic Itô calculus* described in W.S. Kendall (1988). A substantially more detailed treatment (including details of how to obtain the results below and a description and discussion of the necessary extensions to symbolic Itô calculus) will be provided in a later paper. The results are new and will be of interest to workers in the field of statistics of shape, and perhaps also to mathematical physicists. They provide a reinforcement of the view expressed in W.S. Kendall (1988), and further argued in my contribution to the discussion of D.G. Kendall (1989), that computer algebra and the associated equipment now form a powerful tool for probabilists and statisticians, as indeed for the mathematical scientist in general.

Suppose k particles $X_1, \ldots, X_k$ diffuse in Euclidean n-space $\mathbb{R}^n$ according to independent copies of an Ornstein-Uhlenbeck process. Thus $X_1, \ldots, X_k$ obey the system (1.1) of stochastic differential equations

$$d_I X_i = d_I B_i - \frac{\kappa}{2} X_i \, dt \quad \text{for } i = 1, \ldots, k \tag{1.1}$$

in which $B_1, \ldots, B_k$ are independent Brownian motions in $\mathbb{R}^n$ and κ is a non-negative constant, the *Ornstein-Uhlenbeck parameter.* Here and in the following we use the stochastic calculus, so $d_I X_i$ is the *Itô stochastic differential* of the random process X_i. See Rogers and Williams (1987) for an exposition, and also W.S. Kendall (1987, Section 1), for an introduction to the notation used below and some relevant geometric considerations.

Following D.G. Kendall (1977) one may consider the *(Euclidean) shape* formed by the configuration of the k diffusing particles. That is to say, one considers the stochastic evolution of those aspects of the configuration which have nothing to do with its location, orientation, or size. The resulting *diffusion of shape* has a fascinating and beautiful structure, despite the simplicity of the underlying stochastic differential system (1.1). (Strictly speaking the shape performs a diffusion only up to a random time-change; see the comment before equations (2.2), (2.3), and (2.4).) We shall describe this structure by analyzing stochastic differential systems for the stochastic evolution of collections of *shape statistics* — configuration functions depending only on the shape of the configuration in question. These collections of shape statistics will form coordinate systems for the shape diffusion.

Here is a brief summary of some statistical and probabilistic aspects of shape relevant to this paper, and a summary of previous results on shape diffusion. Recall that two configurations each of k points are said to have the same shape if one configuration can be transformed into the other by application of a sequence of translations, rotations, and dilatations. (To avoid degeneracy we stipulate that $k > 2$ and that neither configuration is composed of totally coincident points.) This conception of shape arose from a statistical problem in archaeology (Broadbent 1980; D.G. Kendall and W.S. Kendall 1980) and has been considerably developed over the last decade (see D.G. Kendall 1984, 1986, the reviews of Small 1988, and D.G. Kendall 1989, and the introductory treatment in Chapter 8 of Stoyan et al. 1987). In particular it has been established that the space Σ_n^k of k points in n-space carries a metric which is natural from statistical and probabilistic points of view. The shape spaces Σ_1^k are metrically spheres while Σ_2^k are metrically complex projective spaces. For $n \geq 3$ the shape space Σ_n^3 is metrically a hemisphere. The general shape space does not have such simple geometry and indeed if $k > n + 1$ and $n \geq 3$ then Σ_n^k is not a smooth manifold (D.G. Kendall 1989).

In the case $k = 3$ and $n = 2$ the shape space is a complex projective space of one complex dimension and is therefore isometric to a 2-sphere. The shape of the diffusing triad $X_1X_2X_3$ is actually Brownian motion on this 2-sphere, up to a random time change. This beautiful result (linked to properties of the Hopf fibration) is due to D.G. Kendall (1977) in the case when the Ornstein-Uhlenbeck parameter κ is zero. He also identified the shape diffusion in the cases of Σ_1^4 and Σ_3^3. In W.S. Kendall (1988) an implementation of stochastic calculus in the REDUCE computer algebra language (the symbolic Itô calculus mentioned above) and a description of shape in terms of *homogeneous shape coordinates* were used to identify the shape diffusion of Σ_n^3 for $n \geq 3$ and non-negative κ. Carrying the computer algebra approach further had to await a way of handling vectors of general symbolic dimension n in REDUCE . This has now been developed, and the results described below are the first fruits of this extension.

It should be noted that Carne (1988) has also made a successful study of the shape diffusion, as part of a wider study of the geometry of shape. The explicit calculations given here complement his more algebro-geometric approach. Indeed a direct connection between diffusion theory and Riemannian geometry (as described in Chapter 5 of Ikeda and Watanabe 1981) means one can deduce the shape geometry from knowledge of the shape diffusion. The paper promised above will consider connections both with Carne's results and with unpublished work of D.G. Kendall (manuscript). Carne (1988) and also Le (1988) have obtained the form of a generalized shape diffusion for points on the sphere; in this case it is no longer possible to separate shape from size so the generalization is actually a shape-and-size diffusion.

The contents of the rest of the paper are as follows. Section 2 summarizes the results describing the general shape diffusion in terms of stochastic differential systems. Expressions corresponding to three different coordinate systems are provided: the so-called homogeneous shape coordinates of normalized square side lengths, the coordinates of standardized inner products, and the coordinates corresponding to a singular values decomposition. Section 3 discusses some of the more basic questions concerning the last of these coordinate systems, which provides the most insight of the three into the behaviour of the shape diffusion. The paper concludes with Section 4, which comprises a brief discussion of topics for further work.

I am grateful to T.K. Carne, S.D. Jacka, D.G. Kendall, and Le H.L. for their helpful comments on preliminary and draft versions of this work.

2. Stochastic Differential Systems for Shape

Suppose k particles $X_1, \ldots, X_k$ diffuse in Euclidean n-space $\mathbb{R}^n$ as specified by the stochastic differential system (1.1). As noted above, we stipulate $k > 2$. Consider the *(modified) shape* $\sigma \in \tilde{\Sigma}_n^k$ of the k-tuple $\{X_1, \ldots, X_k\}$. The modified shape is defined using the enlarged symmetry group of rotations, translations, dilatations, *and* reflections (hence the notation $\tilde{\Sigma}_n^k$ rather than Σ_n^k). (Section 3 explains how to carry results over to the full shape space Σ_n^k.) Adapting W.S. Kendall (1988), the (modified) shape σ is usefully parametrized by the *homogeneous shape coordinates* given by the $\binom{k}{2}$ normalized squared side lengths of the k-tuple $\{X_1, \ldots, X_k\}$. The normalization is obtained by dividing by the *size* Σ given in (2.1):

$$\Sigma = \frac{1}{2k} \sum_i \sum_j \|X_i - X_j\|^2. \tag{2.1}$$

Thus the homogeneous shape coordinates are given by $\sigma_{ij} = \|X_i - X_j\|^2/\Sigma$ (so $\sigma_{ii} = 0$ and $\sigma_{ij} = \sigma_{ji}$). They determine the modified shape of the multiplet $\{X_1, \ldots, X_k\}$. Consider the stochastic differential system governing the evolution of Σ and σ. Using a time-change $d\tau = dt/\Sigma$ the system can be summarized by the stochastic differential equations at (2.2), (2.3), and (2.4) below. Note from (2.3) that Σ and σ are infinitesimally uncorrelated; indeed the evolution of σ in the τ time-scale is independent of Σ and the trajectory of Σ forms a sufficient (functional) statistic for the Ornstein-Uhlenbeck parameter κ. A consequence of these observations is that shape σ and size Σ form a *skew-product decomposition* of the process of 'shape-and-size' of the multiplet $\{X_1, \ldots, X_k\}$. In particular the time-change based on size Σ turns the shape σ into a genuine diffusion governed by (2.4) below.

$$\text{Drift}\left(d_I \Sigma\right) = \left\{(k-1)n - \kappa\,\Sigma\right\} dt \tag{2.2a}$$
$$(d_I \Sigma)^2 = 4\,\Sigma\, dt \tag{2.2b}$$

$$\left(d_I \sigma_{ij}\right)\left(d_I \Sigma\right) = 0 \tag{2.3}$$

$$\text{Drift}\left(d_I \sigma_{ij}\right) = \left\{2 - (k-1)\sigma_{ij}\right\} n\, d\tau \tag{2.4a}$$
$$\begin{aligned}(d_I \sigma_{ij})(d_I \sigma_{uv}) = {} & 2\left(\delta_{iu} - \delta_{iv} + \delta_{jv} - \delta_{ju}\right)\left(\sigma_{iv} - \sigma_{iu} + \sigma_{ju} - \sigma_{jv}\right) d\tau \\ & - 4\sigma_{ij}\sigma_{uv}\, d\tau.\end{aligned} \tag{2.4b}$$

Here δ_{ij} is the Kronecker symbol, equal to unity if $i = j$ but otherwise zero. Notation such as Drift $(d_I \Sigma)$ refers to the mean forward infinitesimal increment at a fixed time of Σ, where the mean is the conditional expectation given the σ-field of events determined at the fixed time. The system (2.4) is the stochastic differential system for shape diffusion in *homogeneous shape coordinates.*

Computer algebra proved convenient in finding the above formulae, although their derivation by hand is a straightforward exercise. Indeed all the formulae for stochastic differential systems in this section can be (and have been) checked manually using Itô's lemma and (somewhat laborious) formula manipulation. Finding such formulae for the first time is of course rather harder work. It is in the exploratory phase that the benefits of the computer algebra of symbolic Itô calculus really pay off.

As in the derivation of the other stochastic differential systems described below, the computer algebra procedure for deriving (2.2), (2.3), (2.4) followed closely the method expounded in W.S. Kendall (1988). REDUCE was used in its interactive mode to define expressions for Σ and σ in terms of the Ornstein-Uhlenbeck processes $X_1, \ldots, X_k$. The procedure **d** of symbolic Itô calculus was then applied to derive expressions for $d_I \Sigma$, $d_I \sigma$ in terms of $X_1, \ldots, X_k$ and $d_I X_1, \ldots, d_I X_k$. The known second-order structure of $d_I X_1, \ldots, d_I X_k$ then allowed the determination of expressions for Drift $(d_I \Sigma)$, $(d_I \Sigma)^2$, $(d_I \Sigma)(d_I \sigma_{ij})$, Drift $(d_I \sigma_{ij})$, and $(d_I \sigma_{ij})(d_I \sigma_{uv})$, in terms of $X_1, \ldots, X_k$. Finally the REDUCE package was used to determine equivalent expressions in terms of Σ, σ_{ij} as above.

The advance in technique over W.S. Kendall (1988) lies in the use of REDUCE operators representing the action of summing over dummy variables. This allows the treatment of symbolic dimension. These operators and their associated simplification rules will be described in the future paper promised above.

The results of (2.2), (2.3), (2.4) (summarizing the second-order structure of Σ, σ) formed an intermediary stage in the derivation of results concerning the standardized inner product system below, and these in turn provided a second-order structure by which was derived the system for singular values decomposition. This approach (similar to that employed in the precursor paper of Kendall, 1988) typifies a step-by-step strategy which is important in computer algebra as a means of reducing the (often extreme) length of intermediate expressions. The occurrence of machine-overflow is thereby minimized and (of equal importance) the user finds it easier to see the direction in which the interactive calculations are pointing.

From henceforth we work in the τ-timescale and consider the shape diffusion σ.

The shape-diffusion formulae at (2.4) can be re-expressed in another coordinate system determined by the inner-products of a normalized system of particles representing the shape of the multiplet. These inner-products $\{C_{ij} : i, j = 1, \ldots, k\}$ are defined by the following (in which $\overline{X} = \frac{1}{k}\sum_j X_j$):

$$C_{ij} = \langle X_i - \overline{X}, X_j - \overline{X} \rangle / \Sigma. \tag{2.5}$$

The inner-products satisfy some important relationships:

$$\sigma_{ij} = C_{ii} - 2C_{ij} + C_{jj}, \tag{2.6a}$$

$$\sum_j C_{ij} = 0 \quad \text{for all} \quad i, \tag{2.6b}$$

$$C_{ij} = \frac{1}{k}\left\{\frac{1}{2}\sum_k (\sigma_{ik} + \sigma_{jk} - \sigma_{ij}) - 1\right\}. \tag{2.6c}$$

It may be deduced from the definition of Σ that

$$\sum_i C_{ii} = 1. \tag{2.7}$$

In this new coordinate system the stochastic differential system of (2.4) transforms to the following, the stochastic differential system for shape diffusion in *standardized inner product coordinates*:

$$\text{Drift}\left(d_I C_{ij}\right) = \left\{\delta_{ij} - (k-1)C_{ij} - 1/k\right\} n\, d\tau \tag{2.8a}$$

$$\begin{aligned}(d_I C_{ij})(d_I C_{uv}) = {} & \left(\delta_{iu} - 1/k\right)C_{jv}\, d\tau + \left(\delta_{iv} - 1/k\right)C_{ju}\, d\tau \\ & + \left(\delta_{ju} - 1/k\right)C_{iv}\, d\tau + \left(\delta_{jv} - 1/k\right)C_{iu}\, d\tau \\ & - 4\, C_{ij} C_{uv}\, d\tau.\end{aligned} \tag{2.8b}$$

Formulae (2.8) were derived from (2.4) using computer algebra and then verified (once!) by hand.

Systems (2.4) and (2.8) are not particularly informative about shape diffusion (though it is worth noting that the spatial dimension n enters into the systems only through the drifts). The stochastic matrix $\mathbf{C} = \{C_{ij} : i, j = 1, \ldots, k\}$ is a symmetric non-negative definite matrix and so it is natural to consider its spectral decomposition as providing a further system of coordinates. Consider

$$\mathbf{C} = \mathbf{R} \Lambda \mathbf{R}^T \tag{2.9}$$

where $\mathbf{R} = \{R_{ij} : i, j = 1, \ldots, k\}$ is a stochastic rotation matrix formed from the eigenvectors of $\mathbf{C}$ and $\Lambda = \{\lambda_i \delta_{ij} : i, j = 1, \ldots, k\}$ is a stochastic diagonal matrix formed from the eigenvalues of $\mathbf{C}$. (The matrices $\mathbf{R}$ and Λ are related to a singular values decomposition of a standardized representation of the multiplet $\{X_1, \ldots, X_k\}$.) Suppose that $\mathbf{R}$ is defined by the *Stratonovich* stochastic differential equation

$$d_S \mathbf{R} = \mathbf{R} \, d_S \, \eta \tag{2.10}$$

where $\eta = \{\eta_{ij} : i, j = 1, \ldots, k\}$ is the *rotational noise* for the stochastic rotation process $\mathbf{R}$. The stochastic differential system for shape diffusion in *singular values decomposition coordinates* is a stochastic differential system for Λ and η such that (2.9) yields a set of standardized inner product coordinates with the correct statistics (that is to say, satisfying the standardized inner product stochastic differential system (2.8)). In fact the singular values decomposition system is not uniquely determined by this requirement if rank considerations require more than one eigenvalue to be held fixed at zero, and it transpires that the system exhibits divergence (thus failing to define completely the evolution of $\mathbf{R}$ and Λ) on collision of a pair of eigenvalues neither one being held fixed at zero. Divergence problems will also arise for $\mathbf{R}$ if a moving eigenvalue hits a couple of eigenvalues held fixed at zero. Section 3 shows that these problems do not arise in practice.

A combination of computer algebra and manual calculation (reinforced at certain points by general arguments, and verified by manual calculation) shows that apart from the above considerations the required stochastic differential system must be as follows. Note that because we centralized the configuration (thus allowing the relation (2.6b)) one of the eigenvalues is always zero. We stipulate this to be the first eigenvalue, so $\lambda_1 = 0$ for all time. Indeed considerations of the rank of $\mathbf{C}$ make it clear that a total of at least r eigenvalues must be zero at all times, where $r = \max\{k - n, 1\}$. For convenience we order the eigenvalues in ascending order and stipulate that the first r eigenvalues are to be fixed at zero, so $0 = \lambda_1 = \cdots = \lambda_r \leq$

$\lambda_{r+1} \leq \cdots \leq \lambda_k$. In the sequel we refer to $\lambda_{r+1}, \ldots, \lambda_k$ as the 'moving eigenvalues'. Note finally from (2.7) that $\sum_i \lambda_i = 1$.

$$\text{Drift}\left(d_I \eta_{ij}\right) = 0, \tag{2.11a}$$

$$(d_I \eta_{ij})^2 = -(d_I \eta_{ij})(d_I \eta_{ji}) = \frac{\lambda_i + \lambda_j}{(\lambda_i - \lambda_j)^2}\, d\tau \tag{2.11b}$$

when $\lambda_i \neq \lambda_j$ and neither i nor j equals unity,

$$(d_I \eta_{ij})(d_I \eta_{uv}) = 0 \tag{2.11c}$$

when the conditions for equation (2.11b) do not apply.

$$(d_I \eta_{ij})(d_I \lambda_u) = 0. \tag{2.12}$$

$$\text{Drift}\left(d_I \lambda_i\right) = -\left[\left\{2\Big(\sum_{j:j\neq i} \frac{1}{\lambda_j - \lambda_i}\Big) + (k-1)n\right\}\lambda_i + k - n\right] d\tau \tag{2.13a}$$

when $i > r$ (this drift is zero otherwise),

$$(d_I \lambda_i)^2 = 4\lambda_i(1 - \lambda_i)\, d\tau, \tag{2.13b}$$

$$(d_I \lambda_i)(d_I \lambda_j) = -4\lambda_i \lambda_j\, d\tau \quad \text{when } i \text{ does not equal } j. \tag{2.13c}$$

REDUCE was used to create expressions for the C_{ij} in terms of λ_a and η_{uv}. Symbolic Itô calculus was then employed to find equations for the second order statistics of Λ, η in the particular case $r = 1$. Hence were derived the equations (2.11), (2.12), (2.13). The case of general r could then have been derived from a limiting argument or by modifying the computer algebra manipulations used for $r = 1$. In actual fact the correctness of the stochastic differential system for all r was then checked manually. Thus interactive computer algebra found the form of the solution, which was then verified manually to hold in all cases.

Equations (2.11), (2.12), and (2.13) reveal the structure of the shape diffusion to be that of a *skew-product* in the terminology of Pauvels and Rogers (1988), although it does not quite fall within the scope of the theory described there (since the shape diffusion will not in general be a Riemannian Brownian motion). The skew-product property follows by noting that the stochastic differential system (2.13) for Λ is autonomous and (by (2.12)) infinitesimally uncorrelated with the evolution of $\mathbf{R}$. If on the other hand Λ is conditioned to be held fixed then the stochastic differential system for $\mathbf{R}$ is that of a fixed diffusion with parameters depending on Λ. For (2.6b) and the fact that λ_1 is fixed at zero imply that $\mathbf{U} = \mathbf{R}(0)^{-1}\mathbf{R}$ leaves fixed the unit vector $(1, 0, \ldots, 0)^T$ and so under the Λ-conditioning $\mathbf{U}$ is a left-invariant diffusion on the corresponding subgroup $\mathrm{SO}(k-1) \leq \mathrm{SO}(k)$. In the full-rank case $r = 1$ this conditioned diffusion is actually a Brownian motion with respect to a left-invariant Riemannian metric depending in general on the conditioned value of Λ.

If Λ is not conditioned but left free to diffuse according to (2.12) and (2.13) then in general $\mathbf{U}$ is only a Γ-martingale with respect to a left-invariant connection (see W.S. Kendall 1987 for an explanation of this term). For the left-invariant metric on $\mathrm{SO}(k-1)$ is *not* in general unique (except in the special case of $k=3$ already covered by W.S. Kendall 1988, and the trivial case of $k=2$ which was excluded at the outset of this paper). Consequently except in these special cases $\mathbf{R}$ cannot be expressed as a diffusion on $\mathrm{SO}(k)$ subject to a random time change controlled by Λ. Thus the decomposition is not a skew-product decomposition in this special sense (which is perhaps what is more generally understood by the term 'skew-product'). In geometrical terms the singular values decomposition does not in general decompose the Riemannian metric induced by $\mathbf{R}$ and Λ into a warped product.

3. Answers to Some Basic Questions

In this section two fundamental features of the system (2.11), (2.12), (2.13) are discussed. Only the general lines of proofs are indicated.

3.1. Whether Eigenvalues Collide

The first feature concerns whether the system for the singular values decomposition defines the shape diffusion for all time. As noted above, the stochastic differential system (2.11), (2.12), (2.13) determines the stochastic evolution of shape only up to the first time a pair of the last $k-r$ eigenvalues collide, or (only in the case $r>1$) if λ_{r+1} hits zero. Moreover if $r>1$ then the system is not uniquely determined by the requirement that $\mathbf{C}=\mathbf{R}\,\Lambda\,\mathbf{R}^T$ should satisfy (2.8), as one can introduce extra rotational diffusion on axes corresponding to some pairs of $\{\lambda_2,\ldots,\lambda_r\}$.

The lack of uniqueness presents no problem, since we need only to *synthesize* $\mathbf{C}=\mathbf{R}\,\Lambda\,\mathbf{R}^T$ with the correct statistics. We do not therefore require uniqueness. The system (2.11), (2.12), (2.13) must be minimal in some sense, but we will not pursue this further here.

The question of collision might present a problem. However it can be shown that if initially the 'moving eigenvalues' $\lambda_{r+1},\ldots,\lambda_k$ are distinct then with probability one at no future time will any pair collide. This is established by considering the positive real-valued process

$$\Phi = -\sum_{r+1\leq i<j\leq k}\sum \log(\lambda_j-\lambda_i). \qquad (3.1)$$

The process Φ diverges to infinity precisely when a pair of 'moving eigenvalues' collide. Combination of Itô's lemma, the system (2.13), and a permutation argument for a triple sum produces an argument showing the following:

$$\text{Drift}\left(d_I\Phi\right) = \binom{k-r}{2}\left\{n(k-1)-2\right\}d\tau. \tag{3.2}$$

(A similar but more tedious argument can be applied to evaluate $\left(d\Phi\right)^2$.) This implies that the process

$$\Psi(\tau) = \Phi(\tau) - \binom{k-r}{2}\left\{n(k-1)-2\right\}\tau \tag{3.4}$$

defines a continuous local martingale. As such it may be expressed as a random time-change of real-valued Brownian motion. On any compact time-interval $[0, T]$ the trajectory of Ψ is bounded below by $-\binom{k-r}{2}\{n(k-1)-2\}T$ (by virtue of the positivity of Φ). Consequently it follows that in any given compact time-interval with probability one Ψ and hence Φ must be bounded above by random but finite bounds. (One appeals to the properties of real-valued Brownian motion.)

The above shows that Φ remains finite for all time, and so no pair of 'moving eigenvalues' may collide. We see in Subsection 3.2 that λ_{r+1} will not hit zero if $r > 1$. These arguments show that the system (2.11), (2.12), (2.13) defines the shape diffusion for all time, so long as the initial values of the last $k - r$ eigenvalues are distinct. In effect the shape diffusion is thereby defined in coordinates of the singular values decomposition over all of $\tilde{\Sigma}_n^k$ except on a polar subset

$$\mathcal{P} = \{\lambda_i = \lambda_j \text{ for some pair } i \neq j \text{ with } r < i < j \leq k\}.$$

(A polar subset is one which the shape diffusion never visits after time zero.) This is in close analogy to the way in which the classical expression of Euclidean Brownian motion in polar coordinates (using a Bessel process) breaks down at the origin.

Of course systems (2.8) or (2.4) provide definitions of the shape diffusion holding over all of $\tilde{\Sigma}_n^k$ without exception. Further investigation of the polar subset $\mathcal{P}$ would involve exploitation of the connection between Riemannian geometry and diffusion theory, alluded to in the introduction. This will be discussed in the follow-up paper promised above.

3.2. The Full Shape Diffusion

The second feature concerns the fulfilment of the promise in Section 2 to show how to derive formulae for the full shape diffusion on Σ_n^k. The answer hangs on determining precisely when λ_{r+1}, the 'smallest moving eigenvalue', can ever hit zero.

First note that $\tilde{\Sigma}_n^k = \Sigma_n^k$ in the case $k \leq n$ (since the symmetry group SO(n) can always carry multiplets of n or fewer points into their mirror images). Thus for $k \leq n$ the full shape diffusion is already identified. It remains to discuss the case $k > n$, which will split into the 'critical case' $k = n + 1$ and the case $k > n + 1$.

It is convenient to digress at this point to establish the behaviour of λ_{r+1} in the general case. The 'smallest moving eigenvalue' λ_{r+1} can hit zero if and only if the critical case $r = k - n = 1$ holds (which completes the argument of Subsection 3.1 to show the good behaviour of the system at (2.11), (2.12), (2.13)). This result is proved by comparing λ_{r+1} to a Bessel process of appropriate dimension.

First note that if $n = 1$ then there is nothing to prove, as $r = k - 1$, $\sum_i \lambda_i = 1$, and so $\lambda_{r+1} = \lambda_k = 1$ is constant.

Suppose $n > 1$. Let $T^{(\epsilon)}$ be the first time at which $\lambda_{r+2} - \lambda_{r+1}$ is no larger than ϵ. Assuming the 'moving eigenvalues' are initially distinct, $T^{(\epsilon)}$ is positive for sufficiently small ϵ. By the no-collision result above, $T^{(\epsilon)} \to \infty$ as ϵ tends to zero.

Consider λ_{r+1} in a new time-scale suggested by (2.13b) and defined by $d\tilde{\tau} = (1 - \lambda_{r+1})\, d\tau$. Working up to the random time $\tilde{T}^{(\epsilon)} = \int_0^{T^{(\epsilon)}} (1 - \lambda_{r+1})\, d\tau$ the evolution of λ_{r+1} in the new time-scale is governed by

$$\text{Drift}\left(d_I \lambda_{r+1}\right) = -2H\lambda_{r+1}\, d\tilde{\tau} + \nu\, d\tilde{\tau} \tag{3.5a}$$

$$\left(d_I \lambda_{r+1}\right)^2 = 4\lambda_{r+1}\, d\tilde{\tau} \tag{3.5b}$$

where $\nu = 2r - (k - n)$ and

$$H = \left\{ \left(\sum_{b > r+1} \frac{1}{\lambda_b - \lambda_{r+1}} \right) + \frac{(k-1)n - \nu}{2} \right\} (1 - \lambda_{r+1})^{-1}. \tag{3.6}$$

Note that H is bounded over the time interval $0 < \tilde{\tau} \leq \tilde{T}^{(\epsilon)}$ since over this interval $1 - \lambda_{r+1} > \lambda_b - \lambda_{r+1} > \lambda_{r+2} - \lambda_{r+1} \geq \epsilon$.

From (3.5) it follows that the process $X = \sqrt{\lambda_{r+1}}$ is a Bessel process of dimension ν with superimposed drift $-HX\, d\tilde{\tau}$. But H is bounded up to $\tilde{T}^{(\epsilon)}$ and so the Girsanov change-of-measure theorem implies that X can hit zero if and only if a Bessel process of dimension ν (without superimposed drift) can hit zero. Now it is classical (recalling the expression of Bessel processes as the radial parts of Euclidean Brownian motions) that such hitting of zero is possible if and only if $\nu = 1$. This means the 'smallest moving eigenvalue' λ_{r+1} can hit zero if and only if the critical case $r = k - n = 1$ holds.

Suppose $k > n + 1$, so that λ_{r+1} does not hit zero. Geometrical arguments show that the obvious projection of full shape onto modified shape

$$\pi : \Sigma_n^k - \{\lambda_{r+1} \circ \pi = 0\} \to \tilde{\Sigma}_n^k - \{\lambda_{r+1} = 0\} \tag{3.7}$$

is then a two-to-one map and indeed a covering map. The modified shape diffusion stays away from $\{\lambda_{r+1} = 0\}$ in this case, so it is possible to lift the path of the modified shape diffusion via π to the full shape space Σ_n^k. Thus if the initial point of the full shape diffusion is specified then its evolution can be deduced from that of the modified shape diffusion. So the system (2.11), (2.12), (2.13) does in fact specify the full shape diffusion in the case $k > n+1$ as well as in the case $k \leq n$.

If $k = n+1$ then the operation of taking the signed volume (via a determinant) shows that $\Sigma_n^{k=n+1} - \{\lambda_2 \circ \pi = 0\}$ can be split into two components $\Sigma^{(+)}{}_n^{n+1}$ and $\Sigma^{(-)}{}_n^{n+1}$ (recall that $r+1 = 2$ in this critical case). Moreover the reflection symmetry provides an isomorphism of the full shape diffusion on one component to the full shape diffusion on the other, and the component shape diffusions are isomorphic to the (modified) shape diffusion on $\tilde{\Sigma}_n^{n+1} - \{\lambda_2 = 0\}$. A proper description of the full shape diffusion must explain how the sign of the volume alters when the random process λ_2 visits zero.

In this case the Bessel process argument above shows that λ_2 behaves as a random time change of the square of a real-valued Brownian motion, modified by a locally bounded drift. Define $Y = \pm\sqrt{\lambda_2}$, where the sign is chosen according to the sign of the signed volume of the full shape. Itô calculus and excursion theory can be applied to show

$$\text{Drift}\,(d_I Y) = -\left\{\left(\sum_{j:j>2} \frac{1}{\lambda_j - Y^2}\right) + \frac{n^2-1}{2}\right\} Y\, d\tau, \tag{3.8a}$$

$$\left(d_I Y\right)^2 = (1 - Y^2)\, d\tau, \tag{3.8b}$$

$$\left(d_I \lambda_i\right)\left(d_I Y\right) = -2\lambda_i Y\, d\tau \qquad \text{when } i \neq 2, \tag{3.8c}$$

$$(d_I \eta_{ij})(d_I Y) = 0. \tag{3.8d}$$

If (3.8) is used to replace the corresponding parts of (2.13), (2.12) then we obtain an expression for the full shape diffusion in the case $k = n+1$, using a variation on the coordinates of the singular values decomposition based on η, $0 = \lambda_1$, $Y = \pm\sqrt{\lambda_2}$, and $\lambda_3, \ldots, \lambda_k$.

4. Conclusion

The work above raises a number of questions.

4.1. Relationship to Geometry

As has already been noted, Carne (1988) and D.G. Kendall (manuscript) have considered the Riemannian geometry natural to the Euclidean shape space Σ_n^k. The stochastic differential systems for shape diffusion carry within themselves information about this Riemannian geometry. The shape diffusion can be expressed as Brownian motion on the corresponding manifold modified by a drift. On the other hand the Riemannian metric tensor can be identified from the information summarized in the second-order part of the stochastic differential system for the shape diffusion. For example the level sets in $\tilde{\Sigma}_n^k$ obtained by fixing a value for $\mathbf{R}$ are incomplete but otherwise totally geodesic $(k-r-1)$-dimensional submanifolds of constant positive sectional curvatures $+1$. Indeed the 'eigenvalue map' sending $\sigma \in \tilde{\Sigma}_n^k - \mathcal{P}$ to $(\lambda_{r+1}, \ldots, \lambda_k)$ is a Riemannian submersion of the non-polar part of the modified shape space onto an open fragment of a $(k-r-1)$-sphere of constant positive sectional curvatures $+1$.

An obvious objective is to construct a set of computer algebra procedures to identify various features of the Riemannian geometry from the diffusion characteristics (D.G. Kendall and Le have carried out a similar task, using computer algebra to derive formulae for curvature for various coordinatizations of shape spaces). Account will have to be taken of the need to *complete* the Riemannian geometry to extend over the polar set $\mathcal{P}$ where pairs of moving eigenvalues coincide.

The case of Σ_n^3 for $n \geq 3$ is informative. In this case $\tilde{\Sigma}_n^3 = \Sigma_n^3$, $r = \max\{3-n, 1\} = 1$, $\lambda_1 = 0$ and $\lambda_2 + \lambda_3 = 1$. The eigenvalues provide one degree of freedom in a coordinate space looking like $[0, \frac{1}{2}]$. In the rotational component the only variation is provided by $\eta_{23} = -\eta_{32}$ and so the rotational coordinate space looks like a circle $\mathrm{SO}(1)$. Thus the singular values decomposition is based on a cylinder $[0, \frac{1}{2}] \times \mathrm{SO}(1)$. The singularity set is the circle $\mathcal{P} = \{\lambda_2 = \frac{1}{2}\} \times \mathrm{SO}(1)$ and in fact the associated Riemannian geometry collapses this circle to a point, and gives Σ_n^3 the geometry of a hemisphere (this corresponds to the route followed in W.S. Kendall, 1988). The identification of $\mathcal{P}$ to a point arises from the divergence as $\lambda_3 - \lambda_2$ converges to zero of

$$\begin{aligned}(d_I \eta_{32})^2 &= \frac{\lambda_3 + \lambda_2}{(\lambda_3 - \lambda_2)^2}\, d\tau \\ &= \frac{d\tau}{(\lambda_3 - \lambda_2)^2}. \end{aligned} \tag{4.1}$$

This example is a useful prototype for the way in which the diffusion establishes the geometry; near the polar locus $\mathcal{P}$ the divergent diffusion coefficients of the rotational noise lead to identifications in the system of coordinates of singular values decomposition. Note however in general it is necessary to take account of singularities in the Riemannian geometry.

Thus the task which should be undertaken next is to provide means, using computer algebra, of passing from the diffusion to the geometry and (if possible) taking account of identifications such as above.

4.2. Wishart Matrices

If the multiplet $\{X_1, \ldots, X_k\}$ is composed of independent random points possessing the same multivariate spherically symmetric Gaussian distribution then the resulting distribution of shape corresponds to a certain Wishart distribution normalized to have unit trace. This distribution is the invariant distribution of the shape diffusion if the points of the multiplet diffuse according to (1.1). Corresponding to this, the results above could be obtained as consequences of a stochastic calculus version of Wishart distribution theory. See the work of Bru (1989).

This 'stochastic calculus of multivariate statistical analysis' will provide the next testing ground for symbolic Itô calculus. As a further prospect there is the challenging task of generalizing to the non-spherically-symmetric extension of (1.1). The work of Mardia and Dryden (1989) suggests other exercises connected to non-central Wishart distributions.

4.3. Relevance of the λ_i Coordinates to Shape Theory

The coordinates of singular values decomposition make explicit a natural $\mathrm{SO}(k-1)$ (indeed, $\mathrm{O}(k-1)$) symmetry for Σ_n^k, allowing for certain questions a reduction of dimensionality by considering only the $k-r$ 'moving eigenvalues'. For example the locus of collinear multiplets is invariant under this symmetry, and so natural measures of distance from collinearity will be given by expressions involving only $\lambda_{r+1}, \ldots, \lambda_k$. For this reason the relative simplicity of the system (2.13) is particularly satisfying, and the Riemannian submersion referred to in Subsection 4.1 is of practical importance.

4.4. Matrix Factorization in Stochastic Calculus

We have already noted similarities to the work of Pauvels and Rogers (1988). See also Norris, Rogers, and Williams (1986) and references therein to work of Dynkin, Dyson, McKean, and Orihara on random matrices. Taylor (1988) expounds work of Malliavin and Malliavin which forms a more geometric approach to similar problems for Brownian motion on symmetric spaces. However shape diffusions appear to lack too much symmetry for any of this previous work to apply directly.

4.5. Automatic Reduction of Stochastic Differential Equations

One way to view the work of this paper is as an exploitation of a not entirely evident $\mathrm{O}(k-1)$ symmetry to reduce a stochastic differential system (2.4) or (2.8) to a form (2.11), (2.12), (2.13) involving a reduction in dimensionality. In effect, a stochastic differential system has been partially 'solved'. This raises the enticing prospect of building sets of procedures in REDUCE or another computer algebra package which would search for possible symmetries in a stochastic differential system. Having found a symmetry, this would be exploited to produce a new representation of the system in the manner given above. Sets of REDUCE procedures already exist to perform similar tasks for partial differential equations, so this prospect must be eminently achievable! From this point of view the work of this paper, originally undertaken primarily to further elicit the structure of shape diffusion, becomes a test case suggesting geometric perspectives and algorithms for complex stochastic systems.

Acknowledgement

Part of this work was aided by the provision of a Sun workstation under SERC grant GR/E 39891 while the author was affiliated to Strathclyde University.

References

Ambartzumian, R.V. (1987). *Stochastic and Integral Geometry.* Reidel, Dordrecht/Boston.

Broadbent, S.R. (1980). Simulating the ley-hunter (with discussion). *Journal of the Royal Statistical Society A* 143, 109–140.

Bru, M.F. (1989). Processus de Wishart. *Comptes Rendus de l'Académie des Sciences (Paris)* 308, 29–32.

Carne, T.K. (1988). The geometry of shape spaces. Preprint.

Durrett, R. and Pinsky, M.A. (1988). *The Geometry of Random Motion.* Contemporary Mathematics 73, American Mathematical Society, Providence, R.I.

Ikeda, N. and Watanabe, S. (1981). *Stochastic Differential Equations and Diffusion Processes.* North Holland/Kodansha, Amsterdam/Tokyo.

Kendall, D.G. (manuscript). Ten years late better than never. Unpublished manuscript, circulated Summer 1987.

——— (1977). The diffusion of shape (abstract). *Advances in Applied Probability* 9, 428–430

——— (1984). Shape manifolds, Procrustean metrics, and complex projective spaces. *Bulletin of the London Mathematical Society* 16, 81–121.

——— (1986). Further developments and applications of the statistical theory of shape. *Teorija Verojatnostei* 31, 467–473. Translated as: *Theory of Probability* 31 (1987), 407–412.

——— (1989). A survey of the statistical theory of shape (with discussion). *Statistical Science*, to appear.

Kendall, D.G. and Kendall, W.S. (1980). Alignments in two-dimensional random sets of points. *Advances in Applied Probability* 12, 380–424.

Kendall, W.S. (1987). Stochastic differential geometry: an introduction. In Ambartzumian (1987), reprinted from *Acta Applicandae Mathematicae* 9, 29–60.

——— (1988). Symbolic computation and the diffusion of shapes of triads. *Advances in Applied Probability* 20, 775–797.

Le, H.L. (1988). Unpublished Ph.D. thesis, University of Cambridge.

Mardia, K. and Dryden, I.L. (1989). The statistical analysis of shape data. *Biometrika*, to appear.

Norris, J.R., Rogers, L.C.G., and Williams, D. (1986). Brownian motion of ellipsoids. *Transactions of the American Mathematical Society* 294, 757–765.

Pauvels, E.J. and Rogers, L.C.G. (1988). Skew-product decompositions of Brownian motions. In Durrett and Pinsky (1988, pp. 237–262).

Rogers, L.C.G. and Williams, D. (1987). *Diffusions, Markov Processes, and Martingales*, Volume 2. Wiley, Chichester.

Small, C.G. (1988). Techniques of shape analysis on sets of points. *International Statistical Review* 56, 243–257.

Stoyan, D., Kendall, W.S., and Mecke, J. (1987). *Stochastic Geometry and its Applications.* Wiley/Akademie Verlag, Chichester/Berlin.

Taylor, J.C. (1988). The Iwasawa decomposition and the limiting behaviour of Brownian motion on a symmetric space of non-compact type. In Durrett and Pinsky (1988, pp. 303–332).

Department of Statistics
University of Warwick
Coventry CV4 7AL.

Asymptotics in High Dimensions for Percolation

Harry Kesten[1]

Abstract

We prove that the critical probability for bond or site percolation on $\mathbb{Z}^d$ is asymptotically equal to $1/(2d)$ as $d \to \infty$. If the probability of a bond (respectively site) to be occupied is $\gamma/(2d)$ with $\gamma > 1$, then for the bond model the percolation probability converges as $d \to \infty$ to the strictly positive solution $y(\gamma)$ of the equation $y = 1 - \exp(-\gamma y)$. In the site model the percolation probability is asymptotically equal to $\gamma y(\gamma)/(2d)$ under these conditions. An asymptotic independence property for the random field of sites which belong to the infinite cluster is given.

1. Introduction

Broadbent and Hammersley (1957) created the theory of percolation. Largely because of Hammersley's impetus the subject has grown enormously and is at present a very lively research area in probability and statistical mechanics. Since much of my own research has been inspired by John Hammersley it is a pleasure to dedicate an article on percolation to him in this Festschrift.

Recently Aizenman, Bricmont, and Lebowitz (1987) used the behavior of the critical probability of site percolation in high dimension to obtain some interesting properties of the Ising model. For oriented percolation the asymptotic behavior of the critical probability in high dimension was derived by Cox and Durrett (1983). Here we shall consider the asymptotic properties as $d \to \infty$ of (unoriented) bond and site Bernoulli percolation on $\mathbb{Z}^d$. In (Bernoulli) bond percolation the bonds are occupied (respectively vacant) with probability p (respectively $q := 1 - p$) and all bonds are independent. The corresponding product measure on the configurations of bonds is denoted by P_p. $C(x)$ is the (occupied) *cluster* of x; it is the collection of all points which can be reached from x by an occupied path. (An *occupied path* is a path all of whose edges are occupied.) We write

[1]Research supported by the NSF through a grant to Cornell University.

$\theta(p) = \theta(p, \mathbb{Z}^d, \text{bond})$ for the percolation probability:

$$\theta(p) = P_p\{C(0) \text{ is infinite}\}. \tag{1.1}$$

It is known (Aizenman, Kesten, and Newman 1987; Gandolfi, Grimmett, and Russo 1988) that if $\theta(p) > 0$, then there exists w.p.1 a unique infinite cluster. If $\theta(p) > 0$ we say that *percolation occurs.* Broadbent and Hammersley (1957) and Hammersley (1959) proved that there exists a non-trivial critical probability $p_c = p_c(\mathbb{Z}^d, \text{bond})$ which separates the parameter domains where percolation occurs and where it does not occur. In other words, if we set

$$p_c = \sup\{p : \theta(p) = 0\}, \tag{1.2}$$

then

$$0 < p_c < 1, \ \theta(p) = 0 \text{ if } p < p_c \text{ and } \theta(p) > 0 \text{ if } p > p_c.$$

It is believed (but so far only proven when $d = 2$) that $\theta(p_c) = 0$. All of the preceding has its analogue for site percolation; we merely have to replace 'bond' by 'site' everywhere in the above description of the bond model.

The principal result of this paper gives the asymptotic behavior of p_c. The result is not unexpected, since simple results about branching processes tell us that on a tree with all vertices of degree $2d$, percolation occurs if and only if $p > (2d-1)^{-1}$. Theorem 1 says that asymptotically for large d the critical probability for such a tree and for $\mathbb{Z}^d$ are the same in first order; the circuits which exist on $\mathbb{Z}^d$ play only a small role for large d. Gordon (1988) recently also proved that $2d\,p_c(\mathbb{Z}^d, \text{bond}) \to 1$ as $d \to \infty$ by a rather different method.[2]

THEOREM 1.

$$\frac{1}{2d-1} \le p_c(\mathbb{Z}^d, \text{bond}) \le p_c(\mathbb{Z}^d, \text{site}) \le \frac{1}{2d} + O\left(\frac{(\log\log d)^2}{d \log d}\right). \tag{1.3}$$

Theorem 1 can be used to show that if one takes $p = \gamma/(2d)$, then the random field of the sites which belong to an infinite cluster behaves for large d like an independent random field (with success probability converging to the $y(\gamma)$ of (1.5)). For site percolation we have a similar result after a simple modification of the statement. Such a modification is necessary for the following trivial reason. In the bond model a site is incident to

[2]Note added in proof: It seems that the forthcoming paper of Hara and Slade (1989) implicitly proves that $p_c(\mathbb{Z}^d, \text{bond}) = (2d)^{-1} + O(d^{-2})$. In addition, it has just come to our attention that asymptotic expansions (in powers of $(2d-1)^{-1}$) for $p_c(\mathbb{Z}^d, \text{site})$ and $p_c(\mathbb{Z}^d, \text{bond})$ were given on a non-rigorous basis in Gaunt, Sykes, and Ruskin (1976) and Gaunt and Ruskin (1978).

$2d$ edges, each of which can potentially connect the site to ∞. As we shall see this leads to a strictly positive limit for $\theta(\gamma/(2d), \mathbb{Z}^d, \text{bond})$ when $\gamma > 1$ is fixed. In the case of site percolation our definitions require the site x to be occupied in order for x to be connected to ∞. Consequently $\theta(\gamma/(2d), \mathbb{Z}^d, \text{site}) \le \gamma/(2d)$. In order to obtain a situation comparable to that of the bond model we should ignore the state of x itself or condition on x being occupied. Theorem 2S shows that this indeed leads to a result for the site model which is almost the same as for the bond model.

We should note that Theorems 2B and 2S (and their proofs) express the generally held belief that in high dimensions the system exhibits 'mean field behavior'. E.g. in the bond model this means that around a fixed site x the number of neighbors of x connected to ∞ is close to its expected value $2d\theta(p)$, irrespective of the states of the edges incident to x itself. Once this is accepted it is easy to derive a consistency relation for $\theta(p)$. This is the so called mean field equation; at $p = \gamma/(2d)$ the limit of this equation as $d \to \infty$ is just (1.5).

We write $|A|$ for the number of vertices in the set A.

THEOREM 2B. *In the bond model, when $\gamma > 1$ is fixed,*

$$\lim_{d\to\infty} \theta\left(\frac{\gamma}{2d}, \mathbb{Z}^d, \text{bond}\right) = y(\gamma), \tag{1.4}$$

where $y(\gamma)$ is the unique strictly positive solution of

$$y = 1 - e^{-\gamma y}. \tag{1.5}$$

More generally, for fixed $\gamma > 1$

$$\lim_{d\to\infty} \sup_{A,B} \Big| P_{\gamma/2d}\{\text{all sites in } A \text{ belong to the infinite occupied cluster, but none of the sites in } B \text{ do}\} - \big(y(\gamma)\big)^{|A|}\big(1 - y(\gamma)\big)^{|B|}\Big| = 0. \tag{1.6}$$

The supremum in (1.6) is over all pairs of finite disjoint sets A and B.

THEOREM 2S. *In the site percolation model, when $\gamma > 1$ is fixed,*

$$\lim_{d\to\infty} \frac{2d}{\gamma} \theta\left(\frac{\gamma}{2d}, \mathbb{Z}^d, \text{site}\right) = y(\gamma)$$

(with $y(\gamma)$ as in (1.5)). More generally, for fixed $\gamma > 1$,

$$\lim_{d\to\infty} \sup_{A,B} \Big| P_{\gamma/2d}\{\text{all sites in } A \text{ have a neighbor which belongs to the infinite occupied cluster, but none of the sites in } B \text{ do}\} - \big(y(\gamma)\big)^{|A|}\big(1 - y(\gamma)\big)^{|B|}\Big| = 0. \tag{1.7}$$

The supremum in (1.7) is over the same A, B as in (1.6).

Theorem 1 for bond percolation and Theorem 2B are special cases of similar results for a more general cluster model in which bonds are not independent. These so called Fortuin-Kasteleyn models have an extra parameter Q. For integral $Q \geq 1$ these cluster models have a close relationship with the Potts model with Q colors. In Bricmont, Kesten, Lebowitz, and Schonmann (1989) and Kesten and Schonmann (1989), these models are described in more detail and results corresponding to the above results are proved there for integer Q. For $1 \leq Q \leq 2$ one can even obtain the exact parallels to the above results and we shall give these proofs elsewhere (Kesten 1989). The proof of Theorem 1 has to be given for percolation first and that will be done here. However, Theorem 2 is better treated for all $1 \leq Q \leq 2$ at the same time, and its proof will therefore be deferred to Kesten (1989).

Acknowledgement. The author is indebted to R. Schonmann for suggesting Theorems 2B and 2S and part of their proof.

2. Proof of Theorem 1

The first inequality in (1.3) is one of the earliest results in the subject. It was proven by means of a Peierls argument by Broadbent and Hammersley (1957). The second inequality was proven a number of times; see McDiarmid (1980), Hammersley (1961), and Oxley and Welsh (1979). The only novelty of (1.3) is therefore the last inequality and *for the remainder of this section we shall work with site percolation on $\mathbb{Z}^d$.*

As in Cox and Durrett (1983), which dealt with oriented percolation, we shall basically apply Chebyshev's inequality to the number of occupied paths which connect the origin, $\mathbf{0}$, to points at distance $n-1$ from $\mathbf{0}$ (for n large). Unfortunately, for standard percolation there is less independence among such paths than for oriented percolation, and in order to regain some independence we have to restrict ourselves to certain subclasses of paths which we now define. First, a *path* (of *length* m) on $\mathbb{Z}^d$ is a sequence $v_1, \dots, v_m$ of m vertices of $\mathbb{Z}^d$ such that v_i and v_{i+1} are neighbors. We do not insist that all the v_i are distinct; a path is not necessarily self-avoiding. The ith *step* of the path is the vector $s_i := v_i - v_{i-1}$. e_k will denote the kth unit coordinate vector. We now define for positive integers N and n the following class ($\lfloor a \rfloor$ denotes the largest integer a):

$\mathcal{C}(N,n)$ = collection of paths of length $nN-1$ whose steps s_i satisfy
(a) $s_i \in \{e_k : k > d - \lfloor d/N \rfloor\}$ if $i = jN$ for $j = 1, \dots, n-1$ and
(b) $s_i \in \{\pm e_k : k \leq d - \lfloor d/N \rfloor\}$ if $jN < i < (j+1)N$ for $j = 0, \dots, n-1$
(this also applies to s_1, which we define as v_1). (2.1)

In the sequel we shall make the convention that $v_0 = \mathbf{0}$ and $s_1 = v_1$ for paths in $\mathcal{C}(N, n)$. We note that there are $\lfloor d/N \rfloor$ choices for each of the steps of the form (a) and $2d - 2\lfloor d/N \rfloor$ choices for each of the steps of the form (b). Thus $\#\mathcal{C}(N, n)$, the cardinality of $\mathcal{C}(N, n)$ is

$$(2d - 2\lfloor d/N \rfloor)^{n(N-1)} \lfloor d/N \rfloor^{n-1}. \tag{2.2}$$

Any path $v_1, \ldots, v_{nN-1}$ in $\mathcal{C}(N, n)$ starts at a neighbor of $\mathbf{0}$ and

for $kN \le i < (k+1)N$ the sum of the last $\lfloor d/N \rfloor$ components of v_i equals k. (2.3)

We shall count paths in $\mathcal{C}(N, n)$, but not just occupied paths. Instead we define a stronger property. We attach to each vertex v of $\mathbb{Z}^d$ a sequence of 0–1 valued random variables $Y_1(v), Y_2(v), \ldots$ such that

$$\text{all variables } \{Y_i(v) : i \ge 1, v \in \mathbb{Z}^d\} \text{ are independent} \tag{2.4}$$

and

$$P\{Y_i(v) = 1\} = p \text{ for all } i \text{ and } v. \tag{2.5}$$

If $r = (v_1, \ldots, v_{nN-1})$ is a path of length $nN - 1$ then we define

$$\begin{aligned} k(r, v) &= \text{number of } i \ge 1 \text{ with } v_i \text{ equal to } v \\ &= \text{the number of visits by } r \text{ to } v. \end{aligned}$$

We say that the event $A(r)$ occurs if

$$Y_j(v) = 1 \text{ for } j \le k(r, v) \text{ for all } v. \tag{2.6}$$

Thus if we think of $Y_j(v)$ as the Y value sampled at the jth visit to v, then $A(r)$ occurs if and only if the Y sampled at each visit to a vertex by r is $+1$. Consequently

$$P\{A(r)\} = p^{nN-1} \tag{2.7}$$

for all paths r of length $nN - 1$.

In the proof of Lemma 1 it is explained how the event $A(r)$ is related to r being occupied. In any case we shall be interested in the number of paths r for which $A(r)$ occurs. To estimate the variance of this number we introduce some further quantities. For a pair of paths $r = (v_1, \ldots, v_{nN-1})$, and $r' = (v'_1, \ldots, v'_{nN-1})$, both of length $nN - 1$, we define

$$J(r, r') = \sum_v k(r, v) \wedge k(r', v) \tag{2.8}$$

($a \wedge b$ denotes $\min\{a, b\}$). Thus if r visits v_i at time i, and this is the νth visit to v_i by r, then this visit adds to the count $J(r, r')$ if and only if there exists an index j such that $v'_j = v_i$ and r' visits v_i for the νth time at time j.

Finally we introduce a probability measure on ordered pairs of paths. $\mathbb{P}$ will be the probability measure which picks a pair r,r' from $\mathcal{C}(N, n)$ with all pairs equally likely. Thus, the probability mass assigned to any pair of paths in $\mathcal{C}(N, n)$ is $[\#\mathcal{C}(N, n)]^{-2}$. Actually, at this moment $\mathbb{P}$ depends on nN, but we suppress this dependence in the notation. $\mathbb{E}$ denotes expectation with respect to $\mathbb{P}$.

LEMMA 1. *For any fixed* N,

$$\theta(p, \mathbb{Z}^d, \text{site}) \geq p \limsup_{n\to\infty} [\mathbb{E}\{p^{-J(r,r')}\}]^{-1}. \tag{2.9}$$

PROOF: Choose v occupied if $Y_1(v) = 1$ and vacant if $Y_1(v) = 0$. It is easily seen that under (2.5) the distribution of the occupancy configurations is P_p. Assume now that $A(r)$ occurs for some $r = (v_1, \dots, v_{nN-1}) \in \mathcal{C}(N, n)$. Then by 'loop-removal' we can find an occupied self-avoiding path from v_1 to v_{nN-1}. Loop-removal consists of first finding the last index k such that $v_k = v_1$. We then take out from r the vertices $v_2, \dots, v_k$. We are then left with the path $(v_1, v_{k+1}, \dots, v_{nN-1})$ which visits v_1 only at time 1. Next we find the last index $m \geq k+1$ for which $v_m = v_{k+1}$ and we remove the vertices $v_{k+2}, \dots, v_m$ to obtain the path $(v_1, v_{k+1}, v_{m+1}, \dots, v_{nN-1})$ which visits each of v_1 and v_{k+1} exactly once. We continue this procedure until no vertex is visited more than once. Let $\bar{r}$ be the self-avoiding path which is left over after this procedure. Its first vertex is v_1 and it is easily seen that the last vertex of $\bar{r}$ must be equal to the endpoint of r, v_{nN-1} (even though v_{nN-1} may be visited several times by r, and in the loop-removal procedure the last vertex of $\bar{r}$ may appear as a v_t which equals v_{nN-1}, but with $t < nN - 1$). By (2.3) with $k = n - 1$ this endpoint of $\bar{r}$ is at least at distance $n - 1$ from $\mathbf{0}$ (the distance here is the l^1 distance, not the Euclidean one). Also $\bar{r}$ must be occupied since all its vertices had corresponding $Y_1 = 1$ if $A(r)$ occurred. Thus $A(r)$ implies that there exists an occupied self-avoiding path of length $n - 1$ starting at a neighbor of the origin. As $n \to \infty$ the probability of the last event converges to

$$P_p\{\text{a neighbor of } \mathbf{0} \text{ is connected to } \infty\} = \frac{1}{p}\,\theta(p, \mathbb{Z}^d, \text{site}).$$

Thus (2.9) will follow if we can prove

$$P\{A(r) \text{ occurs for some } r \in \mathcal{C}(N, n)\} \geq [\mathbb{E}\{p^{-J(r,r')}\}]^{-1}. \tag{2.10}$$

However, (2.10) is almost immediate from Schwarz's inequality. Indeed if M denotes the number of r in $\mathcal{C}(N,n)$ for which $A(r)$ occurs, then

$$P\{A(r) \text{ occurs for some } r\} = P\{M > 0\} \geq \frac{(E\{M\})^2}{E\{M^2\}}.$$

Now

$$E\{M\} = \#\mathcal{C}(N,n)p^{nN-1}.$$

(see (2.7)), while

$$\begin{aligned} E\{M^2\} &= \sum_{r,r'} P\{A(r) \text{ and } A(r') \text{ occur}\} \\ &= \sum_{r,r'} p^{2nN-2-J(r,r')} = [\#\mathcal{C}(N,n)]^2 p^{2nN-2}\mathbb{E}\{p^{-J(r,r')}\}. \end{aligned}$$

The second equality here follows from the fact that the number of Y's sampled by r and r' *together* is $2nN-2-J(r,r')$, because J counts precisely the number of times r samples a Y which is also sampled by r'. (2.10) follows from these formulae. ■

To estimate $\mathbb{E}\{p^{-J}\}$ we shall break up J into a sequence of contributions which behave more or less like a Markov chain. Before we do this it is convenient to view the paths $(v_1,\ldots,v_{nN-1})$ and $(v'_1,\ldots,v'_{nN-1})$ as the initial pieces of two infinite paths $r=(v_1,v_2,\ldots)$ and $r'=(v'_1,v'_2,\ldots)$. Accordingly we extend $\mathbb{P}$ to a measure on pairs of infinite paths which are independent under $\mathbb{P}$ and whose ith step is any one of the unit vectors in $\{\pm e_k : k \leq d - \lfloor d/N \rfloor\}$ with probability $(2d - 2\lfloor d/N\rfloor)^{-1}$ when N does not divide i, and whose ith step is any one of $\{e_k : k > d - \lfloor d/N \rfloor\}$ with probability $\lfloor d/N \rfloor^{-1}$ when i is a multiple of N. Here the first step of the path r is v_1 and the first step of r' is v'_1. We also maintain our convention that $v_0 = \mathbf{0}$. One easily checks that the initial pieces of length $nN-1$ of r and r' are independently uniformly distributed over $\mathcal{C}(N,n)$ as with the previous definition of $\mathbb{P}$. If necessary we shall write $J_{nN}(r,r')$ now, instead of our previous $J(r,r')$, to indicate that we are working with the initial pieces of length $nN-1$.

We define the kth *block* of r to be the path $(v_{(k-1)N}, v_{(k-1)N+1},\ldots, v_{kN-1})$. By a slight abuse of notation we shall also say that the time t or the index t occurs in the kth block if $(k-1)N \leq t < kN$. We say that r has a *high density point* in the kth block if there exist t and s in the kth block such that

$$t,s \geq 1,\ |t-s| \geq 2, \text{ and } |v_t - v_s| \leq 1.$$

In particular if v_t is a double point of r in the rth block, then it is also a high density point. However, v_t is also a high density point if one of its neighbors is visited by r at any other time than $t-1$ or $t+1$. Similar definitions hold for r'. By our choice of $\mathbb{P}$, (2.3) still holds for all k so that w.p.1 $v_t = v_s$ can actually occur only if t and s belong to the same block. For the same reason $v_t = v'_s$ can w.p.1 occur only when t and s lie in the same block.

We next define *special indices.* If r has no high density point in the kth block, then $(k-1)N+i$, with $0 \le i < N$, is a special index if and only if

$$v'_{(k-1)N+i} \text{ has not been visited by } r' \text{ at any time } 1 \le t < (k-1)N+i, \text{ and in addition } v'_{(k-1)N+i} = v_{(k-1)N+j} \text{ for some } j. \text{ (If } k=1 \text{ we also require } i,j \ge 1.) \qquad (2.11)$$

We point out that (w.p.1) the occurrence of (2.11) depends on the kth blocks of r and r' only, since the only possible values for t and $(k-1)N+j$ at which r' or r can visit $v_{(k-1)N+i}$ are in the kth block. With each such special index t we associate a contribution $L(t) = L(t;r,r')$ of size 1 to J. Here and in the future we index a contribution L by the special index to which it corresponds. Next, when r has a high density point in the kth block, then there is either no special index in $[(k-1)N, kN)$ or exactly one. The former is the case if there are no t and s in $[(k-1)N \vee 1, kN)$ with $v_t = v'_s$ ($a \vee b$ denotes $\max\{a,b\}$). If there do exist such t and s, then the only special index in the kth block is taken to be $kN-1$ and the corresponding contribution $L(kN-1)$ is defined as

$$L(kN-1) = \text{ number of } s \in [(k-1)N \vee 1, kN) \text{ for which } v'_s \text{ equals a } v_t \text{ in the } k\text{th block of } r.$$

Now let $t(1) < t(2) < \cdots < t(\rho)$ be all the special indices $\le nN-1$ (thus the next special index $t(\rho+1)$ occurs at or after time nN; this defines $\rho = \rho(nN)$). We claim that

$$J_{nN}(r,r') \le \sum_{k=1}^{\rho} L(t(k);r,r'). \qquad (2.12)$$

To prove (2.12) consider the kth block of r. The vertices in this block can be visited only at the times $[(k-1)n, kN)$. If r has no high density points in this block and v is one of the vertices of r in this block then $k(r,v) = 1$. Therefore the only contributions to (2.8) from this block come from v's with $k(r,v) \wedge k(r',v) = 1$. Let v be such a vertex and let t be the smallest index t for which $v'_t = v$. Then t is a special index and the corresponding $L(t) = 1 = k(r,v) \wedge k(r',v)$. Thus all contributions to J from a block without high density points also appear in the right hand side of (2.12). If r has a high density point in the kth block but $v_t \neq v'_s$ for all $t,s \in [(k-1)N, kN)$, then there are no contributions from this block to either side of (2.12). If $v_t = v'_s$ for some t, s, then the contribution to J_{nN} from this block is

$$\sum_{\substack{v \in k\text{th} \\ \text{block of } r}} k(r,v) \wedge k(r',v) \leq \sum_{\substack{v \in k\text{th} \\ \text{block of } r}} k(r',v) = L(kN-1).$$

Thus for a block with high density points the contribution to the right hand side of (2.12) is always at least as large as the one to the left hand side, and (2.12) must hold.

Finally we associate a *type* with each special index and its corresponding contribution L. We make the convention that $t(0) = 0$. For $i \geq 1$ we say that $t(i)$ is of

type 1 if t belongs to a block without high density points, $t(i) - t(i-1) = 1$, and $t(i)$ *is not* a multiple of N,
type 2 if t belongs to a block without high density points, $t(i) - t(i-1) = 1$, but $t(i)$ *is* a multiple of N,
type 3 if t belongs to a block without high density points, and $t(i) - t(i-1) \geq 2$,
type 4 if t belongs to a block with high density points.

It will turn out that the main task is to estimate

$$\mathsf{E}\{p^{-L(t(i))}; i \leq \rho,\ t(i) \text{ is of type } l \mid \mathcal{F}_{i-1}\} \tag{2.13}$$

on the event

$$\{(i-1) \leq \rho,\ t(i-1) \text{ is of type } m\}, \tag{2.14}$$

where

$$\mathcal{F}_j := \text{ the } \sigma\text{-field generated by } t(j) \text{ and } \{v_t : t < \lceil t(j)/N \rceil N\} \cup \{v'_t : t \leq t(j)\}$$

($\lceil a \rceil$ denotes the smallest integer $\geq a$). It may be useful for the reader to skip Lemmas 2–6 at first reading to see how the main line of the argument runs once (2.13) has been estimated.

Note that r and r' are not treated equally in the definition of $\mathcal{F}_j$; we are forced to do this by the asymmetric definition of the special indices which involve first looking at the whole block of r to see whether it contains a high density point, while high density points of r' do not play such a role.

We have defined $\mathcal{F}_j$ in such a way that $L(t(j))$ is measurable with respect to $\mathcal{F}_j$.

To estimate (2.13) we shall need some estimates which are basically known facts about a simple random walk. For the remainder of this section we take $D = d - \lfloor d/N \rfloor$ and $\{S_u\}$ a simple random walk in $\mathbb{Z}^D$ with $S_0 = \mathbf{0}$. K_i will denote some universal constant (independent of d, N and p). Furthermore we restrict p and N to satisfy

$$\frac{1}{2d} \le p \le \frac{2}{2d} \quad \text{and} \quad 8 \le N \le \frac{\log d}{2 \log\log d}. \tag{2.15}$$

LEMMA 2. *Let*

$$\mathcal{G}_t = \sigma\text{-field generated by all } v_i \text{ and by the } v'_q \text{ with } q \le t.$$

and let τ be a stopping time with respect to the $\mathcal{G}_t$. Then for p and N satisfying (2.15) and for any vertex w we have

$$\mathbb{P}\{v'_i = w \text{ for some } \tau \vee (jN - 1) < i < (j+1)N \mid \mathcal{G}_\tau\}$$
$$\le \frac{(1 + K_1 N/d)}{2D} \text{ on the set } \{\tau = s\}, \text{ for any } jN \le s < (j+1)N - 1. \tag{2.16}$$

Moreover

$$\mathbb{P}\{r' \text{ visits the } (j+1)\text{th block of } r \text{ at some time} > \tau \mid \mathcal{G}_\tau\}$$
$$\le \begin{cases} \frac{N}{2D}(1 + K_1 N/d) & \text{on the set } \{\tau \ge jN\} \\ \frac{2N}{d} & \text{on the set } \{\tau < jN\}. \end{cases} \tag{2.17}$$

PROOF: Note that, given r and the event $\{\tau = s\}$, possibly intersected with some other event in $\mathcal{G}_s$, the conditional distribution of the steps s'_i with $i > s$ is still the same as the unconditional distribution under the extended $\mathbb{P}$ as defined above. In particular the steps s'_i for $s < i < (j+1)N$ and i not divisible by N are distributed like the steps of a simple random walk $\{S_u\}$ on $\mathbb{Z}^D$ with $S_0 = \mathbf{0}$. More precisely, this holds for the projection of the s'_i on the span of the first D coordinate vectors. We shall be somewhat cavalier about this and shall not always distinguish between s'_i and this projection. For any vector w in $\mathbb{Z}^d$ or $\mathbb{Z}^D$ we use $w(i)$ to denote the ith component of w and $\overline{w} = (w(1), \dots, w(D))$ for the projection of w on $\mathbb{Z}^D$ if $w \in \mathbb{Z}^d$.

Now it is known for a simple random walk $\{S_u\}$ on $\mathbb{Z}^D$ with $S_0 = \mathbf{0}$ (cf. Kesten 1964, Sect. 3) that

$$\sup_{\overline{w}} P\{S_{2u+1} = \overline{w}\} \le \sup_{\overline{w}} P\{S_{2u} = \overline{w}\} = P\{S_{2u} = \mathbf{0}\}, \tag{2.18}$$

$$\sum_{u=1}^{\infty} P\{S_{2u} = \mathbf{0}\} \le \frac{1}{2D}(1 + K_2 D^{-1}), \quad \sum_{u=2}^{\infty} P\{S_{2u} = \mathbf{0}\} \le K_2 D^{-2}. \tag{2.19}$$

Also, by counting all possibilities (cf. (3.5) in Kesten 1964), one easily obtains

$$\sup_{\overline{w}} P\{S_1 = \overline{w}\} = \frac{1}{2D}, \sup_{\overline{w} \neq \mathbf{0}} P\{S_2 = \overline{w}\} \leq \frac{K_3}{D^2},$$
$$\sup_{\overline{w}} P\{S_3 = \overline{w}\} \leq \frac{K_3}{D^2}. \tag{2.20}$$

It follows from these observations that if $jN \leq s$ and $w = (w(1), \ldots, w(d))$ is such that

$$\sum_{q=1}^{D} \{w(q) - v'_s(q)\} \quad \text{is even,} \tag{2.21}$$

then we have on the set $\{\tau = s\}$

$$\begin{aligned} \mathbb{P}\{v'_i = w \text{ for some } s < i <(j+1)N \mid \mathcal{G}_\tau\} & \\ \leq P\{S_u \text{ visits } \overline{w} - \overline{v}'_s \text{ for some } u > 0\} & \\ = P\{S_u \text{ visits } \overline{w} - \overline{v}'_s \text{ at some even time} > 0\} & \\ \leq \frac{1}{2D}\left(1 + \frac{K_2}{D}\right) \quad \text{(see (2.19)).} & \end{aligned} \tag{2.22}$$

If the sum in (2.21) is odd instead of even then we obtain (2.22) by replacing 'even' by 'odd' in (2.22) and using (2.18), (2.19), as well as the special estimates (2.20) for the terms corresponding to $u = 1$ or 3. This proves (2.16).

Next we note that on $\{\tau = s\}$ with $jN \leq s < (j+1)N$ the first case of (2.17) is immediate from (2.16) since there are only N points in the $(j+1)$th block of r and these can be visited by r' only during the $(j+1)$th block. (2.17) is also clear on $\{\tau \geq (j+1)N\}$ for then the left hand side is zero. In order to obtain (2.17) on $\{\tau = s\}$ when $s < jN$ we observe that the sum of the last $\lfloor d/N \rfloor$ coordinates is the same for all the v_t in the $(j+1)$th block of r (compare (2.3)). The same comment applies to r'. Therefore r' can visit the $(j+1)$th block of r only if the sum of the last $\lfloor d/N \rfloor$ coordinates is the same for v'_{jN} and v_{jN}. Moreover the last $\lfloor d/N \rfloor$ coordinates of v'_{jN} are w.p.1 the same as those of $v'_{(j-1)N} + s'_{jN}$. Thus if we condition on $\mathcal{G}_{\tau \vee (jN-1)}$ then on $\{\tau < jN\}$ the conditional probability that r' visits the $(j+1)$th block of r is bounded by

$$\begin{aligned} \mathbb{P}\{s'_{jN} = \big(0, \ldots, 0, v_{jN}(D+1) - v'_{(j-1)N}(D+1), \ldots & \\ \ldots, v_{jN}(d) - v'_{(j-1)N}(d)\big) \mid \mathcal{G}_{\tau \vee (jN-1)}\} \leq \lfloor d/N \rfloor^{-1}. & \end{aligned}$$

In the last step we used that s'_{jN} takes any given value with probability at most $\lfloor d/N \rfloor^{-1}$ by the definition of $\mathbb{P}$. This implies (2.17) on $\{\tau < jN\}$ as well, since $\mathcal{G}_\tau \subset \mathcal{G}_{\tau \vee (jN-1)}$. ■

We remind the reader of our convention that $t(0) = 0$. If we declare $t(0)$ to be a special index of type 2 then Lemmas 3–6 remain valid even for $i = 1$. In other words, for $i = 1$ the estimates in these lemmas for $m = 2$ apply also to

$$\mathsf{E}\{p^{-L(t(1))}; 1 \le \rho,\ t(1) \text{ is of type } l\}.$$

We leave most of the slight modifications necessary for $i = 1$ to the reader.

LEMMA 3. *Under (2.15) the expression in (2.13) is for $l = 4$ at most*

$$K_4 p^{-N} \left[\frac{N}{2D}\right]^{N+1} \tag{2.23}$$

on the set (2.14) for any $1 \le m \le 4$.

PROOF: First observe that if

$$t(i-1) \text{ occurs in the } k\text{th block,} \tag{2.24}$$

then the next special index can be of type 4 only if it occurs in the jth block for some $j > k$ and if r has a high density point in its jth block (no matter what the type of $t(i-1)$ is). In addition v'_s must equal v_t for some s and t in the jth block for there to be any special index in the jth block. Let us assume for the rest of this proof that (2.24) occurs and let us set

$$\sigma_j = \text{smallest index } s \ge (j-1)N \text{ such that } v'_s \text{ equals some } v_t$$

($\sigma = \infty$ if no such s exists). (If $i = 1$ then we replace kN by 1 in the above definition.) Then on the event (2.24) we have

$$\begin{aligned}\mathsf{E}&\left\{p^{-L(t(i))};\ i \le \rho,\ t(i) \text{ is of type } 4 \mid \mathcal{F}_{i-1}\right\}\\ &\le \sum_{j>k} \sum_{(j-1)N \le \nu < jN} \mathsf{E}\Big\{p^{-L(jN-1)};\ \sigma_j = \nu \text{ and } r \text{ has}\\ &\qquad\qquad \text{a high density point in the } j\text{th block} \mid \mathcal{F}_{i-1}\Big\}.\end{aligned} \tag{2.25}$$

We shall estimate the summands in the right hand side of (2.25) by conditioning on r. First we show that for $(j-1)N \le \nu < jN$

$$\mathsf{E}\{p^{-L(jN-1)} \mid r, \sigma_j = \nu, \mathcal{F}_{i-1}\} \le K_5 p^{-N} \left[\frac{N}{2D}\right]^{N-1}. \tag{2.26}$$

This will be seen to follow from Lemma 2. Indeed, note that $\sigma_j = \nu \in [(j-1)N, jN)$ implies $L(jN-1) \geq 1$, since there is at least the contribution to this L of the visit of r' to r at the time σ_j. For $L(jN-1)$ to be $\geq \lambda + 1$, there must be at least λ further visits by r' to r, necessarily to the jth block of r and during the time interval (σ_j, jN). Thus by the first line of (2.17) (with j replaced by $j-1$)

$$\mathbb{P}\{r' \text{ visits the } j\text{th block of } r \text{ at least } \lambda \text{ times during } (\sigma_j, jN) \mid \mathcal{G}_{\sigma_j}\} \leq \left\{\frac{N}{2D}(1 + K_1 N/d)\right\}^{\lambda}. \quad (2.27)$$

By virtue of (2.27) the left hand side of (2.26) is at most

$$p^{-1} + \sum_{\lambda=1}^{N-1} p^{-\lambda-1}\left\{\frac{N}{2D}(1 + K_1 N/d)\right\}^{\lambda}. \quad (2.28)$$

Note that the upper bound in the sum over λ is $N-1$ because r' cannot visit the jth block of r more than N times. With the choice of p and N restricted by (2.15) the ratio of the geometric series in (2.28) is at least 2 and (2.26) follows.

Substitution of (2.26) into (2.25) now shows that on the event (2.24)

$$\begin{aligned}&\mathbb{E}\{p^{-L(t(i))};\ i \leq \rho,\ t(i) \text{ is of type } 4 \mid \mathcal{F}_{i-1}\}\\ &\quad\leq K_5 p^{-N}\left[\frac{N}{2D}\right]^{N-1} \sum_{j>k} \mathbb{P}\{r \text{ has a high density point in}\\ &\qquad\text{its } j\text{th block and } r' \text{ visits the } j\text{th block of } r \mid \mathcal{F}_{i-1}\}. \quad (2.29)\end{aligned}$$

Note that if $A \in \mathcal{F}_{i-1}$ then $A \cap \{t(i-1) \text{ occurs in the } k\text{th block}\}$ belongs to $\mathcal{H}_{kN-1}$, where

$$\mathcal{H}_t = \sigma\text{-field generated by } \{v_j, v'_j : j \leq t\}.$$

It therefore suffices to estimate the right hand side of (2.29) with $\mathcal{F}_{i-1}$ replaced by $\mathcal{H}_{kN-1}$. Now by estimates entirely analogous to those for (2.16) and (2.27) we have for $j > k$

$$\begin{aligned}&\mathbb{P}\{r \text{ has a high density point in its } j\text{th block} \mid \mathcal{H}_{kN-1}\}\\ &\quad\leq \sum_{(j-1)N \leq t < jN} \Big[\mathbb{P}\{v_s = v_t \text{ for some } t < s < jN \mid \mathcal{H}_{kN-1}\}\\ &\qquad + \sum_{w} \mathbb{P}\{v_s = w \text{ for some } t+2 \leq s < jN \mid \mathcal{H}_{kN-1}\}\Big], \quad (2.30)\end{aligned}$$

where the inner sum over w runs over the $2d$ neighbors of v_t. The first probability in the right hand side is for each fixed t at most

$$\mathbf{E}\Big\{\sup_w \mathbf{P}\{v_s = w \text{ for some } t < s < jN \mid \mathcal{H}_t\} \mid \mathcal{H}_{kN-1}\Big\},$$

which by virtue of (2.16) (with the roles of r and r' interchanged) is at most D^{-1}. As for the second probability in the right hand side of (2.30) note that $v_s = w$ means that $v_s - v_t$ has to be a unit vector, and in fact when t and s lie in the same block this can occur only when $w \in \{\pm e_i : i \le D\}$. Therefore this probability is at most

$$\begin{aligned}\mathbf{E}\Big\{\sup_{i\le D} \mathbf{P}\{v_s - v_t = \pm e_i \text{ for some } t+2 \le s < jN \mid \mathcal{H}_t\} \mid \mathcal{H}_{kN-1}\Big\}&\\ \le \sup_{\overline{w}\ne \mathbf{0}} P\{S_u = \overline{w} \text{ for some } u \ge 2\}&\\ \le K_6 D^{-2}.&\end{aligned}$$

Since t can take at most N values and w at most $2d$ values we obtain that (2.30) is at most K_7N/D. Substituting this into (2.29) we see that its right hand side, with $\mathcal{H}_{kN-1}$ instead of $\mathcal{F}_{i-1}$, is bounded by

$$K_8 p^{-N}\left[\frac{N}{2D}\right]^N \sum_{j>k} \sup_r \mathbf{P}\{r' \text{ visits the } j\text{th block of } r \mid r, \mathcal{H}_{kN-1}\}. \quad (2.31)$$

Since conditioning on r and on $\mathcal{H}_{kN-1}$ is the same as conditioning on $\mathcal{G}_{kN-1}$, the probability in the sum in (2.31) for $j = k+1$ is at most $2N/D$ (by the second line of (2.17)). The remaining sum in (2.31) is bounded by

$$\begin{aligned}\sum_{j\ge k+2} \sup_r \sum_{(j-1)N\le q<jN} \mathbf{P}\{v'_s = v_q \text{ for some}&\\ (j-1)N \le s < jN \mid r, \mathcal{H}_{kN-1}\}&\\ \le \sum_{s\ge(k+1)N} N \sup_w \mathbf{P}\{v'_s = w \mid \mathcal{H}_{kN-1}\}.&\end{aligned} \quad (2.32)$$

In turn the last sum can be estimated by the arguments used in Lemma 2. We can condition on all steps s'_i with i divisible by N in addition to $\mathcal{H}_{kN-1}$. Then $v'_s - v'_{kN-1}$ still contains $s - kN - \lfloor (s-kN+1)/N \rfloor$ simple random walk steps independent of these conditions, so that

$$\sup_w \mathbf{P}\{v'_s = w \mid \mathcal{H}_{kN-1}\} \le \sup_{\overline{w}} P\{S_{s-kN-\lfloor (s-kN+1)/N\rfloor} = \overline{w}\}$$

and (2.32) is therefore at most

$$K_8 N \sum_{u=4}^{\infty} \sup_{\overline{w}} P\{S_u = \overline{w}\} \leq K_9 \frac{N}{D^2}.$$

It follows that (2.31) is bounded by $K_{10} p^{-N}(N/(2D))^{N+1}$. The lemma follows because (2.13) is bounded by the conditional expectation (given $\mathcal{F}_{i-1}$) of (2.31). ■

LEMMA 4. *Under (2.15) the expression (2.13) is for $l = 3$ at most*

$$K_4 \frac{N^2}{pd^2} \tag{2.33}$$

on the set (2.14) for any $1 \leq m \leq 4$.

PROOF: If $t(i)$ is of type 3, then $L(t(i)) = 1$. In addition, $v'_{t(i)}$ cannot have been visited before by r', so that $v'_{t(i)} \neq v'_{t(i-1)}$ (cf. (2.11)). Finally $t(i) \geq t(i-1) + 2$. Therefore

$$\begin{aligned} &\mathsf{E}\{p^{-L(t(i))}; i \leq \rho,\ t(i) \text{ is of type } 3 \mid \mathcal{G}_{t(i-1)}\} \\ &\qquad \leq p^{-1} \mathsf{P}\{v'_s = v_t \text{ for some } s \geq t(i-1) + 2 \text{ and some } t \text{ with} \\ &\qquad\qquad v_t \neq v'_{t(i-1)} \mid \mathcal{G}_{t(i-1)}\}. \end{aligned} \tag{2.34}$$

Next we note that for given r and a time s there are at most N possible v_t which can equal v'_s, since t and s must belong to the same block for this to be possible. The right hand side of (2.34) is therefore bounded by

$$p^{-1} N \sup_{\tau} \sum_{m=2}^{\infty} \sup_{w \neq \mathbf{0}} \mathsf{P}\{v'_{\tau+m} - v'_{\tau} = w\}. \tag{2.35}$$

This sum can be estimated by almost the same method as used for (2.32). First consider the terms with $2 \leq m < 5$. If τ is such that there are no i divisible by N in $(\tau, \tau + m]$, then $v'_{\tau+m} - v'_{\tau}$ has the same distribution as S_m. In particular $v'_{\tau+m} - v'_{\tau} = w$ is possible only if the last $\lfloor d/N \rfloor$ coordinates of w are zero. Also for $\overline{w} \neq \mathbf{0}$

$$P\{S_m = \overline{w}\} \leq K_5 D^{-2}, \tag{2.36}$$

by virtue of (2.18)–(2.20). If $m < 5$ then there may also be exactly one i_0 in $(\tau, \tau+m]$ which is divisible by N. In this case $v'_{\tau+m} - v'_{\tau} = w$ forces w to be the sum of at most $m-1$ vectors from $\{\pm e_k : k \leq D\}$ plus exactly one vector from $\{e_k : D < k \leq d\}$. The step s'_{i_0} has to equal this last vector and the other $m-1$ steps s'_i with $\tau < i \le \tau + m$ have to add up to a vector determined by w. Since the probability of s'_{i_0} having a prescribed value is at most $2N/d$, we obtain that in this case

$$\sup_{w \ne \mathbf{0}} \mathbf{P}\{v'_{\tau+m} - v'_\tau = w\} \le K_5 N/(dD). \tag{2.37}$$

For $m \ge 5$ we simply observe that there are at least $(m - 1 - \lfloor m/N \rfloor) \ge 4$ values of i in $(\tau, \tau + m]$ which are not divisible by N so that

$$\sup_\tau \sup_{w \ne \mathbf{0}} \sum_{m=5}^{\infty} \mathbf{P}\{v'_{\tau+m} - v'_\tau = w\} \le K_6 \sum_{u=2}^{\infty} P\{S_{2u} = \mathbf{0}\} \le K_7 D^{-2} \tag{2.38}$$

(cf. (2.18) and (2.19)). (2.35)–(2.38) show that the right hand side of (2.34) is at most $K_8 p^{-1} N^2 d^{-2}$. Since $\mathcal{F}_{i-1} \subset \mathcal{G}_{t(i-1)}$ this same estimate holds for the expression in (2.13).

For $i = 1$, (2.34) should be replaced by

$$\begin{aligned}
&\mathbf{E}\{p^{-L(t(i))}; 1 \le \rho,\ t(1) \text{ is of type } 3\} \\
&\quad \le p^{-1} \sum_{s=2}^{N-1} \mathbf{P}\{v'_s = \text{some } v_t \text{ with } 1 \le t < N\} + (\text{expression in (2.35)}) \\
&\quad \le p^{-1} \sum_{s=2}^{N-1} \sum_{t=1}^{N-1} \mathbf{P}\{v'_s = v_t = \mathbf{0}\} + p^{-1} N \sum_{s=2}^{N-1} \sup_{w \ne \mathbf{0}} \mathbf{P}\{v'_s = w\} \\
&\qquad\qquad + (\text{expression in (2.35)}) \\
&\quad \le p^{-1} \sum_{s=2}^{N-1} \sum_{t=1}^{N-1} \mathbf{P}\{v'_s = \mathbf{0}\} \mathbf{P}\{v_t = \mathbf{0}\} + K_9 p^{-1} N^2 d^{-2} \\
&\quad \le K_{10} p^{-1} N^2 d^{-2}
\end{aligned} \tag{2.39}$$

(by (2.18)–(2.20)). ∎

For $l = 1$ or 2 our estimate for (2.13) on the set (2.14) does depend on m.

LEMMA 5. *Under (2.15), on the set (2.14) we have*

$$\mathbf{E}\{p^{-L(t(i))}; i \ge \rho,\ t(i) \text{ is of type } 1 \mid \mathcal{F}_{i-1}\} \le \begin{cases} (2pD)^{-1} & \text{if } m = 1 \\ (pD)^{-1} & \text{if } m = 2 \text{ or } 3 \\ 0 & \text{if } m = 4. \end{cases} \tag{2.40}$$

PROOF: As in the last lemma $L(t(i)) = 1$ if $t(i)$ is of type 1. First consider the case $m = 1$, i.e., let $t(i-1)$ be of type 1 as well. Let $t(i-1)$ belong to the kth block. We must then have that $t(i)$ also belongs to the kth block, and in fact $t(i-1)+1 = t(i) < kN$ (since $t(i)$ is not divisible by N). Also $v'_{t(i-1)}$ must equal some v_t with t in the kth block. Since $t(i-1)$ is also of type 1, $v'_{t(i-2)} = v'_{t(i-1)-1}$ is one of the neighbors of v_t and also equals some point of r. Moreover $t(i-1)$ is not divisible by N, so that $t(i-1)-1 = t(i-2) \geq (k-1)N$ also belongs to the kth block. Since r does not have a high density point in the kth block if $t(i-1)$ is of type 1, r does not visit any other neighbors on $\mathbb{Z}^d$ of v_t than v_{t-1} and v_{t+1} during $[(k-1)N, kN)$. One of these is $v'_{t(i-2)}$. But also $v'_{t(i)}$ must be equal to a neighbor of v_t which is visited during the kth block (recall that $v'_{t(i)}$ and $v'_{t(i-2)}$ can only visit points of the kth block of r, by (2.3)). Thus $v'_{t(i)}$ must be either v_{t-1} or v_{t+1}. However, it cannot equal $v'_{t(i-2)}$ because at time $t(i)$, r' must be at a point which it had not visited before (see (2.11)). Since all of r and $v'_{t(i-2)}$ are known when we condition on $\mathcal{F}_{i-1}$, there is only one choice for $v'_{t(i)}$, namely the one point of $v_{t\pm1}$ which is not $v'_{t(i-2)}$. The probability that r' moves to this prescribed site at the $(t(i-1)+1)$th step is $(2D)^{-1}$. This proves the case $m = 1$ of (2.40).

The case $m = 2$ or 3 is very similar, except that there now may be two choices for $v'_{t(i)}$. Again, if $t(i-1)$ belongs to the kth block, then $t(i) = t(i-1)+1$ also belongs to the kth block and $v'_{t(i-1)}$ equals some v_t of the kth block of r, $v'_{t(i)}$ must be one of the neighbors of v_t which are visited by the kth block of r. This allows at most the choices v_{t-1} or v_{t+1} for $v'_{t(i)}$. This takes care of $m = 2$ or 3 when $i \geq 2$.

For $i = 1$ we have by (2.16) (with the roles of r and r' interchanged)

$$\begin{aligned}
&\mathbb{E}\{p^{-L(t(i))}; 1 \leq \rho,\ t(1) \text{ is of type } 1\} \\
&\qquad \leq p^{-1}\mathbb{P}\{v'_1 = v_t \text{ for some } 1 \leq t < N\} \\
&\qquad \leq p^{-1}\sup_w \mathbb{P}\{v_t = w \text{ for some } 1 \leq t < N\} \\
&\qquad \leq (pD)^{-1}.
\end{aligned}$$

Finally, if $m = 4$, then $t(i-1) = kN - 1$ for some k. Then $t(i)$ cannot be of type 1, for this would require on the one hand that $t(i) = t(i-1)+1$, and on the other hand that $t(i)$ is not divisble by N. ■

LEMMA 6. *Under (2.15), on the set (2.14) we have*

$$\mathbb{E}\{p^{-L(t(i))}; i \leq \rho,\ t(i) \text{ is of type } 2 \mid \mathcal{F}_{i-1}\} \leq \begin{cases} 4N(pd)^{-1} & \textit{if } m = 1, 3, \textit{ or } 4 \\ 0 & \textit{if } m = 2. \end{cases} \tag{2.41}$$

PROOF: $t(i)$ can be of type 2 only if $t(i) = kN$ for some k and if $t(i-1) = kN-1$. This rules out that $t(i-1)$ is of type 2, so that the second case of (2.41) is trivial. For $i \geq 2$ and $m = 1, 3$, or 4, on the set $\{t(i-1) = kN-1\}$ the left hand side of (2.41) is bounded by

$$p^{-1}\mathbb{P}\{v_t = v'_{kN} \text{ for some } kN \leq t < (k+1)N \mid \mathcal{H}_{kN-1}\}.$$

This is bounded by $p^{-1}(2N/D)$ by the second case of (2.17). For $i = 1$, $t(0) = kN - 1$ is impossible. ■

We are now ready to carry out the principal estimate for

$$\mathbb{E}\{p^{-J_{nN}(r,r')}\}.$$

By (2.12) this expression is for all n at most

$$\sum_{u=0}^{\infty} \mathbb{E}\left\{p^{-\sum_{k=1}^{u} L(t(k))}; \rho = u\right\}$$

$$\leq 1 + \sum_{u=1}^{\infty} \mathbb{E}\left\{p^{-\sum_{k=1}^{u} L(t(k))}; t(u) < nN\right\}$$

$$\leq 1 + \sum_{u=1}^{\infty} \sum_{\tau} \mathbb{E}\left\{p^{-\sum_{k=1}^{u} L(t(k))}; t(u) < nN,\ t(k) \text{ has type } \tau(k),\ k \leq u\right\}. \tag{2.42}$$

The sum over τ here is over all possible sequences of types $(\tau(1), \dots, \tau(u))$ for $(t(1), \dots, t(u))$. For fixed u and τ the summand here can be written as

$$\mathbb{E}\Big\{p^{-\sum_{k=1}^{u-1} L(t(k))}\mathbb{E}\{p^{-L(t(u))}; u \leq \rho,\ t(u) \text{ is of type } \tau(u) \mid \mathcal{F}_{u-1}\};$$

$$u-1 \leq \rho,\ t(k) \text{ is of type } \tau(k),\ k \leq u-1\Big\}$$

$$\leq \mathbb{E}\Big\{p^{-\sum_{k=1}^{u-1} L(t(k))}\Gamma(\tau(u-1), \tau(u)); u-1 \leq \rho,$$

$$t(k) \text{ is of type } \tau(k),\ k \leq u-1\Big\},$$

where $\Gamma(m, l)$ is an upper bound for (2.13) on the set (2.14). From Lemmas 3–6 we see that we can take for Γ the following matrix:

$$\begin{pmatrix} \frac{1}{2pD} & \frac{4N}{pd} & K_4\frac{N^2}{pd^2} & K_4 p^{-N}\left[\frac{N}{2D}\right]^{N+1} \\ \frac{1}{pD} & 0 & K_4\frac{N^2}{pd^2} & K_4 p^{-N}\left[\frac{N}{2D}\right]^{N+1} \\ \frac{1}{pD} & \frac{4N}{pd} & K_4\frac{N^2}{pd^2} & K_4 p^{-N}\left[\frac{N}{2D}\right]^{N+1} \\ 0 & \frac{4N}{pd} & K_4\frac{N^2}{pd^2} & K_4 p^{-N}\left[\frac{N}{2D}\right]^{N+1} \end{pmatrix}.$$

By iteration of this argument we now obtain

$$\mathbb{E}\left\{p^{-\sum_{k=1}^{u} L(t(k))}; t(u) < nN,\ t(k) \text{ has type } \tau(k),\ k \le u\right\}$$
$$\le \mathbb{E}\{p^{-L(t(1))}; t(1) \text{ has type } \tau(1)\} \prod_{k=1}^{u-1} \Gamma(\tau(k), \tau(k+1)).$$

As pointed out before, the estimates in Lemmas 3–6 with $m = 2$ apply to

$$\mathbb{E}\{p^{-L(t(1))}; 1 \le \rho,\ t(1) \text{ has type } \tau(1)\},$$

so that finally

$$\mathbb{E}\left\{p^{-\sum_{k=1}^{u} L(t(k))};\ t(u) < nN,\ t(k) \text{ has type } \tau(k),\ k \le u\right\}$$
$$\le \Gamma(2, \tau(1)) \prod_{k=1}^{u-1} \Gamma(\tau(k), \tau(k+1)).$$

Substituting this into (2.42) we find

$$\mathbb{E}\{p^{-J_{nN}(\tau,\tau')}\} \le 1 + \sum_{u=1}^{\infty} \sum_{\tau} \Gamma(2, \tau(1)) \prod_{k=1}^{u-1} \Gamma(\tau(k), \tau(k+1)). \tag{2.43}$$

It will not do to take the sum here over *all* sequences $(\tau(1), \dots, \tau(u))$ with values in $\{1, 2, 3, 4\}$ because the largest eigenvalue of the matrix Γ is much bigger than 1 (in fact $\Gamma(1,2)\Gamma(2,1)$ is of order N under the restrictions (2.15) and this will grow with d; see below). However, as we saw in (2.42) we only have to sum over the sequences which are possible sequences of types for $(t(1), \dots, t(u))$. In particular, if $\tau(k) = 2$ for some k, then either all $\tau(j)$ with $k - N < j < k$ equal 1 or one of these $\tau(j)$ equals 3 or 4 and the τ's between $\tau(j)$ and $\tau(k)$ equal 1. We use this to replace Γ in (2.43) by the matrix Δ defined as

$$\begin{pmatrix} \frac{N^{3/(N-1)}}{2pD} & \frac{4}{N^2 pd} & K_4 \frac{N^5}{pd^2} & K_4 p^{-N} \left[\frac{N}{2D}\right]^{N+1} N^3 \\ \frac{N^{3/(N-1)}}{pD} & 0 & K_4 \frac{N^5}{pd^2} & K_4 p^{-N} \left[\frac{N}{2D}\right]^{N+1} N^3 \\ \frac{N^{3/(N-1)}}{pD} & \frac{4}{N^2 pd} & K_4 \frac{N^5}{pd^2} & K_4 p^{-N} \left[\frac{N}{2D}\right]^{N+1} N^3 \\ 0 & \frac{4}{N^2 pd} & K_4 \frac{N^5}{pd^2} & K_4 p^{-N} \left[\frac{N}{2D}\right]^{N+1} N^3 \end{pmatrix}.$$

Δ is obtained from Γ by multiplying the first column by $N^{3/(N-1)}$, and the third and fourth columns by N^3, while dividing the second column by N^3. (2.43) with Γ replaced by Δ is a valid estimate because for each $\tau(k+1) = 2$ for which we lose a factor N^3 in the right hand side of (2.43) we gain a factor of at least N^3 from the $\tau(j)$ which equal 1, 3, or 4 between $\tau(k)$ and the preceding τ which equals 2 (or in all the preceding τ if $\tau(k)$ is the first τ which equals 2).

After the replacement of Γ by Δ we do sum over *all* sequences $(\tau(1), \ldots, \tau(u))$ with values in $\{1,2,3,4\}$ to obtain, uniformly in n,

$$\mathbb{E}\{p^{-J_{nN}(r,r')}\} \leq 1 + \sum_{u=1}^{\infty}\sum_{i=1}^{4} \Delta^u(2,i). \tag{2.44}$$

(1.3) is contained in the following stronger lemma.

LEMMA 7. *The largest eigenvalue of* Δ *is at most*

$$\frac{N^{3/(N-1)}}{2pD} + \frac{12}{pdN^2} + 3K_4\frac{N^5}{pd^2} + 3K_4 p^{-N}\left(\frac{N}{2D}\right)^{N+1} N^3. \tag{2.45}$$

(1.3) holds. Moreover, for fixed $\gamma > 1$,

$$\liminf_{d\to\infty} \theta\left(\frac{\gamma}{2d}, \mathbb{Z}^d, \text{bond}\right) \geq \liminf_{d\to\infty} \frac{2d}{\gamma}\theta\left(\frac{\gamma}{2d}, \mathbb{Z}^d, \text{site}\right) \geq K_5[(\gamma-1)\wedge 1]. \tag{2.46}$$

PROOF: The largest eigenvalue of Δ is the same as the largest eigenvalue of $A^{-1}\Delta A$, where A is the diagonal matrix with entries 1, 3, 3, 3 along the diagonal. $A^{-1}\Delta A$ is obtained from Δ by multiplying the second, third, and fourth columns by 3 and then dividing the corresponding rows by 3. The largest row sum of the resulting matrix occurs in the first row and equals the expression in (2.45). Thus (2.45) is indeed an upper bound for the largest eigenvalue of Δ (Ostrowsky 1973, Theorem 19.1).

For $p = \gamma/(2d)$ with $1 \leq \gamma \leq 2$ and $N = \lfloor(\log d)/(2\log\log d)\rfloor$, (2.45) is bounded above by

$$\frac{d}{\gamma D} + K_6\frac{(\log\log d)^2}{\log d} \leq \frac{1}{\gamma} + K_7\frac{(\log\log d)^2}{\log d}.$$

In particular the largest eigenvalue of Δ is strictly less than one for

$$p = \frac{1}{2d}\left(1 + 2K_7\frac{(\log\log d)^2}{\log d}\right) \tag{2.47}$$

and d large. Thus for large d and p as in (2.47) the right hand side of (2.44) is finite and percolation occurs by Lemma 1. This implies (1.3). Also if we take $p = \gamma/(2d)$ for some fixed $\gamma > 1$, then for large d the right hand side of (2.44) is at most

$$1 + K_8 \sum_{u=1}^{\infty} \max_{i \leq 4} \sum_{j=1}^{4} (A^{-1}\Delta A)^u(i,j)$$

$$\leq 1 + K_9 \sum_{u=1}^{\infty} (\text{expression in (2.45)})^u$$

$$\leq 1 + K_9 \left\{1 - \frac{1}{\gamma} - K_7 \frac{(\log\log d)^2}{\log d}\right\}^{-1}$$

$$\leq K_{10}\gamma(\gamma - 1)^{-1}.$$

The second inequality in (2.46) now follows from Lemma 1. The first inequality can be found in any one of Hammersley (1961), McDiarmid (1980), and Oxley and Welsh (1979). ■

REFERENCES

Aizenman, M., Bricmont, J., and Lebowitz, J. (1987). Percolation of the minority spins in high dimensional Ising models. *Journal of Statistical Physics* 49, 859–865.

Aizenman, M., Kesten, H., and Newman, C.M. (1987). Uniqueness of the infinite cluster and continuity of connectivity functions for short and long range percolation. *Communications in Mathematical Physics* 111, 505–531.

Bricmont, J., Kesten, H., Lebowitz, J., and Schonmann, R. (1989). A note on the large dimensional Ising model. *Communications in Mathematical Physics*, to appear.

Broadbent, S.R. and Hammersley, J.M. (1957). Percolation processes. *Proceedings of the Cambridge Philosophical Society* 53, 629–641 and 642–645.

Cox, J.T. and Durrett, R. (1983). Oriented percolation in dimensions $d \geq 4$: bounds and asymptotic formulas. *Mathematical Proceedings of the Cambridge Philosophical Society* 93, 151–162.

Gandolfi, A., Grimmett, G., and Russo, L. (1988). On the uniqueness of the infinite cluster in the percolation model. *Communications in Mathematical Physics* 114, 549–552.

Gaunt, D.S. and Ruskin, H. (1978). Bond percolation processes in d dimensions. *Journal of Physics A: Mathematical and General* 11, 1369–1380.

Gaunt, D.S., Sykes, M.F., and Ruskin, H. (1976). Percolation processes in d dimensions. *Journal of Physics A: Mathematical and General* 9, 1899–1911.

Gordon, D.M. (1988). Percolation in high dimensions. Preprint.

Hammersley, J.M. (1959). Bornes supérieures de la probabilité critique dans un processus de filtration. In *Le Calcul des Probabilités et ses Applications*, 17–37, CNRS, Paris.

——— (1961). Comparison of atom and bond percolation. *Journal of Mathematical Physics* 2, 728–733.

Hara, T. and Slade, G. (1989). Mean-field critical phenomena for percolation in high dimensions. Preprint.

Kesten, H. (1964). On the number of self-avoiding walks II. *Journal of Mathematical Physics* 5, 1128–1137.

——— (1989). Asymptotics in high dimension for the Fortuin-Kasteleyn cluster model. In Festschrift for T.E. Harris, to appear.

Kesten, H. and Schonmann, R. (1989). Behavior in large dimensions of the Potts and Heisenberg models. *Reviews in Mathematical Physics*, to appear.

McDiarmid, C. (1980). Clutter percolation and random graphs. *Mathematical Programming Study* 13, 17–25.

Ostrowsky, A.M. (1973). *Solutions of Equations in Euclidean and Banach Spaces.* 3rd edition. Academic Press, New York.

Oxley, J.G. and Welsh, D.J.A. (1979). On some percolation results of J.M. Hammersley. *Journal of Applied Probability* 16, 526–540.

Department of Mathematics
Cornell University
Ithaca
New York 14853.

Some Random Collections of Finite Subsets

J.F.C. Kingman

Let $X_1, X_2, \ldots$ be independent random variables having the same continuous distribution function F. For any n, define a family $\mathcal{A}_n$ of subsets of

$$I_n = \{1, 2, \ldots, n\} \tag{1}$$

by the following recipe: a subset A of I_n belongs to $\mathcal{A}_n$ if and only if, whenever $i < j$ and $i, j \in A$,

$$X_i < X_j. \tag{2}$$

The randomness of the X_i means that $\mathcal{A}_n$ is a random family of subsets of I_n, and it is clear that the distribution of $\mathcal{A}_n$ does not depend on F.

Hammersley (1972) studied the problem, proposed by Ulam, of finding at least the asymptotic distribution of the size of the largest set in $\mathcal{A}_n$, the random variable

$$L_n = \max\{|A|; A \in \mathcal{A}_n\}. \tag{3}$$

He showed that there is a constant c such that, with probability one,

$$L_n \sim c\sqrt{n} \tag{4}$$

as $n \to \infty$.

Hammersley conjectured that $c = 2$, but he was only able to prove that

$$\tfrac{1}{2}\pi \leq c \leq e. \tag{5}$$

These bounds are improved in Kingman (1973) to

$$(8/\pi)^{1/2} \leq c \leq \epsilon, \tag{6}$$

where $\epsilon = 2.49\ldots$ can be expressed as

$$\epsilon = \xi^{-1/2}(1-\xi)^{-1/2}, \tag{7}$$

where ξ is the positive root of

$$1 - \xi = e^{-2\xi}. \tag{8}$$

Much later Hammersley's conjecture was proved by Veršik and Kerov (1977), but by complex arguments very specific to the Ulam problem. By contrast, the arguments of Hammersley (1972) and Kingman (1973) are relatively crude, and for this reason apply to other problems. For instance, the upper bound in (5) follows from the obvious fact that, if $A \subset I_n$ has $|A| = r$, then

$$P(A \in \mathcal{A}_n) = \frac{1}{r!}. \tag{9}$$

This implies that the expected number of sets of size r in $\mathcal{A}_n$ is

$$\binom{n}{r}\frac{1}{r!},$$

and this is an upper bound for the probability that there is at least one such set. Thus

$$P(L_n \geq r) \leq \binom{n}{r}\frac{1}{r!}, \tag{10}$$

and Stirling's formula easily shows that this tends to zero as $n, r \to \infty$ in such a way that

$$\liminf rn^{-1/2} > e.$$

The sharpening in (6) is only a little more difficult, and makes use (in a way which will be described below) of the fact that $\mathcal{A}_n$ is *hereditary*: if $A \in \mathcal{A}_n$ and $A' \subset A$, then $A' \in \mathcal{A}_n$.

My interest in these arguments was revived when I encountered, in the context of a genetical problem, another random family with similar properties. Let Y_{ij} $(i, j = 1, 2, \ldots)$ be random variables with a common continuous distribution function F. The Y_{ij} for $i \leq j$ are mutually independent, but the symmetry condition

$$Y_{ji} = Y_{ij} \tag{11}$$

is imposed. The family $\mathcal{A}_n$ is now defined by the requirement that $A \subset I_n$ belongs to $\mathcal{A}_n$ if and only if, for all $i, j \in A$,

$$Y_{ij} \geq \tfrac{1}{2}(Y_{ii} + Y_{jj}). \tag{12}$$

Clearly $\mathcal{A}_n$ is hereditary.

It is shown in Kingman (1988) that, if F corresponds to a uniform distribution, then

$$P(A \in \mathcal{A}_n) \leq \frac{1}{r!} \tag{13}$$

for any A of size r. This allows the arguments of the earlier papers to be carried through, to show that the size of the largest set in $\mathcal{A}_n$ is at most

$\epsilon\sqrt{n}$ for large n. Although crude, this result is of considerable significance in the genetical context. It is however specific to the particular uniform distribution, and the probability

$$P_r(F) = P(A \in \mathcal{A}_n) \tag{14}$$

depends on the choice of F (but not of course on the value of n). For some distributions the bound (13) can be improved; if F corresponds to a negative exponential distribution it is easy to compute that

$$P_r(F) = \left(\frac{2}{r+1}\right)^r \sim \frac{(2\pi r)^{1/2} e^{-1}}{r!}. \tag{15}$$

This ought to make it possible to improve on the coefficient ϵ in the upper bound. On the other hand, there are distributions for which $P_r(F)$ is of larger order than for the uniform distribution, and one may ask whether some cruder upper bound may then be established. Both of these questions are answered by the following theorem.

THEOREM 1. *For each n, let $\mathcal{A}_n$ be a random subset of $I_n = \{1, 2, \ldots, n\}$ having the hereditary property*

$$A \in \mathcal{A}_n, A' \subset A \Rightarrow A \in \mathcal{A}_n. \tag{16}$$

Suppose that, for some constant α, and for sufficiently large n, r,

$$P(A \in \mathcal{A}_n) \leq \alpha^r / r! \tag{17}$$

for every $A \subset I_n$ of size $|A| = r$. Then the size of the largest set in $\mathcal{A}_n$,

$$L_n = \max\{|A|; A \in \mathcal{A}_n\}, \tag{18}$$

satisfies

$$\limsup_{n\to\infty} L_n n^{-1/2} \leq \alpha^{1/2}\epsilon \tag{19}$$

with probability one.

This formulation assumes that the $\mathcal{A}_n$ are all defined on the same probability space. If not, the same argument shows that (19) holds in probability.

PROOF: If $s \leq r \leq n$, the inequality $L_n \geq r$ implies that there is at least one set of size r in $\mathcal{A}_n$. The hereditary property shows that all subsets of this set are in $\mathcal{A}_n$, so that $\mathcal{A}_n$ contains at least $\binom{r}{s}$ sets of size s. But, by (17), the number of sets of size s in $\mathcal{A}_n$ has expectation at most

$$\binom{n}{s}\frac{\alpha^s}{s!},$$

so that

$$\binom{r}{s} P(L_n \geq r) \leq \binom{n}{s} \frac{\alpha^s}{s!}.$$

Hence

$$\log P(L_n \geq r) \leq \log n! - \log s! - \log(n-s)! + s \log \alpha - \log r! + \log(r-s)!.$$

In this inequality let $r, s, n \to \infty$ in such a way that $r \sim \rho n^{1/2}, s \sim \sigma n^{1/2}$ for constants $0 < \sigma < \rho$. Then Stirling's formula yields, after simplification, the inequality

$$\begin{aligned}\log P(L_n \geq r) \leq -\{&\rho \log\rho + \sigma \log \sigma - (\rho - \sigma) \log(\rho - \sigma) \\ &- \sigma \log \alpha - 2\sigma + o(1)\} n^{1/2}.\end{aligned}$$

Hence, by the Borel-Cantelli lemma, $L_n \geq \rho n^{\frac{1}{2}}$ for only finitely many n, so long as

$$\rho \log \rho + \sigma \log \sigma - (\rho - \sigma) \log(\rho - \sigma) - \sigma \log \alpha - 2\sigma > 0. \tag{20}$$

It follows that, with probability one,

$$\limsup L_n n^{-1/2} \leq \rho, \tag{21}$$

so long as $\sigma < \rho$ can be chosen to satisfy (20). Putting $\sigma = \lambda\rho$ for $0 < \lambda < 1$, (20) becomes

$$\log \lambda - (\lambda^{-1} - 1) \log(1 - \lambda) > 2 + \log \alpha - 2 \log \rho. \tag{22}$$

The best choice of λ is that which maximises the left hand side; differentiation gives the equation

$$2\lambda + \log(1 - \lambda) = 0,$$

and comparison with (8) shows that $\lambda = \xi$. With this value of λ, (22) becomes

$$2 + \log \xi + \log(1 - \xi) > 2 + \log \alpha - 2 \log \rho,$$

or

$$\rho > \{\alpha / \xi(1 - \xi)\}^{1/2} = \alpha^{1/2} \epsilon.$$

Thus (21) holds for all $\rho > \alpha^{1/2}\epsilon$, and (19) is proved. ■

For example, if $\mathcal{A}_n$ is defined by (12), and if (for some constant α depending on F)

$$P_r(F) \leq \frac{\alpha^r}{r!} \tag{23}$$

for large r, then the size of the largest set in $\mathcal{A}_n$ is at most

$$(\alpha n)^{1/2}\epsilon \tag{24}$$

for large n. For the uniform distribution $\alpha = 1$, but (15) shows that, for the exponential distribution, (17) holds for any $\alpha > 2e^{-1}$, so that ϵ can be replaced by the smaller constant

$$(2e^{-1})^{1/2}\epsilon = 2.14\ldots .$$

It is an attractive conjecture that, for every continuous distribution F, there is a constant $\beta = \beta(F)$ such that

$$\lim_{r\to\infty} \{P_r(F)r!\}^{1/r} = \beta(F). \tag{25}$$

For any F for which this is true, Theorem 1 shows that, with probability one,

$$\limsup_{n\to\infty} L_n n^{-1/2} \leq \beta(F)^{1/2}\epsilon. \tag{26}$$

Some insight into the way $P_r(F)$ (and thus $\beta(F)$ if it exists) depends on F can be gained by noting that, if the Y_i have distribution function F, the random variables $\phi(Y_i)$, if ϕ is a strictly increasing function, have distribution function

$$G(y) = F\{\phi^{-1}(y)\}. \tag{27}$$

If ϕ is convex, then (12) is implied by

$$\phi(Y_{ij}) > \tfrac{1}{2}\{\phi(Y_{ii}) + \phi(Y_{jj})\},$$

so that

$$P_r(F) > P_r(G).$$

The opposite inequality obtains if ϕ is concave.

This shows in particular that, if F has decreasing density on an interval (as does the exponential distribution), then $P_r(F)$ is less than for the uniform distribution, and so (13) remains valid.

It is natural to ask whether there are non-trivial bounds for $P_r(F)$ which hold for all F. The answer is given by the following theorem.

THEOREM 2. *For any continuous distribution function F, and any $r \geq 2$,*

$$\frac{2^r r!}{(2r)!} < P_r(F) < \frac{2^r}{(r+1)!}, \tag{28}$$

and these bounds are best possible.

The right hand inequality shows that (17) holds, for any F, with $\alpha = 2$. Hence (19) holds universally, if the right hand side is set at

$$2^{1/2}\epsilon = 3.52\ldots .$$

The inequalities (28) also show that, if $\beta(F)$ exists, it satisfies

$$\tfrac{1}{2} \le \beta(F) \le 2. \tag{29}$$

PROOF: Because F is continuous and non-decreasing,

$$\begin{aligned} P_r(F) &= P\Big\{Y_{ij} > \tfrac{1}{2}(Y_{ii} + Y_{jj}) \quad (i, j = 1, 2, \ldots, r)\Big\} \\ &\le P\Big\{Y_{ij} > \min(Y_{ii}, Y_{jj}) \quad (i, j = 1, 2, \ldots, r)\Big\} \\ &= E\Big\{\prod_{i<j}[1 - F(\min(Y_{ii}, Y_{jj}))]\Big\} \\ &= E\Big\{\prod_{i<j}\max(U_i, U_j)\Big\} \end{aligned}$$

where the random variables $U_i = 1 - F(Y_{ii}) \quad (i = 1, 2, \ldots, r)$ are independent and uniformly distributed on $(0, 1)$. It is easy to check by direct integration that this last expectation is $2^r/(r+1)!$, so that

$$P_r(F) \le 2^r/(r+1)!. \tag{30}$$

There is equality in (30) only if F is such that

$$Y_{ij} > \min(Y_{ii}, Y_{jj})$$

for all $i < j \le r$ implies

$$Y_{ij} > \tfrac{1}{2}(Y_{ii} + Y_{jj})$$

a.s. for all $i < j \le r$. This can only happen if, whenever $Y_{(1)} < Y_{(2)} < Y_{(3)}$ are the order statistics of a sample of size 3 from F, then

$$P\{Y_{(2)} > \tfrac{1}{2}(Y_{(1)} + Y_{(3)})\} = 1, \tag{31}$$

and this contradicts the continuity of F.

On the other hand, that (30) is best possible may be seen by considering

$$F(y) = 1 - (1-y)^{1/m} \quad (0 \le y \le 1), \tag{32}$$

where m is a large integer. For this choice of F,

$$\begin{aligned} P_r(F) &= E\left\{\prod_{i<j}\left[1 - F\left(\tfrac{1}{2}(Y_{ii} + Y_{jj})\right)\right]\right\} \\ &= E\left\{\prod_{i<j}\left[1 - \tfrac{1}{2}(Y_{ii} + Y_{jj})\right]^{1/m}\right\} \\ &= E\left\{\prod_{i<j}\left[\tfrac{1}{2}(U_i^m + U_j^m)\right]^{1/m}\right\} \\ &\to E\left\{\prod_{i<j}\max(U_i, U_j)\right\} = \frac{2^r}{(r+1)!} \end{aligned}$$

as $m \to \infty$.

The argument for the lower bound in (28) is exactly similar, starting from

$$P_r(F) \geq P\{Y_{ij} > \max(Y_{ii}, Y_{jj}) \ \ (i, j = 1, 2, \ldots, r)\}.$$

The sharpness is established by taking

$$F(y) = y^{1/m} \quad (0 \leq y \leq 1), \tag{33}$$

and again letting $m \to \infty$. ■

References

Hammersley, J.M. (1972). A few seedlings of research. *Proceedings of the Sixth Berkeley Symposium on Mathematical Statistics and Probability* 1, 345–394.

Kingman, J.F.C. (1973). Subadditive ergodic theory. *Annals of Probability* 1, 883–909.

——— (1988). Typical polymorphisms maintained by selection at a single locus. *Journal of Applied Probability* 25A, 113–125.

Veršik, A.M. and Kerov, S.V. (1977). Asymptotics of the Plancherel measure of the symmetric group and the limiting form of Young tables. *Soviet Mathematics Doklady* 18, 527–531.

Senate House
University of Bristol
Tyndall Avenue
Bristol BS8 1TH.

Probabilistic Analysis of Tree Search

C.J.H. McDiarmid

Abstract

Consider the family tree of an age-dependent branching process, where the branches have costs corresponding to birth times. The first-birth problem of Hammersley (1974) then concerns the cost of an optimal (cheapest) node at depth n. Suppose that we must explore the tree so as to *find* an optimal or nearly optimal node at depth n. We now have a suitable model for analysing the behaviour of tree search algorithms, and we may extend the investigations of Karp and Pearl (1983).

1. Introduction

Many algorithms considered in operations research, computer science and artificial intelligence may be represented as a search or partial search through a rooted tree. Such algorithms typically involve backtracking but try to minimise the time spent doing so. This paper extends work of Karp and Pearl (1983), and gives a probabilistic analysis of backtracking and non-backtracking search algorithms in certain random trees. We thus cast some light on the question of when to backtrack: it seems that backtracking is valuable just for problems with 'dead-ends'.

Let us review briefly the model and results of Karp and Pearl. They consider an infinite rooted tree in which each node has exactly two sons. The branches have independent $0,1$–valued random costs X, with $p = P(X = 0)$. (We have swapped p and $1-p$ from the original paper.) The problem is to find an optimal (cheapest) or nearly optimal path from the root to a node at depth n.

The problem changes nature depending on whether the expected number $m_0 = 2p$ of zero-cost branches leaving a node is greater than 1, equal to 1 or less than 1 (as was suggested in Hammersley 1974, Note 8). When $m_0 > 1$ a simple 'uniform cost' breadth-first search algorithm A1 finds an optimal solution in expected time $O(n)$; and when $m_0 = 1$ this algorithm takes expected time $O(n^2)$. When $m_0 < 1$ any algorithm that is guaranteed to find a solution within a constant factor of optimal must take exponential expected time. However, in this case a 'bounded-lookahead plus partial backtrack' algorithm A2 usually finds a solution close to optimal in linear expected time. This successful performance of the backtracking algorithm A2 for the difficult case when $m_0 < 1$ was taken to suggest that similar heuristics should be of general use for attacking NP-hard problems.

We shall see that with the above search model, a simple non-backtracking bounded-lookahead algorithm A3 performs as successfully as the backtracking algorithm A2. Thus it seems hard to recommend the use of heuristics like A2 on the basis of this search model. Similar comments hold if we allow more general finite random costs on the branches.

However, there is a qualitative difference if we allow nodes to have no sons (or allow branches to have infinite costs) so that there are 'dead-ends'. We extend Karp and Pearl's work by considering search in random trees generated by an age-dependent branching process, in which the mean number of children of an individual is greater than one. The investigation is related to the first-birth (or death) problem, as introduced by J.M. Hammersley (1974) (see also Joffre et al. 1973). This model is discussed further below. Let p_0 be the probability that a node has no sons, and let m_0 be the mean number of zero-cost branches leaving a node (instantaneous births).

Our results concerning algorithms A1 and A2 are natural extensions of Karp and Pearl's results. Thus the breadth-first search algorithm A1 finds an optimal solution in linear expected time if $m_0 > 1$ and in quadratic expected time if $m_0 = 1$. If $m_0 < 1$ then any algorithm with a constant performance guarantee must take exponential expected time, but the backtracking algorithm A2 finds a nearly optimal solution in linear expected time.

However, the performance of the simple non-backtracking bounded lookahead algorithm A3 depends critically on whether $p_0 = 0$ or $p_0 > 0$. Suppose that $m_0 < 1$, so that optimal search is hard. If $p_0 = 0$, so that as in the Karp and Pearl model there are no dead-ends, then algorithm A3 usually finds a nearly optimal solution in linear expected time; that is, it performs as successfully as the backtracking algorithm A2. However, if $p_0 > 0$ then algorithm A3 usually fails to finds a solution. Thus our model suggests that backtracking becomes attractive when there is the possibility of dead-ends.

In the next section we give details concerning the search model and the algorithms A1, A2 and A3, then in Section 3 we present our results, and finally Section 4 contains proofs.

2. Model and Algorithms

We suppose that the tree to be searched is the family tree F of an age-dependent branching process of Crump-Mode type (see Crump and Mode 1968). In this model an initial ancestor is born at time $t = 0$ and then produces children at random throughout his lifetime. If $Z_1(t)$ denotes the number of children born up to time t, then $Z_1(t)$ is an arbitrary counting process, that is a non-negative integer-valued non-decreasing right-continuous random process. We do not insist that $Z_1(0) = 0$. The children of the ancestor form the first generation: from their several birth times they behave like independent copies of their parent. Their children form the second generation, and so on. We let $Z_n(t)$ be the number of individuals born in the nth generation by time t, and let $Z_n(\infty) = \sup_t Z_n(t)$.

We shall always assume that the mean number $m = E[Z_1(\infty)]$ of children of an individual satisfies $m > 1$, so that the extinction probability q satisfies $q < 1$. Let $p_k = P[Z_1(\infty) = k]$. Clearly $q > 0$ if and only if $p_0 > 0$, and these conditions correspond to the existence of 'dead-ends'.

When searching the family tree F we take the cost of a branch to be the difference between the birth times of the child and the parent, and so we take the cost of a node to be the birth time of the corresponding individual. Thus we seek a first born individual in generation n. We shall denote the corresponding optimal cost by $C_n^\star$ (rather than B_n): if $Z_n(\infty) = 0$ then we set $C_n^\star = \infty$. Thus

$$P(C_n^\star = \infty) = P(Z_n(\infty) = 0) = q_n \to q \text{ as } n \to \infty.$$

The interesting case is when the tree to be searched is infinite: we shall often condition on the event S of ultimate survival, and then $C_n^\star$ is finite for all n.

Assumptions. Recall that we assume that the mean family size m satisfies $m > 1$: this is essential. For convenience we shall also assume that m is finite and that lifetimes (branch costs) are bounded. We are thus able to show that certain events of interest fail with exponentially small probabilities. (Truncation arguments as in Kingman (1975) may then be used to obtain 'almost sure' results under weaker assumptions.) Further, a simple translation allows us to assume that small costs can occur, that is $E[Z_1(\delta)] > 0$ for $\delta > 0$.

The distinction between zero and non-zero costs turns out to be important. Let $m_0 = E[Z_1(0)]$ be the expected number of zero-cost branches from a node (instantaneous births to an individual).

We shall discuss the performance of three algorithms, A1, A2 and A3, the first two of which are taken from Karp and Pearl (1983). Each algorithm maintains a subtree T of the family tree F containing the root; and at each step explores some node of T. Here, exploring a node x mean appending to x the next (leftmost) child y of x in F but not yet in T, and observing the cost of the corresponding branch xy; or observing that node x has no more children.

Algorithm A1 is a 'uniform cost' breadth-first search algorithm and will be analysed for the cases $m_0 > 1$ and $m_0 = 1$, when there are many zero-cost branches and search is easy. Algorithm A2 is a hybrid of local and global depth-first search strategies and will be analysed for $m_0 < 1$. Algorithm A3 consists of repeated local optimal searches, and will be analysed also for $m_0 < 1$. Note that A1 is an exact algorithm, whereas A2 and A3 are approximation algorithms or heuristics.

For each algorithm Aj, we let the random cost of the solution found be $C_n^{\mathrm{A}j}$ ($= \infty$ if no solution is found), and the random time taken be $T_n^{\mathrm{A}j}$. We measure time by the number of nodes of the search tree encountered. The three algorithms are as follows.

Algorithm A1: At each step, explore the leftmost node among those active nodes of minimum cost. Here, a node is active if it is in T and may perhaps have further children. The algorithm halts when it would next explore a node at depth n. That node then corresponds to an optimal solution.

Algorithm A2: This algorithm conducts a staged search with backtracking if a local test is failed. It has three parameters: d, L, and α. By an (α, L)-*regular path* we mean a path which consists of segments each of length L and cost at most αL (except that the last segment may have length less than L). The algorithm A2 conducts a depth-first search to find an (α, L)-regular path from a depth d node to a depth n node. If it succeeds in reaching depth n, the algorithm returns the corresponding path as a solution: if it fails, the search is repeated from another depth d node. If all the nodes at depth d fail to root an (α, L)-regular path to a depth n node, the algorithm terminates with failure.

Algorithm A3: This simple bounded-lookahead or 'horizon' heuristic is a staged-search algorithm which avoids backtracking. It has one parameter L. Starting at the root it finds an optimal path to a node at depth L, makes that node the new starting point and repeats.

3. Results

We summarise our results in six theorems. Theorem 3.1 concerns the region where the mean number m_0 of zero-cost branches leaving a node satisfies $m_0 > 1$, Theorem 3.2 concerns $m_0 = 1$ and Theorems 3.3–3.6 concern $m_0 < 1$. When $m_0 \geq 1$ the main distinction is between zero and non-zero costs. Recall that S denotes the event of ultimate survival.

THEOREM 3.1. *Let* $m_0 > 1$.

(a) *The random variable* $C^\star = \lim C_n^\star$ *is finite almost surely on* S, *and indeed there exists* $\delta < 1$ *such that*

$$P(C^\star \geq k \mid S) = O(\delta^k) \quad \text{as } k \to \infty.$$

(b) *The time* T_n^{A1} *taken by algorithm A1 satisfies* $E[T_n^{\mathrm{A1}}] = O(n)$.

Thus, if the family tree is infinite, the optimal cost $C_n^\star$ remains bounded as $n \to \infty$, and algorithm A1 finds an optimal path in linear expected time.

Next we consider the critical case $m_0 = 1$. It is convenient here to restrict attention to a Bellman-Harris age-dependent branching process (see for example Harris 1963). Now children are produced according to a simple Galton-Watson branching process, and branch costs are independent and each distributed like some non-negative random variable X.

THEOREM 3.2. *Consider a Bellman-Harris process with* $m_0 = mP(X = 0) = 1$ *and* $E[Z_1(\infty)^2] < \infty$.
(a) If further $E[Z_1(\infty)^{2+\delta}] < \infty$ *for some* $\delta > 0$, $P(0 < X < 1) = 0$ *and* $P(X = 1) > 0$, *then*

$$C_n^\star / \log\log n \to 1 \quad \textit{almost surely on } S \textit{ as } n \to \infty.$$

(b) The time T_n^{A1} *taken by algorithm A1 satisfies* $E[T_n^{\mathrm{A1}}] = O(n^2)$.

Part (a) shows roughly that if the optimal cost is finite then it is usually close to $\log\log n$: it is a special case of a result of Bramson (1978). Part (b) states that the algorithm A1 finds an optimal path in quadratic expected time.

Our main interest is in the case $m_0 < 1$. The first result for this case shows that we cannot quickly find guaranteed optimal or near optimal solutions, and so it is of interest to analyse heuristic approximation methods. The next result concerns the optimal cost $C_n^\star$ and then we consider the algorithms A2 and A3.

THEOREM 3.3. *Assume that* $m_0 < 1$. *Let* T_n *be the least number of nodes explored in any proof that guarantees a certain path of length* n *to be within a constant factor* β *of optimal. Then there exists* $\eta > 1$ *and* $\delta < 1$ *such that*

$$P(T_n < \eta^n \mid S) = O(\delta^n) \quad \text{as } n \to \infty.$$

THEOREM 3.4. *There is a constant* $\gamma \geq 0$, *defined by equation (4.1) below and satisfying* $\gamma > 0$ *if and only if* $m_0 < 1$, *such that for any* $\epsilon > 0$ *there exists* $\delta < 1$ *with*

$$P\left(\left|\frac{1}{n}C_n^\star - \gamma\right| > \epsilon \middle| S\right) = O(\delta^n) \quad \text{as } n \to \infty.$$

This result shows roughly that if the optimal cost $C_n^\star$ is finite then it is usually close to γn. It is essentially due to Hammersley (1974) and Kingman (1975), see also Kesten (1973), Kingman (1976). We shall find that it follows quite easily from our analysis of the search algorithm A2; see also Biggins (1979).

THEOREM 3.5. *Let $m_0 < 1$, and consider the backtracking algorithm A2. For any $\epsilon > 0$, with appropriate parameters the algorithm runs in linear expected time, and there exists $\delta < 1$ such that*

$$P\big(C_n^{A2} \le (1+\epsilon)C_n^\star\big) = 1 - O(\delta^n) \quad \text{as } n \to \infty.$$

THEOREM 3.6. *Let $m_0 < 1$, and consider the non-backtracking algorithm A3.*

(a) If $p_0 = 0$ then for any $\epsilon > 0$, with appropriate constant lookahead the algorithm runs in linear expected time, and there exists $\delta < 1$ such that

$$P\big(C_n^{A3} \le (1+\epsilon)C_n^\star\big) = 1 - O(\delta^n) \quad \text{as } n \to \infty.$$

(b) If $p_0 > 0$, then for any constant lookahead there exists $\delta < 1$ such that

$$P(C_n^{A3} < \infty) = O(\delta^n) \quad \text{as } n \to \infty.$$

We thus see that the backtracking algorithm A2 is a good heuristic, and so is the non-backtracking algorithm A3 as long as $p_0 = 0$.

Hammersley (1974, Note 8) asked about the concentration of the random variable $C_n^\star$, in particular in the special case considered by Karp and Pearl (1983) when each individual has exactly two children, both born at time 0 or 1. For this case the bounded differences inequality of Hoeffding (1963), Azuma (1967) in the form given in McDiarmid (1989) shows immediately that for any $t \ge 0$,

$$P\left(|C_n^\star - E(C_n^\star)| \ge t\right) \le 2e^{-2t^2/n}.$$

4. Proofs

The first lemma below immediately gives part (a) of Theorem 3.1.

LEMMA 4.1. *Suppose that $m_0 > 1$. Let T be the least depth at which an infinite path of zero-cost branches is rooted, where we let $T = \infty$ if there is no such path. Then there exists $\delta < 1$ such that $P(T > n \mid S) = O(\delta^n)$ as $n \to \infty$.*

PROOF: The zero-cost branches yield a branching process $\tilde{Z}$ with mean $m_0 > 1$ and thus with extinction probability $\tilde{q} < 1$. Suppose that $p_0 + p_1 > 0$. Then $\alpha = f'(q)$ satisfies $0 < \alpha < 1$. So, by a minor extension of Theorem 8.4 in Chapter I of Harris (1963), $f_n(\tilde{q}) = q + O(\alpha^n)$. Here f_n is the generating function for $Z_n(\infty)$. But

$$P(T > n) = \sum_k P\big(Z_n(\infty) = k\big)\tilde{q}^k = f_n(\tilde{q}).$$

Hence

$$q + O(\alpha^n) = P(T > n) = (1-q)P(T > n \mid S) + q,$$

and so $P(T > n \mid S) = O(\alpha^n)$, as required. The case not considered so far is when $p_0 + p_1 = 0$, but then clearly

$$P(T > n \mid S) = P(T > n) \le \tilde{q}^{2^n}. \qquad \blacksquare$$

Now consider a Galton-Watson branching process $\tilde{Z}$. Let D_n be the number of nodes encountered in a depth-first search of the family tree which terminates on reaching a node at depth n or on searching the complete tree, and let $d_n = E[D_n]$.

LEMMA 4.2. *For each* n, $d_n \le n+1$.

PROOF: Let $q_n = P(\tilde{Z}_n = 0)$. Of course $d_0 = 1$. Suppose that d_n is finite. Then by conditioning on $\tilde{Z}_1$ we see that

$$\begin{aligned} d_{n+1} &= 1 + \sum_{k \ge 1} p_k \left(1 + q_n + \cdots + q_n^{k-1}\right) d_n \\ &= 1 + \frac{d_n}{1-q_n} \sum_{k \ge 0} p_k (1 - q_n^k) \\ &= 1 + \frac{d_n}{1-q_n} (1 - f(q_n)) \\ &\le 1 + d_n \end{aligned}$$

since $f(q_n) = q_{n+1} \ge q_n$. $\blacksquare$

We may now prove part (b) of Theorem 3.1. Consider the branching process $\tilde{Z}$ corresponding to the zero cost branches. It has mean $m_0 > 1$ and so it has extinction probability $\tilde{q} < 1$. It now follows from Lemma 4.2 (by Wald's equation) that

$$E[T_n^{\mathrm{A1}}] \le \frac{n+1}{1-\tilde{q}}.$$

This completes our proof of Theorem 3.1. $\blacksquare$

PROOF OF THEOREM 3.2: Part (a) has already been discussed, so consider part (b). Consider again the process $\tilde{Z}$ corresponding to the zero-cost branches. This has mean $E[\tilde{Z}_1] = m_0 = 1$ and variance $\tilde{\sigma}^2 = \sigma^2 p^2 + mp(1-p) < \infty$, where $p = P(X = 0)$. Hence

$$P(\tilde{Z}_n > 0) = \frac{2 + o(1)}{\tilde{\sigma}^2 n}$$

(see for example Athreya and Ney 1972, p. 19). So, arguing as before,

$$E[T_n^{\mathrm{A1}}] \leq \frac{n+1}{P(\tilde{Z}_n > 0)} = O(n^2). \qquad \blacksquare$$

To consider the case $m_0 < 1$ we must investigate the Crump-Mode model in more detail. The key to the analysis is the function $\phi(\theta)$ introduced by Kingman (1975). For $\theta \geq 0$, let

$$\phi(\theta) = E\left[\sum_r e^{-\theta B_{1r}}\right],$$

where the sum is over the birth times B_{1r} of the children r of the initial ancestor. Note that $\phi(0) = E[Z_1(\infty)] = m < \infty$, and so $\phi(\theta) < \infty$ for all $\theta \geq 0$. Next, for $a \geq 0$ let

$$\mu(a) = \inf\left\{e^{\theta a}\phi(\theta) : \theta \geq 0\right\},$$

and define the 'time constant' γ by

$$\gamma = \inf\{a \geq 0 : \mu(a) \geq 1\}. \tag{4.1}$$

The next two lemmas may be found (essentially) in Kingman (1975).

LEMMA 4.3. *The function μ on $[0, \infty)$ is continuous; $\mu(0) = m_0$; and for some $b \geq 0$, μ is strictly increasing on $[0, b]$ and $\mu(a) = m$ for each $a \geq b$.*

LEMMA 4.4. *For any $a \geq 0$,*

$$E[Z_n(an)] = \big(\mu(a) + o(1)\big)^n \quad \text{as } n \to \infty.$$

PROOF OF THEOREM 3.5: Let $0 < \epsilon < \gamma$ and let $\alpha = \gamma + \epsilon$. By Lemmas 4.3 and 4.4 we may choose a constant L such that $E[Z_L(\alpha L)] > 1$. By considering the (α, L)-sons of a depth d node and their (α, L)-sons and so on we obtain a branching process $\hat{Z}$ say. This process has mean $\hat{m} > 1$ and thus has extinction probability $\hat{q} < 1$.

We can bound the expected running time of algorithm A2 as follows. By Lemma 4.2 (and Wald's equation) for each node at depth d, the expected cost of a search to depth n from that node is at most

$$m^{L+1}\big(\lceil (n-d)/L \rceil + 1\big) \leq m^{L+1}(n+1).$$

Hence (by Wald's equation again)

$$E[T_n^{\mathrm{A2}}] \leq \frac{d + m^{L+1}(n+1)}{1-\hat{q}} = O(n).$$

Next consider costs. Let λ be a bound on lifetimes or branch costs, and set $d = d(n) = \lfloor (\epsilon/\lambda)n \rfloor$. If the algorithm A2 succeeds then

$$C_n^{\mathrm{A2}} \leq d\lambda + \lceil (n-d)/L \rceil (\alpha L) \leq (\alpha + \epsilon)n \quad \text{once } d \geq L.$$

But

$$P(C_n^{\mathrm{A2}} = \infty) \leq \sum_k P(Z_d(\infty) = k)\hat{q}^k = f_d(\hat{q}).$$

Also, by Lemmas 4.3, 4.4,

$$P\big(C_n^{\mathrm{A2}} \leq (\gamma - \epsilon)n\big) \leq E\big[Z_n((\gamma - \epsilon)n)\big] = O(\delta_1^n)$$

for some suitable $\delta_1 < 1$. Hence

$$\begin{aligned} P\Big(C_n^{\mathrm{A2}} > &\frac{\gamma + 2\epsilon}{\gamma - \epsilon} C_n^\star\Big) \\ &\leq P\big(C_n^\star \leq (\gamma - \epsilon)n\big) + P\big(\{C_n^{\mathrm{A2}} = \infty\} \setminus \{C_n^\star = \infty\}\big) \\ &\leq O(\delta_1^n) + f_d(\hat{q}) - f_n(0) \\ &= O(\delta_2^n) \quad \text{for suitable } \delta_2 < 1. \end{aligned}$$

■

We do not need to prove Theorem 3.4 here, since it is implicit in Kingman (1975), but note that we have actually done so above. We may adapt the proof idea above to yield a variant of the 'Chernoff theorem' of Biggins (1979), which we shall use to prove Theorem 3.3.

LEMMA 4.5. *If* $1 < \mu(a)$ *then for any* $1 < \eta < \mu(a)$ *there exists* $\delta < 1$ *such that*

$$P\big(Z_n(an) < \eta^n \mid S\big) = O(\delta^n) \quad \text{as } n \to \infty.$$

PROOF: By Lemma 4.3 we may choose $b < a$ such that $\mu(b) > \eta$. Then by Lemma 4.4 we may choose $\epsilon > 0$ sufficiently small and L sufficiently large that $\epsilon\lambda + b < a$ and $(\hat{m} - \epsilon)^{(1-2\epsilon)/L} > \eta$, where $\hat{m} = E[Z_L(bL)]$.

Consider the branching process $\hat{Z}$, with mean $\hat{m}$, formed by taking (b, L)-sons and their sons and so on. By a theorem of Seneta and Hyde (see for example Athreya and Ney 1972, Theorem 3, p. 30)

$$P\big(\hat{Z}_k < (\hat{m} - \epsilon)^k\big) \to \hat{q} \quad \text{as } k \to \infty.$$

By considering these processes rooted at the nodes at depth $d + i$ where $d = \lfloor \epsilon n \rfloor$ and $0 \leq i < L$, we see that

$$\begin{aligned} P\Big(Z_{d+i+kL}\big((d+L)\lambda + kbL\big) < (\hat{m} - \epsilon)^k\Big) &\leq f_{d+i}(\hat{q} + o(1)) \\ &= q + O(\delta_1^k) \end{aligned}$$

for suitable $\delta_1 < 1$. But, if $n = d + i + kL$ where $0 \leq i < L$, then $an \geq (d+L)\lambda + kbL$ and $(\hat{m} - \epsilon)^k \geq \eta^n$ for n sufficiently large. Hence

$$P(Z_n(an) < \eta^n) \leq q + O(\delta_2^n).$$

Thus

$$P\big(Z_n(an) < \eta^n \mid S\big)(1-q) + q_n \leq q + O(\delta_2^n),$$

and the result follows, since $q_n = q + O(\delta_3^n)$ for some suitable δ_3 (< 1). ■

PROOF OF THEOREM 3.3: Let $0 < \epsilon < \gamma$, let $\beta' = \beta(\gamma+\epsilon)/(\gamma-\epsilon)$ and let $k = k(n) = \lfloor n/\beta' \rfloor$. The key observation is that if $C_n^\star > (\gamma - \epsilon)n$ then

$$T_n \geq Z_k\big((\gamma-\epsilon)n/\beta\big) \geq Z_k\big((\gamma+\epsilon)k\big);$$

for each node counted by $Z_k\big((\gamma-\epsilon)n/\beta\big)$ has cost less than $C_n^\star/\beta$ and so must be explored. Now let $1 < \eta < \mu(\gamma+\epsilon)$. Then

$$\begin{aligned} P(T_n < \eta^k \mid S) &\leq P(C_n^\star \leq (\gamma-\epsilon)n \mid S) + P\left(Z_k\big((\gamma+\epsilon)k\big) < \eta^k \mid S\right) \\ &= O(\delta^k) \end{aligned}$$

for suitable $\delta < 1$, by Lemmas 4.4 and 4.5. ■

PROOF OF THEOREM 3.6:
(a) Let $p_0 = 0$. By Theorem 3.4 we may choose L so that $E[C_L^\star/L] < (1+\epsilon)\gamma$. Now C_n^{A3} is bounded above by $\lceil n/L \rceil$ independent copies of $C_L^\star/L$, and we are done.
(b) Observe that

$$P(C_n^{\mathrm{A3}} = \infty) \geq 1 - (1-p_0)^{n/L}.$$ ■

Acknowledgement

I would like to acknowledge stimulating conversations on tree search with Greg Provan.

References

Athreya, K.B. and Ney, P.E. (1972). *Branching Processes.* Springer-Verlag, Berlin.

Azuma, K. (1967). Weighted sums of certain dependent random variables. *Tôkuku Mathematical Journal* 19, 357–367.

Biggins, J.D. (1976). The first- and last-birth problems for a multitype age-dependent branching process. *Advances in Applied Probability* 8, 446–459.

——— (1977). Martingale convergence in the branching random walk. *Journal of Applied Probability* 14, 25–37.

——— (1979). Chernoff's theorem in the branching random walk. *Journal of Applied Probability* 14, 630–636.

Bramson, M.D. (1978). Minimal displacement of branching random walk. *Zeitschrift für Wahrscheinlichkeitstheorie und verwandte Gebiete* 45, 89–108.

Chernoff, H. (1952). A measure of asymptotic efficiency for tests of a hypothesis based on the sum of observations. *Annals of Mathematical Statistics* 23, 493–507.

Crump, K.S. and Mode, C.J. (1968). A general age-dependent branching process I. *Journal of Mathematical Analysis and Applications* 24, 494–508.

Grimmett, G.R. (1985). Large deviations in subadditive processes and first-passage percolation. In *Particle Systems, Random Media, and Large Deviations*, ed. R. Durrett, Contemporary Mathematics 41, 175–194.

Hammersley, J.M. (1974). Postulates for subadditive processes. *Annals of Probability* 2, 652–680.

Harris, T.E. (1963). *The Theory of Branching Processes.* Springer-Verlag, Berlin.

Hoeffding, W. (1963). Probability inequalities for sums of bounded random variables. *Journal of the American Statistical Association* 58, 13–30.

Joffre, A., LeCam, L. and Neveu, J. (1973). Sur la loi des grands nombres pour des variables aléatoires de Bernoulli attachées à un arbre dyadique. *Compte Rendus de l'Académie des Sciences, Paris A* 277, 963–964.

Karp, R.M. (1976). The probabilistic analysis of some combinatorial search algorithms. In *Algorithms and Complexity*, ed. J.F. Traub, Academic Press, New York, 1–19.

Karp, R.M. and Pearl J. (1983). Searching for an optimal path in a tree with random costs. *Artificial Intelligence* 21, 99–116.

Kesten, H. (1973). Contribution to the discussion of Kingman (1973).

Kingman, J.F.C. (1973). Subadditive ergodic theory. *Annals of Probability* 1, 883–909.

——— (1975). The first birth problem for an age-dependent branching process. *Annals of Probability* 3, 790–801.

——— (1976). Subadditive processes. In *Ecole d'Eté de Probabilités de Saint Flour V-1975*, ed. P.-L. Hennequin, Lecture Notes in Mathematics no. 539, Springer-Verlag, Berlin.

McDiarmid, C.J.H. (1989). On the method of bounded differences. Manuscript.

Department of Statistics
Oxford University
Oxford OX1 3TG.

Probability Densities for Some One-Dimensional Problems in Statistical Mechanics

J.S. Rowlinson

1. Introduction

The statistical mechanics of the real three-dimensional world is generally too difficult to allow us to obtain analytic results for the structural and other properties of matter. A few problems, most of them variants of the Ising model, can be solved for two-dimensional systems, but it is only for one-dimensional systems that analytic results can be obtained in profusion (Lieb and Mattis 1966). There is a heavy price to pay for this simplification since many features of the real world, such as transitions between the different phases of matter, are absent in one-dimensional systems with intermolecular forces of realistic range. If, however, we are interested in the behaviour of fluids near planar solid walls or in slits between parallel walls (e.g. Kjellander and Sarman 1988) then we wish to know only how the properties change with distance in the one direction perpendicular to the wall(s), since the system will generally be translationally invariant in the two directions parallel to the wall(s). The behaviour of one-dimensional systems mimics these three-dimensional systems reasonably realistically. Here we consider several such systems, in different thermodynamic ensembles, in order to obtain explicit expressions for the probability densities for the number of molecules in a system of fixed length or for the length in a system of fixed number of molecules. Most previous work on one-dimensional systems has considered only the so-called thermodynamic limit, that is the limit of infinite size but finite non-zero density. Here the emphasis is on finite systems although comparisons are made with infinite systems. The analogy of such systems with the counting rates of Geiger counters is also explored. Such devices have for a long time aroused the interest of those working on probability theory (Kosten 1943; Malmquist 1947; Feller 1948; Hammersley 1953; Albert and Nelson 1953).

2. Thermodynamic Ensembles

The most primitive thermodynamic ensemble is a totally closed system of fixed number of molecules of each species, N_α, of fixed volume, V, and of fixed energy, U. Its characteristic function is the entropy $S(\mathbf{N}, V, U)$. Here $\mathbf{N}$ denotes the set of N_α. More useful are the two canonical ensembles of fixed temperature, T, one of constant volume and the other of constant pressure, for which the characteristic functions are the Helmholtz and Gibbs free energies, $A(\mathbf{N}, V, T)$ and $G(\mathbf{N}, p, T)$. The pressure fluctuates about its mean value in the constant volume ensemble, and vice versa. In the grand canonical ensemble the activity ζ_α (or its logarithm, the ratio of chemical potential to temperature, $\mu_\alpha/kT = \ln \zeta_\alpha$) is specified for each species but the numbers of molecules fluctuate; the characteristic function is the grand potential, $\Omega(\zeta, V, T)$. In the thermodynamic limit this is equal to $-pV$ in a homogeneous system, but in a finite inhomogeneous system it can also contain terms proportional to area, length etc. (see below). The reduced variance of the number of molecules in a one-component grand ensemble, $(\overline{N^2} - \overline{N}^2)/\overline{N}^2$, is proportional to N^{-1}, but is nevertheless of physical interest. Thus if k is Boltzmann's constant we have in the thermodynamic limit

$$\frac{\overline{N^2} - \overline{N}^2}{\overline{N}^2} = -\frac{kT}{V^2}\left(\frac{\partial V}{\partial p}\right)_{T,\overline{N}}, \tag{2.1}$$

and so the variance is proportional to the compressibility.

The grand ensemble of fixed pressure is not properly specified since all its variables, ζ, p and T, are intensive and so its size is undetermined. Moreover its characteristic function would be zero by virtue of the Gibbs-Duhem equation which requires that

$$\sum_\alpha N_\alpha d(\ln \zeta_\alpha) - V dp + S dT = 0. \tag{2.2}$$

Each characteristic function is related to the energy $\mathcal{U}(\mathbf{r}^N)$ of N molecules at positions $\mathbf{r}^N$ through the appropriate partition function, Z. Thus

$$\begin{aligned} A &= -kT \ln Z(\mathbf{N}, V, T), \\ G &= -kT \ln Z(\mathbf{N}, p, T), \\ \Omega &= -kT \ln Z(\zeta, V, T); \end{aligned} \tag{2.3}$$

$$\begin{aligned} Z(\mathbf{N}, V, T) &= n_0^N \Big(\prod_\alpha N_\alpha!\Big)^{-1} \int_V d\mathbf{r}^N e^{-\mathcal{U}(\mathbf{r}^N)/kT}, \\ Z(\mathbf{N}, p, T) &= n_0 \int_0^\infty dV\ Z(\mathbf{N}, V, T) e^{-pV/kT}, \\ Z(\zeta, V, T) &= \sum_{N=0} Z(\mathbf{N}, V, T) \prod_\alpha \zeta_\alpha^{N_\alpha}. \end{aligned} \tag{2.4}$$

The density n_0 is introduced to ensure that all the partition functions are dimensionless. We shall see that its size does not affect the value of observable thermodynamic properties. The sum in $Z(\zeta, V, T)$ is over all values of N_α, N_β etc. The Laplace transform that defines $Z(\mathbf{N}, V, T)$ was used in the treatment of one-dimensional systems by Takahashi (1942) and, more recently, by Percus (1982), Bishop and Boonstra (1983) and Finn and Monson (1988).

Since $\Omega = -pV$ for a homogeneous system, it follows that $Z(\zeta, V, T)$ diverges exponentially with volume, that is, as $\exp(pV/kT)$. So we can introduce a function $Z(\zeta, P, T)$ by the equation

$$Z(\zeta, P, T) = n_0 \int_0^\infty dV \; Z(\zeta, V, T) e^{-PV/kT}, \tag{2.5}$$

which becomes infinite as P approaches p. From the nature of this approach one can deduce the behaviour of a one-dimensional system in the thermodynamic limit (Rushbrooke and Ursell 1948; Longuet-Higgins 1958). The important result, for a system of pair-wise additive intermolecular potentials is

$$\Delta(\zeta, p, T) \equiv \begin{vmatrix} \zeta_a \eta_{aa} - 1 & \zeta_a \eta_{ab} & \cdots \\ \zeta_b \eta_{ba} & \zeta_b \eta_{bb} - 1 & \cdots \\ \vdots & \vdots & \ddots \end{vmatrix} = 0, \tag{2.6}$$

where

$$\eta_{\alpha\beta} = n_0 \int_0^\infty dr \; e^{-u_{\alpha\beta}(r)/kT} e^{-pr/kT}, \tag{2.7}$$

where $u_{\alpha\beta}(r)$ is the energy of a pair of molecules on a line with α to the left of β, and separated by a distance r. If the line forms a circle then we speak of clockwise and anti-clockwise interactions. The canonical partition function for the line is a $(N+1)$-fold convolution, and for the circle a N-fold convolution of the Boltzmann factors $\exp[-u_{\alpha\beta}(r)/kT]$. The extra factor in the first case arises from the interaction of the molecules with the fixed boundaries. Hence $Z(\mathbf{N}, p, T)$ is the $(N+1)$- or N-fold product of the appropriate factors $\eta_{\alpha\beta}$, and, from the relation of Z to the Gibbs free energy, it follows that for a pure substance η^{-1} is the activity ζ. For a binary system with $\eta_{ab} = \eta_{ba}$, we have

$$\Delta = 1 - \zeta_a \eta_{aa} - \zeta_b \eta_{bb} + \zeta_a \zeta_b (\eta_{aa} \eta_{bb} - \eta_{ab}^2) = 0. \tag{2.8}$$

There is an important class of systems for which the last term vanishes. For example, if we have a mixture of hard rods of lengths ρ_a and ρ_b then

$$u_{aa}(r) = \infty \;\; \text{if } r < \rho_a, \quad u_{bb} = \infty \;\; \text{if } r < \rho_b, \tag{2.9}$$

and

$$u_{ab}(r) = \infty \quad \text{if } r < \tfrac{1}{2}(\rho_a + \rho_b),$$

and all are zero otherwise. For such a system

$$\eta_{ab} = \pi^{-1} \exp\left[-\tfrac{1}{2}\pi n_0(\rho_a + \rho_b)\right] = (\eta_{aa}\eta_{bb})^{1/2}, \quad \pi = p/n_0kT. \tag{2.10}$$

If there are c components then

$$\Delta = 1 - \pi^{-1} \sum_{\alpha=1}^{c} \zeta_\alpha e^{-\pi\rho_\alpha} = 0. \tag{2.11}$$

This equation and that for the densities, n_α,

$$n_\alpha = -(\partial\Delta/\partial \ln \zeta_\alpha)/(\partial\Delta/\partial\pi), \tag{2.12}$$

have the solutions

$$\begin{aligned} n_0\pi &= \sum_\alpha n_\alpha \bigg/ \left(1 - \sum_\alpha \rho_\alpha n_\alpha\right), \\ \zeta_\alpha &= \pi n_\alpha e^{\pi\rho_\alpha} \bigg/ \sum_\alpha n_\alpha. \end{aligned} \tag{2.13}$$

Another system of interest is the penetrable sphere model (Widom and Rowlinson 1970; Hammersley, Lewis and Rowlinson 1975; Rowlinson 1980) the primitive form of which is a binary mixture with

$$\begin{aligned} u_{aa}(r) = u_{bb}(r) = 0 \quad &\text{for all } r, \\ u_{ab}(r) = \infty \quad \text{if } r < \rho/2 \quad &\text{and zero otherwise.} \end{aligned} \tag{2.14}$$

It is now convenient to choose $n_0 = \rho^{-1}$, so that n is a dimensionless density. The equation $\Delta = 0$ takes the form

$$(\pi - \zeta_a)(\pi - \zeta_b) = \zeta_a\zeta_b e^{-\pi}, \tag{2.15}$$

and

$$\pi = \frac{n_a\zeta_a - n_b\zeta_b}{n_a - n_b}, \tag{2.16}$$

$$n_a^{-1} = \pi^{-1} + \frac{(1 + \pi - \zeta_b)}{(\pi - \zeta_b)(1 + e^{\pi}(\pi - \zeta_b)/\zeta_b)}, \tag{2.17}$$

together with a corresponding equation for n_b^{-1} in terms of ζ_a.

These results are for the primitive or two-component version of the penetrable sphere model. The name itself belongs to the transcribed or one-component version which is obtained by integrating over all positions of the molecules of one species (say b) in the grand partition function. We have then a system with an intermolecular energy that is now negative (or attractive) for all configurations. It is given by

$$\mathcal{U}(\mathbf{r}^N) = \mathcal{W}(\mathbf{r}^N) - N, \tag{2.18}$$

where $\mathcal{W}(\mathbf{r}^N)$ is the total volume covered by N freely penetrating spheres of unit volume (or rods of unit length, etc.) at positions denoted by $\mathbf{r}^N$. If the molecules are well separated then $\mathcal{U}$ is zero, but if they are close it is negative, and it has a minimum value of $(1 - N)$ when all the molecules are at the same position.

3. Finite Systems

The results above are either already known or are simple extensions of what is known. Of more current interest are finite systems for which the properties calculated from the several thermodynamic ensembles are significantly different. It is now difficult to handle a multi-component system of hard rods with any generality since the maximum number of molecules that can be accommodated on a line of finite length is a discontinuous function of composition. A case for which there is a simple explicit solution is a binary mixture of rods of unit length (species a) and points of zero length (species b), the centres of the former of which are confined to a line of length L_a, and the latter to the same line, but to a length L_b of it, symmetrically chosen with respect to L_a. If the boundaries are formed from fixed molecules, of either species, then $L_b = L_a + 1$, but we can equally well consider the slightly more general case $L_b \geq L_a + 1$. We have now

$$Z(N_a, N_b, L_a, L_b) = (L_a - N_a + 1)^{N_a}(L_b - N_a)^{N_b}/N_a!\,N_b!, \qquad L_a \geq N_a - 1. \tag{3.1}$$

The partition function is not a function of temperature, because of the nature of the potential energy of a system of hard rods, and so this variable is omitted. The grand partition function is

$$Z(\zeta_a, \zeta_b, L_a, L_b) = \sum_{N_a=0}^{[L_a+1]} \sum_{N_b=0}^{\infty} \frac{[\zeta_a(L_a - N_a + 1)]^{N_a}[\zeta_b(L_b - N_a)]^{N_b}}{N_a!\,N_b!} \tag{3.2}$$

where $[L_a + 1]$ is the integral part of $L_a + 1$. The summation over N_b can be made at once to give

$$Z(\zeta_a, \zeta_b, L_a, L_b) = e^{\zeta_b L_b} \sum_{N_a=0}^{[L_a+1]} \frac{[\zeta_a e^{-\zeta_b}(L_a - N_a + 1)]^{N_a}}{N_a!}. \tag{3.3}$$

The sum is now the partition function for a one-component system of rods of unit length, on a line of length L_a, at an activity of $\zeta_a e^{-\zeta_b}$. This we know already (Robledo and Rowlinson 1986).

$$Z(\zeta_a, \zeta_b, L_a, L_b) = e^{\zeta_b L_b} Z(\zeta_a e^{-\zeta_b}, L_a)$$
$$= n_a^* \left(\frac{1-n_a}{n_a}\right) \exp\left[\zeta_b L_b + \frac{n_a(L_a+1)}{1-n_a}\right], \qquad (3.4)$$

where n_a is the density of this pure substance in an infinite system at an activity $\zeta_a e^{-\zeta_b}$, and n_a^* is the density of the same substance at the same activity in a semi-infinite system at a distance L_a+1 from the one boundary. If L_a is greater than about 4 then n_a^* is almost indistinguishable from n_a, but it is very different in smaller systems.

It follows from equation (3.4) that in the thermodynamic limit we have

$$\zeta_a e^{-\zeta_b} = \left(\frac{n_a}{1-n_a}\right) \exp\left(\frac{n_a}{1-n_a}\right), \quad \zeta_b = \frac{n_b}{1-n_a}, \qquad (3.5)$$

or

$$\zeta_a = \left(\frac{n_a}{1-n_a}\right) \exp\left(\frac{n_a+n_b}{1-n_a}\right). \qquad (3.6)$$

It follows from equation (2.13), that ζ_a is therefore also the activity of component a in the original mixture (in the thermodynamic limit) at densities n_a and n_b. Thus n_a is both the density of a in the original system at an activity ζ_a, and that of substance a in the hypothetical one-component system at an activity $\zeta_a e^{-\zeta_b}$.

In the finite system

$$Z(\zeta_a, \zeta_b, L_a, L_b) = \frac{n_a^*(1-n_a)}{n_a} \exp\left[\frac{n_a(L_a+1)+n_b L_b}{1-n_a}\right], \qquad (3.7)$$

and

$$\pi = \frac{n_a+n_b}{1-n_a} = \zeta_a e^{-\pi} + \zeta_b, \qquad (3.8)$$

where π, n_a, and n_b are the pressure-to-temperature ratio and the densities of the infinite system at the same activities, ζ_a and ζ_b, as the finite system. The negative of the grand potential, $-\Omega$, is equal to pL, where $L \sim L_a, L_b$, together with two 'end-effects' that are the one-dimensional analogues of the surface tension, γ, of a fluid against a wall. We can re-write equation (3.7) as

$$-\Omega(\zeta_a, \zeta_b, L_a, L_b)/kT = \pi L - 2\gamma/kT + O(L)^{-1}, \qquad (3.9)$$

where π is given by equation (3.8) and

$$2\gamma/kT = \frac{n_a(L_a - L + 1) + n_b(L_b - L)}{(1-n_a)} + \ln(1-n_a), \qquad (3.10)$$

and where the difference between n_a^* and n_a has been lost in the term of $O(L)^{-1}$. Clearly if γ is defined as an 'excess' or surface contribution to the grand potential then it is not determined precisely, since L is arbitrary, a conclusion that has been drawn also for more general cases (McQuarrie and Rowlinson 1987). If $\zeta_b = 0$ then the natural choice is $L = L_a$, when

$$2\gamma/kT = \frac{n_a}{1-n_a} + \ln(1-n_a) < 0, \tag{3.11}$$

a result that has been obtained also by Henderson (1983) by a different route. If $\zeta_a = 0$ then we choose $L = L_b$ to give $\gamma = 0$. There is no obviously natural choice for a mixture.

4. Probability Densities

We now obtain the probability densities for grand and constant-pressure ensembles, which may be written $\mathcal{P}_V(N)$ and $\mathcal{P}_N(V)$ respectively since the first is the probability of finding N molecules in a system of fixed V (at specified ζ and T), and the second is the probability density of finding V to lie between V and $V+dV$ in a system of fixed N (at specified π and T). In the thermodynamic limit we can show that these functions have the same algebraic form. (We cannot say that they are the same function, since the first is a discrete function of N and the second a continuous function of V.)

The general expressions are, for a system of one component,

$$\mathcal{P}_{V,T,\zeta}(N) = \frac{Z(N,V,T)\zeta^N}{\sum_{N=0}^{\infty} Z(N,V,T)\zeta^N} = \frac{Z(N,V,T)\zeta^N}{Z(\zeta,V,T)}, \tag{4.1}$$

$$\mathcal{P}_{N,T,\pi}(V) = \frac{Z(N,V,T)e^{-pV/kT}}{\int_0^{\infty} dV\, Z(n,V,T)e^{-pV/kT}} = \frac{n_0 Z(N,V,T)e^{-pV/kT}}{Z(N,\pi,T)}, \tag{4.2}$$

where n_0 is the density introduced in the definition of the partition functions in equation (2.4). In the thermodynamic limit the partition functions are related to the thermodynamic functions by equation (2.3), so, with an abbreviation of the subscripts,

$$\frac{\mathcal{P}_N(V)}{\mathcal{P}_V(N)} = n_0 \exp[-(pV + N\mu)/kT] \exp[(G - \Omega)/kT] = n_0. \tag{4.3}$$

Here $\zeta = \exp(\mu/kT)$ has been written out in full to show its relation to the function $G = N\mu$ in the thermodynamic limit. Thus $\mathcal{P}_N(V)$ and $\mathcal{P}_V(N)$ are the same thing, their ratio being a fixed density (see below). We shall see that this simple relation between the two probability densities is only slightly modified in the finite one-dimensional system.

The probability, in the binary system discussed above, of there being N_a molecules (rods) and N_b molecules (points) on a line of lengths L_a, L_b is proportional to the appropriate term in the grand partition function, as in equation (4.1).

$$\mathcal{P}_{\zeta_a,\zeta_b,L_a,L_b}(N_a, N_b) = \frac{[\zeta_a(L_a - N_a + 1)]^{N_a}[\zeta_b(L_b - N_a)]^{N_b}}{N_a!\, N_b!\, Z(\zeta_a, \zeta_b, L_a, L_b)}, \qquad N_a \leq [L_a + 1]. \tag{4.4}$$

It follows that the conditional probability of there being N_b molecules, if the number of species a is fixed at N_a, is a Poisson distribution,

$$\mathcal{P}(N_b \mid N_a) = \frac{[\zeta_b(L_b - N_a)]^{N_b}}{N_b!} e^{-\zeta_b(L_b - N_a)}. \tag{4.5}$$

Since

$$\mathcal{P}(N_a, N_b) = \mathcal{P}(N_b \mid N_a) \cdot \mathcal{P}(N_a), \tag{4.6}$$

it follows that the absolute probability of there being N_a molecules in the system is

$$\begin{aligned}\mathcal{P}(N_a) &= \frac{e^{\zeta_b L_b}[\zeta_a e^{-\zeta_b}(L_a - N_a + 1)]^{N_a}}{N_a!\, Z(\zeta_a, \zeta_b, L_a, L_b)} \\ &= \frac{(\pi - \zeta_b)^{N_a+1}(L_a - N_a + 1)^{N_a}}{n_a^* N_a!} \exp[-(\pi - \zeta_b)(L_a - N_a + 1)], \\ &\qquad N_a \leq [L_a + 1],\end{aligned} \tag{4.7}$$

where π is given by equation (3.8). This probability is independent of L_b since it has been assumed that $L_b \geq L_a + 1$.

If ζ_b is zero this reduces to the probability of finding N rods of unit length on a line of length L when the activity is ζ and where the pressure in an infinite system of the same activity is given by π;

$$\mathcal{P}_L(N) = \frac{\pi^{N+1}(L - N + 1)^N}{n^* N!} \exp[-\pi(L - N + 1)], \quad N \leq [L + 1], \tag{4.8}$$

where

$$\pi = \frac{n}{(1 - n)} \quad \text{and} \quad \zeta = \pi e^{\pi}. \tag{4.9}$$

I cannot obtain the absolute probability $\mathcal{P}(N_b)$ in closed form but its lower cumulants can be found. We have, for fixed ζ_a, ζ_b, L_a, L_b,

$$\overline{N_b} = \zeta_b(L_b - \overline{N_a}) \tag{4.10}$$

$$\overline{N_b^2} - \overline{N_b}^2 = \zeta_b(L_b - \overline{N_a}) + \zeta_b^2(\overline{N_a^2} - \overline{N_a}^2). \tag{4.11}$$

Thus the variance of N_b comprises two terms, the first of which is the variance of a Poisson distribution of N_b for a fixed value of N_a, and the second of which is the additional fluctuation in N_b caused by the fluctuations in N_a. The higher cumulants separate similarly into an intrinsic term and a contribution from the fluctuations in N_a. The covariance is negative;

$$\overline{N_a N_b} - \overline{N_a}\,\overline{N_b} = -\zeta_b(\overline{N_a^2} - \overline{N_a}^2). \tag{4.12}$$

The more rods there are in the system, the fewer points, and vice versa.

To obtain the distribution of the length of the system in the constant pressure ensemble we choose $L_b = L_a + 1$ so that $L_a = -1$ corresponds formally to $N_a = N_b = 0$. We have then

$$\begin{aligned} Z(N_a, N_b, \pi) &= \int_{N_a-1}^{\infty} dL_a \, \frac{(L_a - N_a + 1)^{N_a+N_b}}{N_a!\, N_b!} e^{-\pi L_a}, \\ &= \frac{(N_a + N_b)!}{N_a!\, N_b!} \pi^{-(N_a+N_b+1)} \exp[-\pi(N_a - 1)]. \end{aligned} \tag{4.13}$$

(Since our rods have 'unit' length we take $n_0 = 1$.) Hence

$$\mathcal{P}_{N_a,N_b}(L_a) = \frac{\pi^{N_a+N_b+1}}{(N_a + N_b)!}(L_a - N_a + 1)^{N_a+N_b} \exp[-\pi(L_a - N_a + 1)], \quad L_a \geq N_a - 1, \tag{4.14}$$

which is a gamma distribution. If $N_a = 0$ then this is the probability distribution for a line of length $L_b = L_a + 1$ with N_b molecules of a perfect gas;

$$\mathcal{P}_{N_b}(L_b) = \frac{\pi^{N_b+1}}{N_b!} L_b^{N_b} e^{\pi L_b}. \tag{4.15}$$

If $N_b = 0$, and if we abbreviate N_a and L_a to N and L, then

$$\mathcal{P}_N(L) = \frac{\pi^{N+1}}{N!}(L - N + 1)^N e^{-\pi(L-N+1)}, \quad L \geq N + 1. \tag{4.16}$$

So we have a simple relation between $\mathcal{P}_L(N)$, the discrete distribution in the grand ensemble, and $\mathcal{P}_N(L)$, the continuous distribution function in the constant pressure ensemble, namely

$$n^* \mathcal{P}_L(N) = \mathcal{P}_N(L). \tag{4.17}$$

We recall, however, that the density n^*, which first appeared in equation (3.4), is the local density at a distance $L + 1$ from the wall in a semi-infinite grand ensemble of the same activity or pressure, $\zeta = \pi e^{\pi}$. This density is, therefore, a function of L and of ζ (or π), and equation (4.17) does not have the simplicity of equation (4.3), where n_0 is a constant. In the thermodynamic limit n^* in equation (4.17) does become constant and is equal to the actual density at activity ζ. The normalisations of $\mathcal{P}_N(L)$ and $\mathcal{P}_L(N)$ are

$$\int_{N-1}^{\infty} dL\, \mathcal{P}_N(L) = \sum_{N=0}^{[L+1]} \mathcal{P}_L(N) = 1. \qquad (4.18)$$

The local densities, $n(l)$, at distance l from a wall are known for the grand ensemble (Robledo and Rowlinson 1986) and for the constant pressure ensemble (Finn and Monson 1988), in the latter case both analytically and by Monte Carlo computer simulation. Figure 1 shows the Monte Carlo results, the exact curve, the mean $\overline{L}$, and the square root of the variance $(\overline{L^2} - \overline{L}^2)^{1/2}$ calculated from $\mathcal{P}_N(L)$. It is seen that the local density is highest in contact with the fixed wall, $n(0)$, and then oscillates about $n = \pi/(1+\pi)$, its thermodynamic limit. The density falls to zero for lengths much beyond $\overline{L}$; that is, the 'floating' piston or wall that exerts a constant force on the finite system rarely makes excursions much beyond $L = \overline{L}$. For each instantaneous position of the piston we should expect to see a symmetric distribution of the local density, $n(l)$, about $l = L/2$, and so sharp peaks against both walls. The right-hand peak is, however, smoothed out by the fluctuations in the position of the piston, but the underlying symmetry is revealed by a calculation of the mean position of the centre of mass of the fluid. We have

$$n_N(l) = \sum_{q=0} \frac{(l-q)^2}{q!} \pi^{q+1} e^{-\pi(l-q)}, \qquad (4.19)$$

where $n_N(l)$ is the local density at distance l from the left-hand (fixed) wall, and where the upper limit of the sum is $\min(N-1, [l])$. As π becomes infinite the distribution becomes a sum of delta-functions

$$n_N(l) = \sum_{q=0} \delta(l-q). \qquad (4.20)$$

In general,

$$\bar{l} = N^{-1} \int_0^{\infty} dl\, l\, n_N(l) = \frac{1}{2}\left(N - 1 + \frac{N+1}{\pi}\right) = \frac{1}{2}\overline{L}; \qquad (4.21)$$

that is, the mean of l is half of the mean of the total length.

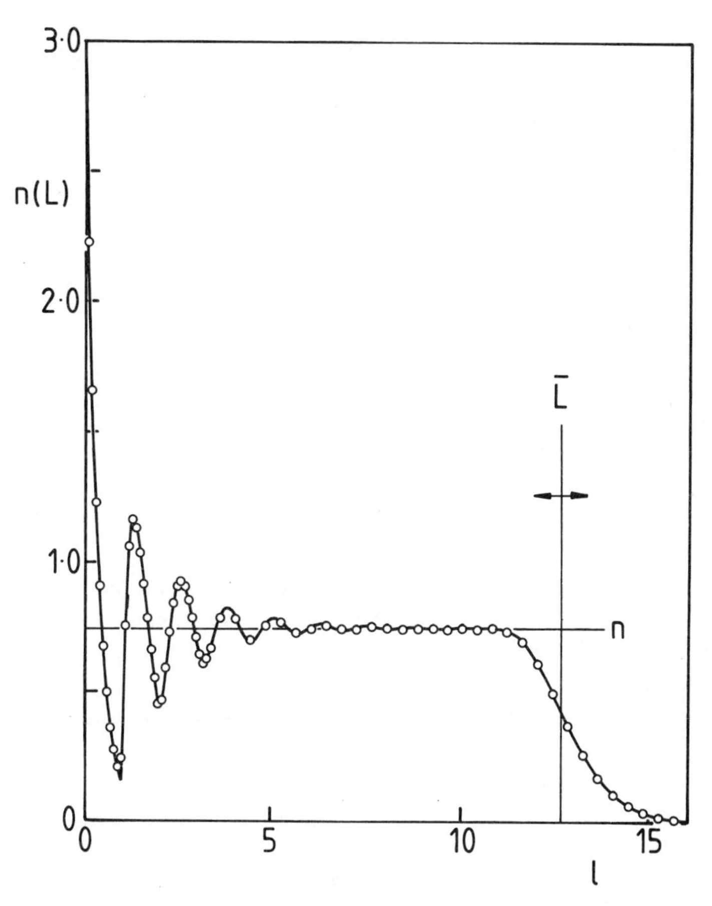

FIG. 1. The density in a constant pressure ensemble with $N = 10$ and $\pi = 3$. The points are the results of a Monte Carlo simulation by Finn and Monson (1988) and the line through them is the calculated density. The horizontal line, n, is the density in the thermodynamic limit, and the vertical line, $\overline{L}$, the mean length of the assembly. The square root of the variance is shown by the short line between the arrowheads.

The difference between $\overline{N}$ at fixed L in the grand ensemble and $\overline{L}$ at fixed N in the constant-pressure ensemble is seen by writing the equations for these quantities in the form

$$\overline{N} = L\left(\frac{\pi}{1+\pi}\right) + \left(\frac{\pi}{1+\pi}\right)^2, \quad N = \overline{L}\left(\frac{\pi}{1+\pi}\right) - \left(\frac{1-\pi}{1+\pi}\right). \qquad (4.22)$$

The difference is of the order unity and so negligible in the thermodynamic limit.

Since L has a gamma distribution it exhibits the property of convolution, which can be given a physical interpretation. We have, from equation (4.16),

$$\mathcal{P}_{M+N+1}(L) = \int_M^{L-N} dl\, \mathcal{P}_M(l-1)\mathcal{P}_N(L-l-1). \tag{4.23}$$

Let us number the molecules from the left in a system of $(M + N + 1)$ molecules, and single out the $(M + 1)$th molecule. We can regard this as an internal piston that divides the system into two parts, with M molecules to the left of it and N to the right. The equation above then tells us that, at a fixed external pressure, the probability that the total length is L the convolution of the probability that the length of the left part of the system is $(l - 1)$ and the right part $(L - l - 1)$. The internal piston can occupy any position from $l = M$ to $l = L - N$, and it excludes other molecules from a length of two units, thus giving a total length to the system of $(l - 1) + (L - l - 1) + 2 = L$.

The one-dimensional system of hard rods has an obvious analogy with a Type I Geiger counter. In this instrument the emissions of a radioactive source are counted, but after each successful count the instrument is dead for a fixed time, and any emissions in that interval go unrecorded. If the emission is a Poisson process of rate π (in units of reciprocal dead time) then the rate at which counts are recorded is $n = \pi/(1 + \pi)$. Thus in the language of statistical physics, the rate of emission is the pressure π and the rate of counting is the density in the thermodynamic limit, n. If the dead-time is zero then these rates are the same — the counter 'obeys Boyle's Law'.

The recorded counts do not have a Poisson distribution, as is shown above. From these results we have for the variance of a finite system

$$\overline{N^2} - \overline{N}^2 = n(1-n)^2(L+2n) = \pi(1+\pi)^{-3}\left(L + \frac{2\pi}{1+\pi}\right). \tag{4.24}$$

In the limit of L becoming infinite this is the expression for the variance of a Type I counter given by Feller (1971).

5. Penetrable Sphere Model

There are no obviously simple boundary conditions for the transcribed version of this model, so the finite system is best studied by placing the molecules on the circumference of a ring. The reciprocal of the activity η

is given by

$$\eta = \int_0^\infty dr\, e^{-\pi r} e^{\theta[1-\min(1,r)]} \tag{5.1}$$

$$= \frac{\pi e^\theta + \theta e^{-\pi}}{\pi(\pi+\theta)} = \zeta^{-1}, \tag{5.2}$$

where θ is a dimensionless measure of the reciprocal temperature, and π is, as before, the dimensionless ratio of pressure to temperature. The constant-pressure partition function is

$$Z(N,\pi,\theta) = \eta^N. \tag{5.3}$$

In principle we could obtain the constant-volume partition function by inverting the Laplace transform $Z(N,\pi,\theta)$. Since η has a pole only at $\pi = 0$, and since the contribution to $Z(N,L,\theta)$ of the first molecule is L, we have

$$Z(N,L,\theta) = \frac{L}{N!}\left(\frac{\partial^{N-1}}{\partial \pi^{N-1}}\left[\pi^N \eta^N e^{\pi L}\right]\right)_{\pi=0}. \tag{5.4}$$

This route is practicable for a system of hard rods for which it gives

$$Z(N,L) = \frac{L(L-N)^N}{N!}, \tag{5.5}$$

but not for the penetrable sphere model. We can, however, use the constant pressure ensemble to obtain the moments of the length of the system. If the length per molecule is denoted $\lambda \equiv L/N$, then

$$\begin{aligned}\eta\overline{\lambda^n} &= \int_0^\infty d\lambda\, \lambda^n \exp[-\pi\lambda + \theta(1-\min(1,\lambda))] \\ &= \frac{n!\, e^\theta}{(\pi+\theta)^{n+1}} + \frac{n!}{e^\pi}\sum_{i=0}^{n}\frac{1}{i!}\left[\pi^{i-n+1} - (\pi+\theta)^{i-n+1}\right].\end{aligned} \tag{5.6}$$

The cumulants of the length are obtained by differentiating the Gibbs free energy,

$$C_n(\lambda) = (-)^n \left(\frac{\partial^n \ln \eta}{\partial \pi^n}\right)_\theta. \tag{5.7}$$

We have

$$\overline{\lambda} = \frac{\overline{L}}{N} = \frac{1}{\pi+\theta} + \frac{1+\pi}{\pi(1+\pi e^{\pi+\theta}/\theta)}, \tag{5.8}$$

$$\overline{\lambda^2} - \overline{\lambda}^2 = \frac{\overline{L^2} - \overline{L}^2}{N} = \frac{1}{(\pi+\theta)^2} + \frac{\theta + 2\pi e^{\theta+2\pi}(1+\pi+\pi^2/2)}{\pi^2(1+\pi e^{\pi+\theta}/\theta)^2}. \tag{5.9}$$

The limits of these expressions as $\theta \to 0$ give the mean-field results,

$$\overline{\lambda} = \frac{1}{\pi}\left[1 + \frac{\theta}{\pi}\left(e^{-\pi}(1+\pi) - 1\right)\right] + O(\theta^2), \tag{5.10}$$

$$\overline{\lambda^2} - \overline{\lambda}^2 = \frac{1}{\pi^2}\left[1 + \frac{2\theta}{\pi}\left(e^{-\pi}(1+\pi+\pi^2/2) - 1\right)\right] + O(\theta^2). \tag{5.11}$$

The complexity of these results does not make it seem likely that there is any simple expression for $\mathcal{P}_N(\lambda)$ for this system.

A Type II Geiger counter has a dead time that runs for a fixed interval after each emission, not after each count. There is, therefore, an analogy with the one-dimensional penetrable sphere model in its mean-field or random distribution limit. Each emission is a molecule and now the analogue of the total dead time is the covered length $\mathcal{W}$ of equation (2.18). The count rate of the instrument is the analogue of the number of blocks into which $\mathcal{W}$ is divided, since each block, followed by a covered interval (dead time) of variable length is one recorded count. The ratio of the number of ends of blocks (and so twice the number of blocks) to the sum of total lengths of the blocks is an analogue of the ratio of the covered area to the covered volume in a three-dimensional system. So we can use Widom's (1971) result for the mean areas of systems of arbitrary dimensionality to show that an emission rate of n gives a count rate of ne^{-n}. This result was obtained by other means by Levert and Scheen (1943) and by Kosten (1943). More generally, from Widom's result and the equations above, the mean density of blocks, $\overline{\nu}$, in a one-dimensional system is

$$\overline{\nu} = \frac{\zeta\pi(\pi+\theta)}{\pi^2 e^{\pi} + \zeta\theta(1+\pi)}. \tag{5.12}$$

The random or mean-field limit of $\overline{\nu} = \pi e^{-\pi} = ne^{-n}$, is attained at $\theta = 0$. However high the density or pressure, $\overline{\nu}$ remains non-zero; that is, the blocks never coalesce, or there is no percolation limit. A Type II counter never records once and then stays dead for all time, however high the rate of emission. Even when the penetrable sphere model has the additional incentive of an attractive intermolecular potential ($\theta > 0$) an infinite number of molecules never forms a single cluster in a one-dimensional system.

Thus the analogy of the hard-rod problem with a Type I counter requires that the count rate is the number of molecules and the emission rate is the pressure, while that between the penetrable sphere model and a Type II counter requires that the emission rate is the number of molecules and the count rate is the number of blocks into which the penetrating molecules are grouped. If, to choose a neutral symbol, the emission rate is r, in units of the reciprocal of the dead time caused by an isolated emission, then a

Type I instrument has a count rate of $r(1+r)^{-1}$ and a Type II instrument of re^{-r}. The Type I instrument is the more efficient.

I thank Dr. P.A. Monson for sending me his results before publication.

REFERENCES

Albert, G.E. and Nelson, L. (1953). Contributions to the statistical theory of counter data. *Annals of Mathematical Statistics* 24, 9–22.

Bishop, M. and Boonstra, M.A. (1983). Exact partition functions for some one-dimensional models via the isobaric ensemble. *American Journal of Physics* 51, 564–566.

Feller, W. (1948). On probability problems in the theory of counters, in *Studies and Essays presented to R. Courant on his 60th birthday, January 8 1948*, 105–115, ed. K.O. Friedrichs, O.E. Neugebauer and J.J. Stoker, Interscience, New York.

——— (1971). *An Introduction to Probability Theory and Its Applications*, Volume 2. 2nd edition, 372–373. John Wiley and Sons, New York.

Finn, J.E. and Monson, P.A. (1988). Adsorption equilibria in an isobaric ensemble. *Molecular Physics* 65, 1345–1361.

Hammersley, J.M. (1953). On counters with random dead time. *Proceedings of the Cambridge Philosophical Society* 49, 623–637.

Hammersley, J.M., Lewis, J.W.E. and Rowlinson, J.S. (1975). Relationships between the multinomial and Poisson models of stochastic processes, and between the canonical and grand canonical ensembles in statistical mechanics, with illustrations and Monte Carlo methods for the penetrable sphere model of liquid-vapour equilibrium. *Sankhyā A* 37, 457–491.

Henderson, J.R. (1983). Statistical mechanics of fluids at spherical structureless walls. *Molecular Physics* 50, 741–762.

Kjellander, R. and Sarman, S. (1988). On the statistical mechanics of inhomogeneous fluids in narrow slits. An application to a hard-sphere fluid between hard walls. *Chemical Physics Letters* 149, 102–108.

Kosten, L. (1943). On the frequency distribution of the number of discharges counted by a Geiger-Müller counter in a constant interval. *Physica* 10, 749–756.

Levert, C. and Scheen, W.L. (1943). Probability fluctuations of discharges in a Geiger-Müller counter produced by cosmic radiation. *Physica* 10, 225–238.

Lieb, E.H. and Mattis, C.C. (1966). *Mathematical Physics in One Dimension. Exactly Soluble Models of Interacting Particles.* Academic Press, New York.

Longuet-Higgins, H.C. (1958). One-dimensional multicomponent mixtures. *Molecular Physics* 1, 83–92.

McQuarrie, D.A. and Rowlinson, J.S. (1987). The virial expansion of the grand potential at spherical and planar walls. *Molecular Physics* 60, 977–989.

Malmquist, S. (1947). A statistical problem connected with the counting of radioactive particles. *Annals of Mathematical Statistics* 18, 255–264.

Percus, J.K. (1982). Non-uniform fluids. In *The Liquid State of Matter*, 31–140, ed. E.W. Montroll and J.L. Lebowitz, North-Holland, Amsterdam.

Robledo, A. and Rowlinson, J.S. (1986). The distribution of hard rods on a line of finite length. *Molecular Physics* 58, 711–721.

Rowlinson, J.S. (1980). Penetrable sphere models of liquid-vapor equilibrium. *Advances in Chemical Physics* 41, 1–57.

Rushbrooke, G.S. and Ursell, H.D. (1948). One-dimensional regular assemblies. *Proceedings of the Cambridge Philosophical Society* 44, 263–271.

Takahashi, H. (1942). Eine einfache Methode zur Behandlung der statistischen Mechanik eindimensionaler Substanzen. *Proceedings of the Physical-Mathematical Society of Japan* 24, 60–62. Reprinted in English in Lieb and Mattis (1966, pp. 25–27).

Widom, B. (1971). Geometrical aspects of the penetrable-sphere model. *Journal of Chemical Physics* 54, 3950–3957.

Widom, B. and Rowlinson, J.S. (1970). New model for the study of liquid-vapor phase transitions. *Journal of Chemical Physics* 52, 1670–1684.

Physical Chemistry Laboratory
University of Oxford
South Parks Road
Oxford OX1 3QZ.

Seedlings in the Theory of Shortest Paths

J. Michael Steele

Abstract

This article explores three developments that arise from the fundamental theorem of Beardwood, Halton, and Hammersley on the asymptotic behavior of the shortest path through n random points. The first development concerns the role of martingales in the theory of shortest paths, especially their role in large deviation inequalities. The second development concerns the use of Lipschitz spacefilling curves to obtain analytical bounds in the theory of the TSP, and it provides some bounds that refine those of Bartholdi and Platzman on the worst case performance of the spacefilling heuristic for the TSP. The final topic addresses the relationship between Karp's partitioning heuristic and the BHH theorem.

1. Introduction

In 1959 Beardwood, Halton, and Hammersley established the following theorem:

If $X_i, 1 \leq i < \infty$ are independent identically distributed random variables with bounded support in $\mathbb{R}^d$, then the length L_n under the usual Euclidean metric of the shortest path through the points $\{X_1, X_2, \ldots, X_n\}$ satisfies

$$n^{-(d-1)/d} L_n \to c_d \int_{\mathbb{R}^d} f(x)^{(d-1)/d}\, dx \quad \textit{almost surely.} \tag{1.1}$$

Here, $f(x)$ is the density of the absolutely continuous part of the distribution of the X_i.

This result has proved fruitful in most of the ways that are open to a mathematical discovery. In particular, it has lead to interesting applications, provoked useful generalizations and inspired new techniques of analysis. The intention of this article is to review and contribute to three developments associated with the Beardwood, Halton, Hammersley theorem.

The first development concerns the extent to which (1.1) can be complemented by large deviation results. This exploration leads us to consider some basic results of large deviations for martingales, particularly Azuma's inequality, for which we give two proofs. While exploring the relationship of the TSP martingale theory, we also examine the demands it places on results like the square function inequality of Burkholder, bounds on Hermite moments, and related ideas. In the course of the review we give new proofs of two inequalities of Rhee and Talagrand, and we examine essentially all of the available information concerning the tail of the probability distribution of L_n.

We next address the use of spacefilling curves in the analytical theory of the TSP. Such techniques are relatively new, but their simplicity and generality suggests that their use will grow. The fact that underlies this development is the existence of measure preserving transformations from $[0,1]$ onto $[0,1]^d$ that are Lipschitz of order $1/d$. A basic objective of Section 4 is to review the background of a problem of Platzman and Bartholdi on the ratio of the length of the tour provided by the spacefilling heuristic and the length of an optimal tour is bounded independently of n.

The third development concerns the role of (1.1) in Karp's polynomial time partitioning algorithm for the TSP. This topic is addressed briefly, but two results are reviewed that will make clear how one can show the effectiveness of Karp's algorithm without resort to the full force of (1.1).

In the concluding section, we discuss some open problems and promising research directions. Finally there are two appendices that stand somewhat apart from our basic themes. The first of these gives S. Lalley's previously unpublished proof of the Beardwood, Halton, Hammersley Theorem in $d = 2$ for random variables with the uniform distribution on $[0,1]^2$. This proof uses minimal machinery and illustrates a technique that is applicable to many related problems. The second appendix develops an inequality for martingales that R.E.A.C. Paley introduced for Walsh functions. Paley's old argument is examined for the suggestions it provides about how one might pursue large deviation inequalities for L_n without paying the price of bounds on L^∞ norms as demanded by Azuma's inequality.

2. Martingale Bounds for the TSP

For $X_i, 1 \le i < \infty$, independent and uniformly distributed in $[0,1]^d$, the length L_n of the shortest path through $\{X_1, X_2, \ldots, X_n\}$ is a random variable that we can show to be tightly concentrated about its mean. In $d = 2$, for example, we know that Var L_n is bounded independently of n. This fact is proved in Steele (1981b) by means of the jackknife inequality of Efron and Stein (1981), but one can provide a proof that offers considerably more potential for further development by following Rhee and Talagrand (1987) and introducing martingale arguments.

If $\mathbf{F}_k$ is the σ-field generated by $\{X_1, X_2, \ldots, X_k\}$ and

$$d_i = E(L_n \mid \mathbf{F}_i) - E(L_n \mid \mathbf{F}_{i-1}), \tag{2.1}$$

then $d_i, 1 \leq i \leq n$, is a sequence of martingale differences that satisfy

$$L_n - EL_n = \sum_{i=1}^{n} d_i. \tag{2.2}$$

This well-known representation is available for any integrable random variable, but there are features that make it particularly effective for L_n. The most central of these is that the d_i can be related to the change that takes place in L_n as one of the X_i is changed. In this respect, the analysis of L_n by means of martingale differences comes to rely on calculations that are quite close to those that made the jackknife inequality effective. By working out the details of the L^p theory associated with the martingale representation, we are led to some of the basic themes of martingale theory.

For each $1 \leq i \leq n$, let $L_n^{(i)}$ denote the length of the shortest path through $S_i = \{X_1, X_2, \ldots, X_{i-1}, \hat{X}_i, X_{i+1}, \ldots, X_n\}$ where the random variables $\{\hat{X}_i : 1 \leq i \leq n\}$ are independent, uniformly distributed and also independent of the variables in the set $S = \{X_i : 1 \leq i \leq n\}$. Since $E(L_n^{(i)} \mid \mathbf{F}_i) = E(L_n \mid \mathbf{F}_{i-1})$, we have the key observation that

$$d_i = E(L_n - L_n^{(i)} \mid \mathbf{F}_i). \tag{2.3}$$

Since one can build a path through S_i by following the minimal path through S and making a detour from X_j to $\hat{X}_i$ and back for some $j \neq i$, we have

$$L_n^{(i)} - L_n \leq 2 \min_{j:j\neq i} |\hat{X}_i - X_j|. \tag{2.4a}$$

By the same reasoning but starting from the optimal tour for S_i we have

$$L_n - L_n^{(i)} \leq 2 \min_{j:j\neq i} |X_i - X_j|, \tag{2.4b}$$

and, moreover, the right hand sides of (2.4a) and (2.4b) both have the same distribution. Next we note that simple geometric considerations as applied in Steele (1981b) give us a bound on the tail of these distributions:

$$P(\min_{j:j\neq i} |\hat{X}_i - X_j| \geq t) \leq Ae^{-Bnt^d}, \quad t > 0, \tag{2.5}$$

where $A = A_d$ and $B = B_d$ are constants that depend on d (but not on n or t). From (2.3), (2.4a,b), (2.5), and Jensen's inequality, we therefore find for any $p \geq 1$ that

$$E|d_i|^p \leq 4^p p d^{-1} A (nB)^{-p/d} \Gamma(p/d). \tag{2.6}$$

Finally, in terms of L^p norms, we find from Stirling's formula that

$$\|d_i\|_p \le C_1(p/n)^{1/d} \tag{2.7}$$

where C_1 is a constant that depends only on $d \ge 2$.

Inequality (2.7) is a basic one for the theory of the traveling salesman problem. In particular, since the d_i are orthogonal random variables, we find that if we restrict attention to $p = 2$ and $d = 2$, then (2.7) completes the proof of the rather surprising uniform bound Var $L_n \le 2A_2B_2^{-1}$ mentioned earlier.

For large p inequality (2.7) is not as effective as one would hope since from (2.4a,b) it is already immediate that the norms $\|d_i\|_\infty$ are bounded by $2d^{1/2}$. Still, by applying the argument used in (2.6) to the conditional probabilities (2.3), we can get a sharper bound on the $\|d_i\|_\infty$. In particular, if we relax the bounds (2.4a,b) to

$$|L_n^{(i)} - L_n| \le 2 \min_{j:j>i} |\hat{X}_i - X_j| + 2 \min_{j:j>i} |X_i - X_j|,$$

then from (2.3) we find

$$|d_i| \le 2E\{\min_{j:j>i} |\hat{X}_i - X_j|\} + 2E\left\{\min_{j:j>i} |X_i - X_j| \big| X_i\right\}. \tag{2.8}$$

Using $\|d_i\|_\infty \le 2d^{1/2}$ to deal with $i = n$, we thus can find a constant C_2 that depends only on $d \ge 2$, so for all $1 \le i \le n$ we have

$$\|d_i\|_\infty \le C_2(n - i + 1)^{-1/d}. \tag{2.9}$$

The beauty of (2.9) is that it permits us to use traditional martingale techniques to obtain reasonably sharp large deviation inequalities on $L_n - EL_n$. To develop one such inequality we first note that for any $y \ge 0$,

$$e^{xy} \le \cosh y + x \sinh y \quad \text{for all } |x| \le 1, \tag{2.10}$$

because (2.10) trivially holds for $x \in \{-1, 0, 1\}$, e^{xy} is convex, and the right hand side is linear in x. If we now let $x = d_i/\|d_i\|_\infty$ and $y = t\|d_i\|_\infty$, we find for $1 \le k \le n$ that

$$\exp\left(t\sum_{i=1}^k d_i\right) \le \prod_{i=1}^k \left(\cosh t\|d_i\|_\infty + \frac{d_i(\sinh t\|d_i\|_\infty)}{\|d_i\|_\infty}\right).$$

Taking expectations and using the fact that the d_i are martingale differences gives us

$$E\exp\left(t\sum_{i=1}^k d_i\right) \le \prod_{i=1}^k \cosh(t\|d_i\|_\infty)$$

$$\le \exp\left(\frac{t^2}{2}\sum_{i=1}^k \|d_i\|_\infty^2\right), \tag{2.11}$$

where in the last inequality we used the elementary bound $\cosh x \leq e^{x^2/2}$. From (2.11) and the fact that the right hand bound is an even function of t, we find for all $t \geq 0$ that

$$P\left(\left|\sum_{i=1}^{k} d_i\right| \geq \lambda\right) \leq 2e^{-\lambda t}\exp\left(\frac{t^2}{2}\sum_{i=1}^{k}\|d_i\|_\infty^2\right), \tag{2.12}$$

so letting $t = \lambda\left(\sum_{i=1}^{k}\|d_i\|_\infty^2\right)^{-1}$ we conclude

$$P\left(\left|\sum_{i=1}^{k} d_i\right| \geq \lambda\right) \leq 2\exp\left(-\lambda^2 \Big/ \left(2\sum_{i=1}^{k}\|d_i\|_\infty^2\right)\right). \tag{2.13}$$

This inequality is valid for any martingale difference sequence $\{d_i\}$, and it is due to Azuma (1967). When we apply (2.13) to our particular $\{d_i\}$ satisfying (2.1) we find a theorem which was established in the case for $d = 2$ in Rhee and Talagrand (1987).

THEOREM 2.1. *There is a constant C_3 depending only on d such that for all $n \geq 1$ and $\lambda > 0$ we have*

$$P(|L_n - EL_n| \geq \lambda) \leq \begin{cases} 2\exp(-C_3\lambda^2/\log n) & \text{if } d = 2 \\ 2\exp(-C_3\lambda^2 n^{(2-d)/d}) & \text{if } d \geq 3. \end{cases} \tag{2.14}$$

The technique used to obtain Azuma's inequality (2.13) is apparently quite crude, and one might hope to do better in several ways. One natural idea is to try to generalize (2.10) to

$$e^{xy} \leq xf(y) + g(y), \quad |x| \leq 1, \; y \geq 0, \tag{2.15}$$

for f and g that might be more effective than sinh and cosh. To see why this idea does not succeed, we let $x = \pm 1$ in (2.15) and add the two resulting inequalities. We find that (2.15) forces the bound $\cosh y \leq g(y)$, and thus no inequality like (2.15) serves us any better than that used in the argument leading to (2.13).

A second seedling concerning Azuma's technique and the TSP comes from viewing (2.10) as a separation of variables for the bivariate function e^{xy}. A classical approach to such separation might call on the generating function for Hermite polynomials:

$$G(x, y) = e^{2xy - y^2} = \sum_{n=0}^{\infty} \frac{H_n(x)y^n}{n!}. \tag{2.16}$$

This approach has not been developed very far, but it seems rich enough to deserve a brief digression. Because of the basic orthogonality relation

$$\int_{-\infty}^{\infty} H_m(x)H_n(x)e^{-x^2}\,dx = \sqrt{\pi}2^n n!\,\delta_{mn},$$

it is not difficult to give a condition on the L^2 norm of $H_n(L_n - EL_n)$ that implies a large deviation inequality of Gaussian type. In fact it suffices to assume that

$$EH_n^2(Z) \leq A^n n!$$

for some constant A.

Before closing this digression on separation of variables in e^{xy}, we should note that (2.16) is closely related to (2.10); in particular from (2.16) we easily find expressions for $\sinh y$ and $\cosh y$ in terms of odd and even Hermite polynomial (see e.g. Section 8.957 of Gradshteyn and Ryzhik 1963). Still, because of special properties of Hermite polynomials such as their recursion relation, one might expect some progress through Lemma 2.1.

Returning to the direct exploration of large deviation inequalities, we should note their easy application to moments. Thus, we multiply (2.13) by $p\lambda^{p-1}$ and integrate over $[0,\infty)$ to find for $p \geq 1$ that

$$E\left|\sum_{i=1}^{k} d_i\right|^p \leq 2p\Gamma(p/2)\left\{2\sum_{i=1}^{k}\|d_i\|_\infty^2\right\}^{p/2},$$

or, in terms of norms,

$$\left\|\sum_{i=1}^{k} d_i\right\|_p \leq C_4 p^{1/2}\left(\sum_{i=1}^{k}\|d_i\|_\infty^2\right)^{1/2}, \tag{2.17}$$

where C_4 is a universal constant which does not even depend on d. When (2.17) is specialized to $\{d_i\}$ satisfying (2.3), we find from (2.9) that for all $n \geq 1$

$$\|L_n - EL_n\|_p \leq \begin{cases} C_5 p^{1/2}(\log n)^{1/2} & \text{if } d = 2, \\ C_5 p^{1/2} n^{(d-2)/(2d)} & \text{if } d \geq 3. \end{cases} \tag{2.18}$$

Inequality (2.8) can be obtained in another way that also provides an interesting proof of Azuma's inequality. The key idea comes from work of Jakubowski and Kwapień (1979), and, in our context, the main point is that if we let $r_k(s)$ be the kth Rademacher function (i.e. $r_k(s) = \text{sign}(\sin 2^k \pi s)$, $0 \leq s \leq 1$) then

$$f(\omega, s) = \prod_{k=1}^{n}\left(1 + \frac{r_k(s)d_k(\omega)}{\|d_k\|_\infty}\right) \tag{2.19}$$

is a density function with respect to the product measure $ds\,dP$. The identities that make (2.19) effective are

$$\sum_{k=1}^{n} \frac{a_k d_k(\omega)}{\|d_k\|_\infty} = \int_0^1 \sum_{k=1}^{n} a_k r_k(s) f(\omega, s)\, ds \tag{2.20a}$$

and

$$1 = \int f(\omega, s)\, dP. \tag{2.20b}$$

The proof of (2.20a) just requires expanding (2.19) and using the fact that the Rademacher functions $\{r_k(s)\}_{1 \le k \le n}$ have mean zero and variance 1. Similarly, (2.20b) follows from expanding (2.19) and using the martingale property. Since $F(\omega, s) \ge 0$ we also see from (2.20b) that $f(\omega, s)$ must indeed be a density with respect to $ds\,dP$.

To get our second proof of Azuma's inequality we first apply Jensen's inequality in (2.20a), integrate with respect to P, and change order of integration:

$$\begin{aligned} E\exp\left(t\sum_{k=1}^{n} a_k d_k \Big/ \|d_k\|_\infty\right) &\le E\left(\int_0^1 \exp\left(t\sum_{k=1}^{n} a_k r_k(s)\right) f(\omega, s)\, ds\right) \\ &= \int_0^\infty \exp\left(t\sum_{k=1}^{n} a_k r_k(s)\right) ds \\ &= \prod_{k=1}^{n} \cosh(t a_k). \end{aligned} \tag{2.21}$$

In the second line of (2.21) we used (2.20b), and in the last we used the fact that the r_k are Bernoulli random variables. If we now let $a_k = \|d_k\|_\infty$ in (2.21), we find that (2.21) reduces to the same bound as (2.11), so one can complete the proof of Azuma's inequality just as before.

The direct application of the Jakubowski-Kwapień representation (2.20) also provides a route to L^p bounds on $\sum_{i=1}^{n} d_i$. Letting $a_k = \|d_k\|_\infty$ in (2.20a) we have

$$\sum_{k=1}^{n} d_k = \int_0^1 \left(\sum_{k=1}^{n} \|d_k\|_\infty r_k(s)\right) f(\omega, s)\, ds, \tag{2.22}$$

so if we raise both sides to the pth power, apply Jensen's inequality on the right and then use (2.20b), we have

$$E\left|\sum_{k=1}^{n} d_k\right|^p \le \int_0^1 \left|\sum \|d_k\|_\infty r_k(s)\right|^p ds. \tag{2.23}$$

Since the $\{r_k\}$ are independent Bernoulli random variables, we can apply Khintchine's inequality (Chow and Teicher 1978 or Haagerup 1982) to obtain

$$E\left|\sum_{k=1}^{n} d_k\right|^p \le \left(\frac{p+1}{2}\right)^{p/2} \left(\sum_{k=1}^{n} \|d_k\|_\infty^2\right)^{p/2}. \tag{2.24}$$

Comparison of (2.24) with (2.17) shows that (2.24) is not an essential improvement. Still, the approach via the representations seem to be a bit better, at least it simplified tracking the constant. An intriguing feature of both approaches is the appearance of the sum of squares of the L_∞ norms. Possibly this quantity is really rooted in the large deviation problem, but more likely, it is a coincidental artifact of the approaches. In the next section we systematically pursue the relationship of moments and large deviations in the context of the TSP. By introducing a few additional martingale tools, we can extract almost all of the information available on the tails of behavior of L_n.

3. Large Deviations and Moment Inequalities

We begin with a lemma that must be classical. It reminds us that the hunt for large deviation inequalities of Gaussian type can be conducted by pursuing appropriate L^p bounds. The interest in this observation comes from the fact that for some variables closely connected with L_n those bounds are easily proved.

LEMMA 3.1. *For any random variable Z, a necessary and sufficient condition that*

$$P(|Z| \ge t) \le Ae^{-Bt^2}, \quad t \ge 0, \tag{3.1}$$

for some constants $A > 0$ and $B > 0$ is that for all $p \ge 1$

$$\|Z\|_p \le Cp^{1/2} \tag{3.2}$$

for some constant C.

PROOF: If (3.1) holds, we multiply by pt^{p-1} and integrate as in (2.6) to obtain (3.2). For the converse, we just note by (3.2) and Markov's inequality that

$$P(|Z| \ge t) \le \frac{1}{t^p} C^p p^{p/2} = e^{p\log C + (1/2)p\log p - p\log t},$$

so, choosing p such that $\log p = 2(\log t - \log C) - 1$, or $p = t^2C^{-2}e^{-1}$, yields (3.1) with $A = 1$ and $B = (2C^2e)^{-1}$. ■

A central theme in the theory of martingales is that for any martingale difference sequence $\{Y_i, 1 \le i \le n\}$ the square function,

$$S_n = \left(\sum_{i=1}^{n} Y_i^2\right)^{1/2}, \tag{3.3}$$

and the maximal function,

$$M_n^* = \sup_{1 \le k \le n} \left|\sum_{i=1}^{k} Y_i\right|,$$

share many properties with the underlying martingale

$$M_k = \sum_{i=1}^{k} Y_i, \quad 1 \le k \le n.$$

In particular, the inequalities of Doob and Burkholder tell us, among other things, that if any one of S_n, M_n^*, or M_n is in L^p for some $1 < p < \infty$ then all three are in L^p. The comparability of the moments S_n and M_n is particularly interesting for the theory of the TSP in $\mathbb{R}^d$ because, as we see in the next lemma, the L^p-norm of S_n can be bounded with enough precision to yield powerful large deviation inequalities. In fact, for $d = 2$ the resulting L^p bound is good enough to guarantee a large deviation inequality of Gaussian type.

LEMMA 3.2. *For the TSP martingale summands d_i of (2.1), we have for even integers $p \ge 2$ and any set $S \subset \{1, 2, \ldots, n\}$ that*

$$\left\|\left(\sum_{i\in S} d_i^2\right)^{1/2}\right\|_p \le C_1 p^{1/d} |S|^{1/2} n^{-1/d}, \tag{3.4}$$

where C_1 is the same constant as given in (2.7) and $|S|$ is the cardinality of S.

PROOF: We first expand and apply the generalized Hölder inequality:

$$\begin{aligned} E\left(\sum_{i\in S} d_i^2\right)^p &= \sum_{i_1\in S}\sum_{i_2\in S}\cdots\sum_{i_p\in S} E d_{i_1}^2 d_{i_2}^2 \ldots d_{i_p}^2 \\ &\le \sum_{i_1\in S}\sum_{i_2\in S}\cdots\sum_{i_p\in S} (E d_{i_1}^{2p})^{1/p} (E d_{i_2}^{2p})^{1/p} \ldots (E d_{i_p}^{2p})^{1/p}. \end{aligned}$$

Next, using the bound from (2.7), together with $\|d_i\|_{2p} \le C_1 (2p/n)^{1/d}$ or $E d_i^{2p} \le C_1^{2p} (2p/n)^{2p/d}$, we find

$$E\left(\sum_{i\in S} d_i^2\right)^p \le |S|^p C_1^{2p} (2p/n)^{2p/d},$$

and hence for even integers p we conclude

$$\left\| \left(\sum_{i=1}^{n} d_i^2 \right)^{1/2} \right\|_p \le C_1 p^{1/d} |S|^{1/2} n^{-1/d}. \qquad \blacksquare$$

This bound is of particular interest for $d = 2$ and $S = \{1, 2, \ldots, n\}$, since it is then of the form required in Lemma 3.1, i.e.

$$\left\| \left(\sum_{i=1}^{n} d_i^2 \right)^{1/2} \right\|_p \le C p^{1/2}. \tag{3.5}$$

Thus for $d = 2$ the square function associated with the TSP martingale differences of (2.1) satisfies a large deviation inequality of Gaussian type (3.1).

One hope raised by (3.4) and (3.5) is that of extracting a Gaussian type large deviation inequality for L_n from that available for the square function $\left(\sum d_i^2\right)^{1/2}$ associated with L_n. To assess this possibility we first recall the square function inequalities of Burkholder (1966, 1973):

For $1 < p < \infty$ and any sequence of martingale differences Y_i with associated square function S_n defined by (3.3), we have

$$(18 p^{1/2} q)^{-1} \|S_n\|_p \le \left\| \sum_{i=1}^{n} Y_i \right\|_p \le 18 q^{1/2} p \|S_n\|_p \tag{3.6}$$

where $1/p + 1/q = 1$.

To see how (3.6) relates to the inequalities considered earlier, we note that we always have

$$|S_n|^p \le \left(\sum_{i=1}^{n} \|d_i\|_\infty^2 \right)^{p/2}, \tag{3.7}$$

so, in particular, the second inequality of (3.3) gives us a bound like (2.16) which expresses the L^p version of Azuma's inequality. In this instance there is a critical difference in that the factor $p^{1/2}$ is inflated to p. Since large deviation results depend on the L^p inequalities for large p, this change in the constant is a major concern.

Still, when $d = 2$ we can use Lemma 3.2 to get good bounds on the tail probabilities of $L_n - EL_n$. We will give two illustrations of this approach. The first consists of showing that the moment generating function of $L_n - EL_n$ can be bounded independently of n.

To begin we note that for $|td_i| < 1$, the Taylor expansion of $\log(1 + td_i)$ gives us

$$\prod_{i=1}^{n} (1 + td_i) = \exp\left(\sum_{k=1}^{\infty} (-1)^{k+1} \beta_k t^k / k \right) \tag{3.8}$$

where $\beta_k = d_1^k + d_2^k + \cdots + d_n^k$. We next note for $k \geq 3$ that

$$|\beta_k| \leq \sum_{j=1}^{n} \|d_j\|_\infty^k \leq \sum_{j=1}^{n} C_1^k (n-j+1)^{k/2}$$
$$\leq C_1^k \sum_{j=1}^{\infty} j^{-3/2} = C_1^k \zeta(3/2) \tag{3.9}$$

where $\zeta(s) = \sum_{j=1}^{\infty} j^{-s}$, so from (3.8) and (3.9) we find

$$\exp\left(\beta_1 t - \tfrac{1}{2}\beta_2 t^2\right) \leq \prod_{i=1}^{n}(1 + td_i)\exp\left(\zeta(3/2)\sum_{k=3}^{\infty} C_1^k t^k / k\right).$$

After taking expectations, we see

$$E\exp\left(\beta_1 t - \tfrac{1}{2}\beta_2 t^2\right) \leq \exp\left(\zeta(3/2)\sum_{k=3}^{\infty} C_1^k t^k / k\right) \equiv \phi(t), \tag{3.10}$$

so writing $\exp(\beta_1 t) = \exp(\beta_1 t - \beta_2 t^2)\exp(\beta_2 t^2)$ and applying Schwarz's inequality gives

$$E\exp(\beta_1 t) \leq \phi(2t)^{1/2}\left(E\exp(2\beta_2 t^2)\right)^{1/2}. \tag{3.11}$$

By (3.5) and Lemma 3.1 we know there is a constant $A > 0$ not depending on n such that $P(\beta_2^{1/2} > t) \leq Ae^{-t^2/A}$; hence, we have for $|t| < A^{-1}$ that

$$E\exp(t\beta_2) \leq \frac{1}{1 - At} \tag{3.12}$$

and the bound (3.11) does not depend upon n.

The uniform bound on the moment generating function given by (3.11) and (3.12) naturally give a large deviation bound. For reference purposes we record the following consequence of (3.11) and (3.12) that was first obtained in Rhee and Talagrand (1988a) by different means.

PROPOSITION 3.1. *For $d = 2$, there is a constant C such that for all $n \geq 2$ and $t > 0$*

$$P(|L_n - EL_n| \geq t) \leq Ce^{-Ct}. \tag{3.13}$$

A stronger result than (3.13) can be obtained by the use of Burkholders inequality. In fact, the following theorem seems to be about as much as one can obtain without going beyond the information on the TSP that is incorporated in (2.7) and (2.9).

THEOREM 3.1. *For $d = 2$, there is a constant C such that for all $n \geq 1$ and $p \geq 1$,*

$$\|L_n - EL_n\|_p \leq Cp^{1/2}(\log p)^{1/2}. \tag{3.14}$$

PROOF: We rely on the martingale representation (2.3) and split the representing sum into two terms,

$$\|L_n - EL_n\|_p = \left\|\sum_{i=1}^{n} d_i\right\|_p \leq \left\|\sum_{i\leq \alpha n} d_i\right\|_p + \left\|\sum_{i>\alpha n} d_i\right\|_p \tag{3.15}$$

for any $0 < \alpha < 1$. To the first summand we apply (2.17), the L^p version of Azuma's inequality, and to the second we apply Burkholder's second inequality to find

$$\|L_n - EL_n\|_p \leq C_4 p^{1/2}\Big(\sum_{i\leq \alpha n} \|d_i\|_\infty^2\Big)^{1/2} + 18pq^{1/2}\left\|\Big(\sum_{i>\alpha n} d_i^2\Big)^{1/2}\right\|_p. \tag{3.16}$$

Now we apply (2.9) to the first sum and (3.4) to the second,

$$\begin{aligned}\|L_n - &EL_n\|_p \\ &\leq C_1C_4 p^{1/2}\Big(\sum_{i\leq \alpha n}(n-i+1)^{-1}\Big)^{1/2} + 18pq^{1/2}C_1p^{1/2}(1-\alpha)^{1/2} \\ &\leq C_6 p^{1/2}\big(\log 1/(1-\alpha)\big)^{1/2} + C_6p^{3/2}q^{1/2}(1-\alpha)^{1/2}. \end{aligned} \tag{3.17}$$

When we let $(1-\alpha)^{1/2} = p^{-1}$, we find (3.14). ■

COROLLARY. *There is a constant B such that for $d = 2$ we have*

$$P(|L_n - EL_n| \geq t) \leq 2e^{-Bt^2/\log(1+t)} \tag{3.18}$$

for all $t \geq 0$.

The proof of (3.18) from (3.14) follows just as in Lemma 3.1. This time the proper choice of p is $t^2/(C\log t)$ where C is the constant of (3.8).

Inequality (3.18) was also first established in Rhee and Talagrand (1988a). Their proof grew out of the idea of interpolating between the $d = 2$ case of (2.14) where the tails have quadratic exponential behavior that depends on n, and on (3.13), where the bound is independent of n but is linear exponential. Rhee and Talagrand (1988a) bring these two bounds together to prove (3.18) by use of interpolation results from Bergh and Lofstrom (1976). The present proof via (3.8) is simpler than that of

Rhee and Talagrand, at least so far it relies on methods that are familiar to probabilists. Still, even now, the Burkholder inequalities might not be regarded as completely commonplace tools, and the proof of (3.18) is not yet elementary.

The quest that has been traced here, the derivation of a Gaussian type large deviation bound for $L_n - EL_n$, has very recently come to fruition through Rhee and Talagrand (1988b). By combining their basic martingale approach with a bare-handed investigation of the geometry of an n-sample from $[0,1]^2$, they show that one can indeed remove the logarithmic factor from (3.18). The resulting inequality for the TSP in $d = 2$ stands as both the natural end to a line of investigation and as a hard challenge. What can one say for $d \geq 2$? What other functionals permit a comparable analysis?

4. Analytical Bounds from Spacefilling Curves

For many problems concerning combinatorial optimization in $\mathbb{R}^d$ one can obtain useful bounds by appealing to the existence of a map ϕ from $[0,1]$ onto $[0,1]^d$ that is Lip α with $\alpha = 1/d$, i.e. $|\phi(s) - \phi(t)| \leq c|s-t|^{1/d}$ for a constant c and all $0 \leq s \leq t \leq 1$. Moreover, Milne (1980) established that one can further require ϕ to be measure preserving, and from our perspective, the benefit of that fact is that it lets us use spacefilling curve techniques to get probabilistic inequalities, at least in the case of uniformly distributed random variables.

For our first example we again consider the traveling salesman problem in $\mathbb{R}^d$, but this time we take the cost of travel from x to y to be $|x-y|^p$, the pth power of the Euclidean distance. If $S = \{x_1, x_2, \ldots, x_n\}$ is a set of n points in $[0,1]^d$, how can we bound $\tilde{L}(S)$, the length of the shortest tour through the points of S under this metric, i.e. how can we bound

$$\tilde{L}(S) = \min_{\sigma} \sum_{i=1}^{n-1} |x_{\sigma(i)} - x_{\sigma(i+1)}|^p \tag{4.1}$$

where the minimum is over all cyclic permutations?

Since ϕ is a surjection, each $x_i \in S \subset [0,1]^d$ has a pre-image $y_i \in [0,1]$. If we choose a cyclic permutation σ so that $y_{\sigma(1)} \leq y_{\sigma(2)} \leq \cdots \leq y_{\sigma(n)}$, then a heuristic tour of the $\{x_i\}$ can be formed by visiting them in the order of the $\{y_i\}$. For this heuristic we find

$$\begin{aligned}\tilde{L}(S) &\leq \sum_{i=1}^{n-1} \left|\phi(y_{\sigma(i)}) - \phi(y_{\sigma(i+1)})\right|^p \\ &\leq c^p \sum_{i=1}^{n-1} \left|y_{\sigma(i)} - y_{\sigma(i+1)}\right|^{p/d} \\ &\leq c^p n^{(d-p)/d} \end{aligned} \tag{4.2}$$

where we applied Hölder's inequality and the fact that $\sum |y_{\sigma(i)} - y_{\sigma(i+1)}|$ is bounded by 1. The key idea of (4.2), i.e. building a path through $\{x_n, 1 \leq i \leq n\}$ by visiting the points in the linear ordering of the $\{y_i, 1 \leq i \leq n\}$, is called the *spacefilling heuristic.* For application to the TSP, this idea was first proposed by Bartholdi and Platzman (1982) and independently by D.H. Fremlin (see e.g. Fremlin 1982). Both for heuristic algorithms and analytic bounds, the idea of using a spacefilling map to exploit the linear ordering, or simple geometry, of $[0,1]$ has many natural applications, and the breadth of these variations can be seen by consulting the survey by Bartholdi and Platzman (1988), the papers by Glass (1985) and Imai (1986), or the recent thesis by Bertsimas (1988).

For $p = 1$, inequality (4.2) recaptures the familiar $O(n^{(d-1)/d})$ bound, but for $p = d$ it provides new information by providing a $O(1)$ bound. In contrast, one only obtains the weaker inequality

$$\tilde{L}(S) \leq c \log n. \tag{4.3}$$

by classical arguments that rest on the fact that any set of n points in $[0,1]^d$ contains a pair within $cn^{-1/d}$ of each other.

The argument used for sharper bound (4.2) was also applied in Steele (1988) to show that the sum of the dth powers of the lengths of the edges of a minimal spanning tree of n points in $[0,1]^d$ can be bounded independent of n. For $d = 2$ the uniform boundedness of the sum of squares of the edge lengths had been established earlier by Gilbert and Pollak (1968), but their delicate geometric argument has no natural analogue for $d > 2$. In contrast, the bound provided by the spacefilling heuristic works pleasantly in all $d \geq 2$. For the spacefilling heuristic applied to the TSP the most interesting problems concern the ratio of the length of the tour produced by the spacefilling curve to the length of the optimal tour. In $\mathbb{R}^2$ Platzman and Bartholdi (1988) provided a bound of order $O(\log n)$, and they conjectured that there is a uniform bound on the ratio. Bertsimas and Grigni (1989) settled the conjecture by giving an example that shows the ratio can be as bad as $c \log n$. The following special case of work in Steele (1989) complements the results of Platzman and Bartholdi (1988) in a way that may be useful in algorithmic applications. The proof does not require any detailed properties of the spacefilling curve in order to provide ratio bounds, except that the curve is measure preserving and is as smooth as feasible.

THEOREM 4.1. *Let ϕ be a measure preserving transformation of $[0,1]$ onto $[0,1]^2$ that is Lipschitz of order $\alpha = 1/2$, i.e.*

$$|\phi(x) - \phi(y)| \leq c|x - y|^{1/2} \tag{4.4}$$

for some c and all $x, y \in [0,1]$. If H_n is the length of the path through the points $\{x_1, x_2, \ldots, x_n\} \subset [0,1]^2$ that is constructed using the spacefilling *heuristic based on ϕ, then for $n \geq 2$*

$$H_n \leq L_n\{1 + 2c^2 \log(m/\bar{e})\} + \pi c^2 m, \tag{4.5a}$$

where L_n is the length of the optimal path through $\{x_1, x_2, \ldots, x_n\}$, m is the length of the longest edge in the heuristic path, and $\bar{e}$ is the average length of the edges in the optimal path.

COROLLARY.

$$H_n \leq \big(1 + \pi c^2 + 2c^2 \log n\big)L_n. \tag{4.5b}$$

PROOF: We suppose the heuristic tour visits the points in the order $x_1, x_2, \ldots, x_n$, i.e. we suppose there are $t_i \in [0, 1]$ such that $t_1 \leq t_2 \leq \cdots \leq t_n$ with $x_i = \phi(t_i)$. For $\lambda > 0$ we introduce two basic subsets of $\{1, 2, \ldots, n-1\}$ by

$$U(\lambda) = \{i : |t_{i+1} - t_i| > \lambda, 1 \leq i < n\}$$

and

$$V(\lambda) = \{i : |\phi(t_{i+1}) - \phi(t_i)| > \lambda, 1 \leq i < n\}.$$

For $i \in V(\lambda)$ inequality (4.4) implies

$$c|t_i - t_{i+1}|^{1/2} \geq |\phi(t_{i+1}) - \phi(t_i)| \geq \lambda$$

so $i \in V(\lambda)$ implies $i \in U(\lambda^2/c^2)$, i.e.

$$V(\lambda) \subset U(c^{-2}\lambda^2). \tag{4.6}$$

If $g(\lambda)$ is the cardinality of $V(\lambda)$, we also have

$$H_n = \int_0^m g(\lambda)\, d\lambda, \tag{4.7}$$

where $m = \max_{1 \leq i < n} |\phi(t_{i+1}) - \phi(t_i)|$, so our goal is now to use (4.6) to bound $g(\lambda)$.

For $i \in U(\lambda)$ the intervals $[t_i, t_i + \lambda]$ are non-intersecting, so if we set

$$A_i = \phi([t_i, t_i + \lambda])$$

then since ϕ preserves measure, each A_i has Lebesgue measure $\lambda = \mu(A_i)$ and $A_i \cap A_j$ has measure zero for any pair $i \neq j$, $i, j \in U(\lambda)$.

We let $D(x, C) \subset [0, 1]^2$ denote the set of all points within distance x of the curve C, and let T_n be an optimal tour of $\{x_1, x_2, \ldots, x_n\}$ with length L_n. By (4.4) and the fact that each x_i is somewhere on the path, we have for each $i \in S$ that

$$A_i \subset D(c\lambda^{1/2}, T_n). \tag{4.8}$$

Naiman's inequality on the volume of tubes (e.g. Naiman 1986, or the easier version of the basic result given in Johnstone and Siegmund 1989) tells us that for any rectifiable curve C of length L one has

$$\mu\big(D(x,C)\big) \leq 2xL + \pi x^2, \tag{4.9}$$

for all $x \geq 0$. If $f(\lambda)$ denotes the cardinality $U(\lambda)$ we then have by (4.8) and (4.9) that

$$\lambda f(\lambda) = \mu\bigg(\bigcup_{i\in U(\lambda)} A_i\bigg) \leq \mu\big(D(c\lambda^{1/2}, T_n)\big) \leq 2c\lambda^{1/2}L_n + \pi c^2\lambda,$$

so

$$f(\lambda) \leq 2c\lambda^{-1/2}L_n + \pi c^2. \tag{4.10}$$

By (4.6) and (4.10) we find our basic bound

$$g(\lambda) \leq 2c^2 L_n \lambda^{-1} + \pi c^2. \tag{4.11}$$

For any $0 < \alpha < m$, we can apply the trivial bound $g(\lambda) \leq n-1$ for $\lambda \in [0,\alpha]$ and apply (4.11) for $\lambda \in [\alpha, m]$; so, when we integrate in (4.7), we find

$$H_n \leq \alpha(n-1) + 2c^2 L_n \log(m/\alpha) + \pi c^2(m-\alpha). \tag{4.12}$$

Finally, since $L_n \leq H_n \leq (n-1)m$ we have for $\alpha = L_n/(n-1) = \bar{e}$ that $\alpha \in [0,m]$, so we can let $\alpha = \bar{e}$ in (4.12) to find (4.5a). To see that (4.5b) follows from (4.5a) we just invoke the very crude bound $m \leq L_n$ and $\bar{e} = L_n/(n-1)$. ■

The argument used in the proof of Theorem 4.1 uses several ideas from Bartholdi and Platzman (1988), and it makes progress mainly by being systematic in the exploitation of the bound (4.9).

The next section deals more directly with the geometry and topology of spacefilling curves.

5. Schoenberg's Map and Smoother Maps

Section 4 made use of smooth spacefilling curves, but it did not provide concrete examples. This section engages the problem of constructing spacefilling curves, especially curves that are as smooth as possible and that preserve Lebesgue measure. It also points out a topological barrier to the sharpening of Theorem 4.1.

We begin by considering a method of Schoenberg (1938) that gives perhaps the shortest classical example of a continuous map from $[0,1]$ onto

$[0,1]^2$. Schoenberg's map is not as smooth as we need, but it points the way to a map that is both simpler and smoother. We first define a real valued even function f of period 2 by taking $f(t)=0$ in $(0,1/3)$, $f(t)=1$ in $(1/3,1)$, and making $f(t)$ linear in $(1/3,2/3)$. We then define Schoenberg's spacefilling curve by the explicit formulas

$$x(t) = \frac{1}{2}f(t) + \frac{1}{2^2}f(3^2t) + \frac{1}{2^3}f(3^4t) + \cdots \tag{5.1a}$$

and

$$y(t) = \frac{1}{2}f(3t) + \frac{1}{2^2}f(3^3t) + \frac{1}{2^3}f(3^5t) + \cdots . \tag{5.1b}$$

To prove the map $t \to (x(t), y(t))$ is surjective, we first note that if $\{a_k\}$ is any infinite sequence of 0's and 1's, then a typical point in the Cantor set $\mathbf{C} \subset [0,1]$ can be written uniquely as

$$t_0 = \frac{2a_0}{3} + \frac{2a_1}{3^2} + \frac{2a_2}{3^3} + \dots . \tag{5.2}$$

By straightforward, but tedious, bounds one can also show that f can be used to extract the kth term in the ternary expansion of t_0, specifically

$$f(3^k t_0) = a_k. \tag{5.3}$$

Now, given any $(x_0, y_0) \in [0,1]^2$, we can use the binary expansion of x_0 and y_0 together with the explicit formulas (5.1) and (5.2) to write down a point in $\mathbf{C}$ that ϕ maps to (x_0, y_0), so ϕ is a surjection of $[0,1]$ onto $[0,1]^2$.

One important aspect of the explicit formulas (5.1) and (5.2) is their computational feasibility. Not only do we know that for every point (x_0, y_0) of $[0,1]^2$ that there *exists* a point of $[0,1]$ that maps onto (x_0, y_0), but we can also quickly compute a point $t \in \mathbf{C}$ such that $\phi(t) = (x_0, y_0)$.

Now we need to assess the smoothness of Schoenberg's map $\phi(t) = (x(t), y(t))$. By uniform convergence, we see ϕ is continuous on $[0,1]$. In fact, it is easy to show there is an α so that ϕ is in Lip α, and we can even determine the best value of α. First, just consider $x(t)$ and note that f satisfies the two naive bounds $|f(s) - f(t)| \le 3|s-t|$ and $|f(t)| \le 1$. Thus we have for any $n \ge 1$ that

$$|x(s) - x(t)| \le 3\sum_{k=1}^{n} 2^{-k}|3^{2k-2}s - 3^{2k-2}t| + 2\sum_{k=n+1}^{\infty} 2^{-k}, \tag{5.4}$$

so for all integers n we have

$$|x(s) - x(t)| = O\big(|s-t|(9/2)^n + 2^{-n}\big). \tag{5.5}$$

Finally, by choosing n to be the integer nearest $-(\log_2|s-t|)/\log_2 9$, we find $|x(s)-x(t)| = O(|s-t|^\alpha)$ where $\alpha = (2\log_2 3)^{-1}$.

Having achieved an α for which $f \in \text{Lip } \alpha$, we will show that $f \notin \text{Lip } \alpha'$ for any $\alpha' > \alpha$ by using some elementary facts about Hausdorff dimension. In fact, we use the result of Hausdorff (1919) that the dimension of the Cantor ternary set equals $\log 2/\log 3$. If we let $N(\epsilon)$ be the least number of intervals $\{I_i\}$ of length 2ϵ, $0 < \epsilon < 1$, that cover the Cantor set $\mathbf{C}$, then in terms of $N(\epsilon)$, the fact that $\mathbf{C}$ has Hausdorff dimension $1/\log_2 3$ tells us that for any $\delta > 0$ there are constants A and B such that $A\epsilon^{-\beta-\delta} > N(\epsilon) > B\epsilon^{-\beta+\delta}$, where $\beta = 1/\log_2 3$.

Now suppose ψ is any map of $\mathbf{C}$ onto $[0,1]^2$, and suppose that ψ is also Lip α'. If λ denotes Lebesgue measure in $\mathbb{R}^2$, then since the compact set $\psi(\mathbf{C})$ covers $[0,1]^2$, and since we have a collection of $N(\epsilon)$ intervals $\{I_i\}$ of length 2ϵ that cover $\mathbf{C}$, we have

$$1 \le \lambda(\psi(\mathbf{C})) \le \sum_{i=1}^{N(\epsilon)} \lambda(\psi(I_i)) \le N(\epsilon)\pi(c\epsilon^{\alpha'})^2 = O(N(\epsilon)\epsilon^{2\alpha'}). \tag{5.6}$$

From (5.6) and the arbitrariness of $\delta > 0$, we conclude that $\beta \ge 2\alpha'$, i.e. $\alpha \le 1/(2\log_2 3)$ for any Lip α map of the Cantor set onto $[0,1]^2$. We have thus established that Schoenberg's spacefilling curve is precisely of smoothness type Lip α with $\alpha = 1/(2\log_2 3)$.

Although Schoenberg's mapping is a rich source of insight, one has to put in considerable modification in order to attain the maximal level of smoothness that one can have. Still the Lip 1/2 measure preserving property is shared by several of the classical spacefilling curves, particularly those due to Hilbert and Lebesgue. For a proof of these features of the classical curves as well as some remarkable analytical applications of spacefilling curves, one can consult Milne (1980). Also, to show one cannot find a map smoother than Lip 1/2 from $[0,1]$ onto $[0,1]^2$, we just use the fact that the Hausdorff dimension of $[0,1]$ is 1 and repeat the argument given for the lower bound of smoothness for Schoenberg's map.

There is nothing more we need to say about the construction of smooth spacefilling curves, but there are some final issues concerning the spacefilling heuristic and the topology of $[0,1]^2$. The bound on the ratio H_n/L_n that was given in Section 4 really relied on bounding the ratio H_n^*/L_n where

$$H_n^* = \sum_{i=1}^{n-1} |t_{i+1} - t_i|^{1/2} \tag{5.7}$$

and $\phi(t_i) = x_i, 1 \le i \le n$. To see a subtlety in this process, we first recall that the dimension theorem of general topology tells us that there is no continuous bijection between $[0,1]$ and $[0,1]^2$ (see e.g. Dugundji 1970, p.

359). Thus, every continuous surjection must have a double point. The investigation of multiple points was pursued further by Pólya (1913) who gave a spacefilling curve with multiple points with multiplicity bounded by three. This explicit line of investigation was completed by Hurewicz (1933) who showed that any surjection of $[0,1]$ onto $[0,1]^2$ must have a triple point. These facts can be used to show that bounding of H_n^* can be slippery.

For example, suppose (x,y) is the triple point guaranteed by Hurewicz and therefore suppose we have $t_1 < t_2 < t_3$ with $\phi(t_i) = (x,y)$. Now, if $s_{1,j} < s_{2,j} < s_{3,j}$ and $s_{i,j} \to t_i$ as $j \to \infty$ for each $i \in \{1,2,3\}$ we have that $L_3 = L\big(\phi(s_{1,j}), \phi(s_{2,j}), \phi(s_{3,j})\big) \to 0$ as $j \to \infty$. On the other hand,

$$H_3^* = \sum_{1 \le i \le 3} |s_{i,j} - s_{i+1,j}|^{1/2} \ge \tfrac{1}{2}|t_3 - t_1| \tag{5.8}$$

for all sufficiently large j. We thus have that H_3^*/L_3 can be made arbitrarily large, and, at first blush, this fact might seem to cast doubt on (4.5a) or (4.5b). There is no contradiction between (5.8) and the earlier bounds, but (5.8) nicely shows that one cannot rely too heavily on H_n^* for a detailed understanding of H_n.

6. Karp's Partitioning Algorithm

The Euclidean traveling salesman problem is the task of *computing* the shortest path through a set of points in $\mathbb{R}^d$. As a computational challenge, the TSP has become an essential test problem for combinatorial optimization, and, as one can see by considering the range of techniques in *The Traveling Salesman Problem: A Guided Tour of Combinatorial Optimization* (Lawler, et al. 1985), the TSP has provided the inspiration for some of the most fundamental developments in the field.

One such development took place when Karp (1976, 1977) used the Beardwood, Halton, Hammersley theorem to show how a simple partitioning algorithm yields a solution to the TSP that is (1) computable in polynomial time and (2) asymptotically optimal in an appropriate probabilistic sense. In this section, we will review Karp's basic idea and make a point that deserves to be more widely known. The asymptotic optimality of Karp's algorithm can be obtained independently of the Beardwood, Halton, Hammersley theorem. In fact, we will see that one can justify Karp's algorithm with results that are considerably less refined than the BHH theorem.

The simplest version of Karp's algorithm addresses the case of the uniform distribution, and, for ease of exposition, we will keep to that case. Let $X_i, 1 \le i < \infty$, be independent random variables with the uniform distribution in $[0,1]^d$, and suppose k_n is a sequence of integers that grows more slowly than $n^{1/d}$. Karp's method for obtaining a path through the points of $S = \{X_1, X_2, \ldots, X_n\}$ is as follows:

(1) Partition $[0,1]^d$ into k_n^d congruent subcubes $\{Q_j\}$.

(2) For each $j, 1 \leq j \leq k_n^d$, calculate an optimal path P_j through the points $S \cap Q_j$.

(3) Join the endpoints of the P_j to form a heuristic path H through all the points of S.

This description is incomplete until we specify k_n, provide a method for finding the paths P_j, and spell out how the P_j are joined to form H. As it happens, virtually any reasonable choices will suffice. For example, we can calculate the P_j by complete enumeration of the possible orders of visiting the points of $Q_j \cap S$, or we can use dynamic programming. Either of these methods will be fast enough to yield a polynomial time algorithm if k_n is chosen appropriately. Thus, for the moment, our concern is just with the effectiveness of the algorithm.

Still, we need to pick a specific rule concerning the connection of the partial paths. Thus, for each $1 \leq i \leq k_n^d$, we label the two end points of the partial path of $S \cap Q_i$ by a_i and b_i, and we connect b_i to a_{i+1} where the Q_i have been ordered lexicographically according to the vertex within each square that is lexicographically minimal. With these procedures assumed, one can show the following:

THEOREM 6.1. *If L_n^K denotes the length of the path produced by Karp's method and if k_n is any unbounded increasing sequence such that $n/k_n^d \to \infty$, then for any $\epsilon > 0$ we have*

$$\sum_{n=1}^{\infty} P\left\{\frac{L_n^K}{L_n} \geq 1+\epsilon\right\} < \infty. \tag{6.1}$$

The proof of (6.1) follows from the next two lemmas. The first guarantees that under the hypotheses on $\{X_i\}$ that L_n cannot be too short.

LEMMA 6.1. *There exist constants $A > 0$ and $0 < \rho < 1$ such that for all $n \geq 1$, we have*

$$P\left\{L_n < An^{(d-1)/d}\right\} \leq \rho^n. \tag{6.2}$$

The proof of (6.2) is easily achieved by dividing $[0,1]^d$ into n subcubes of volume $1/n$, applying standard occupancy results, and a little geometry. The second lemma is more challenging.

LEMMA 6.2. *There is an r_n depending only on n and k_n such that for all n,*

$$L_n \leq L_n^K \leq L_n + r_n \tag{6.3}$$

and

$$r_n \le c\left\{n^{(d-2)/(d-1)}k_n^{1/(d-1)} + k_n^{d-1}\right\} \tag{6.4}$$

where the constant c depends on d and the sequence $\{k_n\}$.

Remark. When $k_n = o(n^{1/d})$ then (6.4) says that $r_n = o(n^{(d-1)/d})$. Since this result holds everywhere, (6.4) and (6.2) yield (6.1).

It is relatively easy to sketch a proof of (6.4). Let $\{F_i\}$ be the set of faces of the k_n^d subcubes $\{Q_j\}$. We will use the optimal path P and some additional edges in order to bound L_n^K. If e is an edge of the optimal path, P, we associate with e a set of points that we will call *pierce points.* If e is interior to some Q_j, then the set of pierce points created by e is just the null set. On the other hand, if $e = (a, b)$ where $a \in Q$ and $b \in Q'$ and Q and Q' are distinct subcubes, then e will create a set of two pierce points. In particular, if F and F' are the faces of Q and Q' that intersect the line from a to b, then $p = e \cap F$ and $p' = e \cap F'$ are called the pierce points associated with a and b, respectively.

We will now build a set of points that may have rather large cardinality, but that can be proved to lie on a relatively short path. First, note the set of pierce points has cardinality bounded by n since each X_i is associated with at most one pierce point. Next, to each face F of each cube Q, we associate a set S_F consisting of (1) its 2^{d-1} subfaces of dimension zero (i.e. its vertices) and (2) its set of pierce points. For each F the set S_F is contained in a $d-1$ dimensional cube of edge length k_n^{-1}, so, by classical bounds (e.g. Few 1955), there is a tour through the points of S_F of length bounded by $ck_n^{-1}|S_F|^{(d-2)/(d-1)}$, where $|S_F|$ denotes the cardinality of S_F. Now consider the union of all of the tours through S_F for all F together with the optimal path P. This set of edges has the property that for each j it contains a path that is contained in Q_j and goes through all the points of S that are contained in Q_j.

We finally see that L_n^K can be bounded by three terms: (1) the length of the edges in the optimal tour, (2) the sum of the edges needed to tour S_F for all F, and (3) the cost of the edges required in Step 3 of Karp's heuristic. We thus see that

$$L_n^K \le L_n + c\sum_F k_n^{-1}|S_F|^{(d-2)/(d-1)} + ck_n^{d-1}. \tag{6.5}$$

The bound on r_n given in (6.4) now follows from (6.5) by Hölder's inequality, the fact that the sum of the $|S_F|$ is $O(n)$, and the fact that there are $O(k_n^d)$ faces of the cubes $\{Q_i\}$.

To some extent, the preceding sketch follows the lines of Halton and Terada (1982) which one can consult for additional details. From the present perspective, the main point of interest is that one requires so little probability theory. All one needs is the elementary occupancy theory in Lemma 6.1.

7. Concluding Remarks

In 1959 the work of Beardwood, Halton and Hammersley was a singular event in the sense that prior to that date and for many years subsequent one finds no comparable work relating probability theory and combinatorial optimization. The power and beauty of the Beardwood, Halton, Hammersley theorem were immediately present, but considerable time needed to elapse before wide appreciation was possible. The key step in the process toward that appreciation is the work of Karp (1976). By connecting the asymptotic result of Beardwood, Halton, and Hammersley with the possibility of effective algorithms, Karp created an eager audience for both the original work and for results that complement it. In Karp and Steele (1985) and the recent thesis of Bertsimas (1988), one can find a review of that development. This article also provides a review, but here the focus is narrowed to the developing roles of martingale theory and of spacefilling curves.

The field of martingale inequalities is so rich that the applications in Sections 2 and 3 only offer a hint of future possibilities. Connections between martingale theory and problems like the TSP can be counted on to develop vigorously in the next few years.

Among the concrete problems that may, or may not, be attacked via martingales, the one that stands out most concerns the completion of our understanding of the tails of $L_n - EL_n$ in $d = 2$. More broadly we would like to understand the ways in which $L_n - EL_n$ behaves like a Gaussian random variable. In particular, we would like to know if $L_n - EL_n$ converges in distribution to a Gaussian limit.

The force behind applications of the spacefilling heuristic is not as great as that behind martingale theory, but one can still expect vigorous activity. The strong interest in the geometry of fractals provides one motivation, but the fact that the heuristic is easily coded also helps. Even though the conjecture of Bartholdi and Platzman is formally settled by the example of Bertsimas and Grigni (1989), many questions remain. As suggested in Section 5, one can expect some more negative results. Nevertheless, one may be able to provide further positive results like Theorem 4.1 that are of use in practical problems.

John Hammersley coined the inviting phrase 'seedlings of research', and throughout his work one finds a generous willingness to reveal interesting ideas that still have room to grow. The intention of this article has been to try to live up to that tradition while engaging the shortest path through many points.

Acknowledgement

I wish to thank S. Lalley for permitting me to include his proof of the critical case of the Beardwood, Halton, Hammersley theorem. I also thank I. Karatzas and L. Platzman for their comments on parts of this article, and I thank D. Bertsimas, D.H. Fremlin, I. Johnstone, W.-S. Rhee, and D. Siegmund for making available their unpublished work. This research was supported in part by National Science Foundation Grant # DMS-8812868.

References

Azuma, K. (1967). Weighted sums of certain dependent random variables. *Tôhoku Mathematical Journal* 19, 359–367.

Bartholdi, J.J. and Platzman, L.K. (1982). An $O(n \log n)$ planar travelling salesman heuristic based on spacefilling curves. *Operations Research Letters* I (4), 121–125.

——— (1988). Heuristics based on spacefilling curves for combinatorial problems in Euclidean space. *Management Science* 34, 291–305.

Beardwood, J., Halton, J.H., and Hammersley, J. (1959). The shortest path through many points. *Proceedings of the Cambridge Philosophical Society* 55, 299–327.

Bergh, J. and Lofstrom, J. (1976). *Interpolation Spaces: An Introduction.* Springer-Verlag, New York.

Bertsimas, D. (1988). *Probabilistic Combinatorial Optimization Problems.* Ph. D. Thesis, Operations Research Center, Massachusetts Institute of Technology, Cambridge, Massachusetts.

Bertsimas, D. and Grigni, M. (1989). On the spacefilling curve heuristic for the Euclidean traveling salesman problem. Technical Report, Operations Research Center, Massachusetts Institute of Technology.

Bially, T. (1969). Spacefilling curves: their generation and their application to bandwidth reduction. *IEEE Transactions on Information Theory* II-15, 658–664.

Burkholder, D.L. (1966). Martingale transforms. *Annals of Mathematical Statistics* 37, 1494–1504.

——— (1973). Distribution function inequalities for martingales: The 1971 Wald Memorial Lectures. *Annals of Probability* 1, 19–42.

Chow, Y.S. and Teicher, H. (1978). *Probability Theory: Independence, Interchangeability, Martingales.* Springer-Verlag, New York.

Dugundji, J. (1970). *Topology.* Allyn and Bacon, Boston.

Efron, B. and Stein, C. (1981). The Jackknife estimate of variance. *Annals of Statistics* 9, 586–596.

Few, L. (1955). The shortest path and the shortest road through n points in a region. *Mathematika* 2, 141–144.

Fremlin, D.H. (1982). On the traveling salesman problem. University of Essex, Note of 6 May 1982, unpublished.

Gilbert, E.N. and Pollak, H.O. (1968). Random minimal trees. *SIAM Journal of Applied Mathematics* 16, 376–387.

Glass, M. (1985). The application of spacefilling curves to some VLSI design problems. Unpublished manuscript, Fermi National Accelerator Laboratory, Batavia, Illinois.

Gradshteyn, I.S. and Ryzhik, I.M. (1963). *Tables of Integrals, Sums, Series and Products.* 4th edn., Fiznatgizdat, Moscow.

Haagerup, U. (1982). The best constants in the Khintchine inequality. *Studia Mathematica* 52, 231–283.

Halton, J.H. and Terada, R. (1982). A fast algorithm for the Euclidean traveling salesman problem, optimal with probability one. *SIAM Journal of Computing* 11, 28–46.

Hausdorff, F. (1919). Dimension und äußeres Maß. *Mathematische Annalen* 79, 157–179.

Hurewicz, W. (1933). Über dimensionser höhende steitige Abbildungen. *Journal für Reine und Angewandte Mathematik* 169, 71–78.

Imai, H. (1986). Worst-case analysis for planar matching and tour heuristics with bucketing techniques and spacefilling curves. *Journal of the Operations Research Society of Japan* 29, 43–67.

Jakubowski, J. and Kwapień, S. (1979). On multiplicative systems of functions. *Bulletin de L'Académie Polonaise des Sciences* 27, 689–694.

Johnstone, I. and Siegmund, D. (1989). On Hotelling's formula for the volume of tubes and Naiman's inequality. *Annals of Statistics* 16, to appear.

Kakutani, S. (1986). *Collected Works.* Ed. R.R. Kallman, Birkhäuser, Boston.

Karp, R.M. (1976). The probabilistic analysis of some new combinatorial search algorithms. In *Algorithms and Complexity: New Directions and Recent Results*, ed. J. Traub, Academic Press, New York, 1–19.

——— (1977). Probabilistic analysis of partitioning algorithms for the traveling salesman problem in the plane. *Mathematics of Operations Research* 2, 209–224.

Karp, R.M. and Steele, J.M. (1985). Probabilistic analysis of heuristics. In *The Traveling Salesman Problem: A Guided Tour of Combinatorial Optimization*, eds. E.L. Lawler et al., John Wiley and Sons, New York, 181–206.

Lalley, S. (1984). Personal communication.

Lawler, E.L., et al. (1985). *The Traveling Salesman Problem: A Guided Tour of Combinatorial Optimization.* Editors, John Wiley and Sons, New York.

Milne, S.C. (1980). Peano curves and smoothness of functions. *Advances in Mathematics* 35, 129–157.

Naiman, D.Q. (1986). Conservative confidence bounds in curvilinear regression. *Annals of Statistics* 14, 896–906.

Paley, R.E.A.C. (1932). A remarkable series of orthogonal functions (I). *Proceedings of the London Mathematical Society* 34, 241–264.

Platzman, L.K. and Bartholdi, J.J. (1988). Spacefilling curves and the planar traveling salesman problem. *Journal of the Association of Computing Machinery*, to appear.

Pólya, G. (1913). Über eine Peanosche Kurve. *Bulletin de l'Académie de Sciences, Cracovie A* 305–313 (also *George Pólya's Collected Works* 3, 1–8,

eds. J. Hersch and G.-C. Rota, MIT Press, Cambridge, Massachusetts).
Rhee, W.T. and Talagrand, M. (1987). Martingale inequalities and NP-complete problems. *Mathematics of Operations Research* 12, 177–181.
——— (1988a). Martingale inequalities, interpolation, and NP-complete problems. *Mathematics of Operations Research*, to appear.
——— (1988b). A sharp deviation for the stochastic traveling salesman problem. *Annals of Probability*, to appear.
Schoenberg, I.J. (1938). On the Peano curve of Lebesgue. *Bulletin of the American Mathematical Society* 44, 519.
Steele, J.M. (1981a). Subadditive Euclidean functionals and non-linear growth in geometric probability. *Annals of Probability* 9, 365–376.
——— (1981b). Complete convergence of short paths and Karp's algorithm for the TSP. *Mathematics of Operations Research* 6, 374–378.
——— (1988). Probabilistic and worst case analyses of some classical problems of combinatorial optimization in Euclidean space. Technical Report, Program in Statistics and Operations Research, Princeton University.
——— (1989). Efficacy of spacefilling heuristics in Euclidean combinatorial optimization. Technical Report, Program in Statistics and Operations Research, Princeton University.

Appendix I. Lalley's Proof of BHH

S. Lalley (1984) provided a remarkable proof of the most interesting case of the Beardwood, Halton, Hammersley theorem. Lalley's previously unpublished proof wins the prize for using minimal machinery. Moreover, his proof serves as a model of the power of similarity arguments and provides a line of attack that is applicable to many other functionals.

Let $U_1, U_2, \ldots$ be independent random variables each having the uniform distribution on $[0,1]^2$, and let L_n be the length of the shortest path through $U_1, U_2, \ldots, U_n$. Observe that L_n is nondecreasing in n. We are to prove that $n^{-1/2}L_n \to C$ a.s. for a constant $C \in (0,\infty)$. For this it suffices to prove that if $N(t)$, $t \geq 0$, is a Poisson process with rate 1, then as $t \to \infty$ we have

$$t^{-1/2}L_{N(t)} \to C \quad \text{almost surely.} \tag{AI.1}$$

Partition the square $[0,1]^2$ into squares $Q_1, Q_2, \ldots, Q_{m^2}$ of side m^{-1}, and define $\lambda_t^m(Q_i)$ to be the length of the shortest path through $\{U_1, U_2, \ldots, U_{N(t)}\} \cap Q_i$. It is easy to see that for each $t > 0$ and each $m = 1, 2, \ldots$, the random variables $\lambda_t^m(Q_1), \lambda_t^m(Q_2), \ldots, \lambda_t^m(Q_{m^2})$ are independent and identically distributed. Moreover, $m\lambda_{m^2t}^m(Q_i)$ has the same distribution as $L_{N(t)}$. Finally, we note $\mathrm{Var}(L_{N(t)}) < \infty$ for each $t \geq 0$ as one can see from the trivial bound $L_{N(t)} \leq 2^{1/2}N(t)$.

LEMMA 1. *For each $t > 0$ and each $m = 1, 2, \ldots,$ we have*

$$-6m + \sum_{i=1}^{m^2} \lambda_t^m(Q_i) \leq L_{N(t)} \leq m\sqrt{5} + \sum_{i=1}^{m^2} \lambda_t^m(Q_i). \tag{AI.2}$$

PROOF: To prove the right inequality we only need to obtain a path through $U_1, U_2, \ldots, U_{N(t)}$. For $1 \leq i \leq m^2$ we first find the shortest path through $\{U_1, U_2, \ldots, U_{N(t)}\} \cap Q_i$ then knit these m^2 paths together by joining endpoints in adjacent squares ordered in snake raster order. Since points in adjacent squares are not separated by a distance greater than $\sqrt{5}\, m^{-1}$, the resulting path has length no greater than $m\sqrt{5} + \sum_{i=1}^{m^2} \lambda_t^m(Q_i)$, establishing the right hand inequality. One should note that it does not hurt this bound if some of the sets $\{U_1, U_2, \ldots, U_n\} \cap Q_i$ are empty.

To prove the left hand inequality consider the shortest path γ through $U_1, U_2, \ldots, U_{N(t)}$. If the two endpoints of γ do not lie in $\bigcup_i \partial Q_i$, extend the path γ so that the endpoints of the extended path $\bar{\gamma}$ lie in $\bigcup_i \partial Q_i$; this can be done in such a way that the length of $\bar{\gamma}$ is bounded by $|\gamma| + 2/m$, where $|\gamma|$ denotes the length of γ. Fix a square Q_i. The intersection $Q_i \cap \bar{\gamma}$ consists of a finite number of paths $\gamma_1, \gamma_2, \ldots, \gamma_k$ in Q_i each having its endpoints on ∂Q_i. Clearly, each point in $\{U_1, U_2, \ldots, U_{N(t)}\} \cap Q_i$ lies in $\bigcup_j \gamma_j$. The paths $\gamma_1, \gamma_2, \ldots, \gamma_k$ may be joined together by cutting and pasting and adding arcs $\beta_1, \ldots, \beta_{k-1}$ on ∂Q_i in such a way that no point in ∂Q_i lies on more than one of $\beta_1, \ldots, \beta_{k-1}$. Consequently, $\lambda_t^m(Q_i) \leq |\gamma_j| + 4/m$. Summing over $i = 1, \ldots, m^2$ yields the left hand inequality. ∎

LEMMA 2. *For each $t > 0$,*

$$\lim_{m \to \infty} m^{-1} \sum_{i=1}^{m^2} \lambda_{m^2 t}^m(Q_i) = EL_{N(t)} \quad \textit{almost surely.}$$

PROOF: This does not quite follow from the strong law of large numbers. But, since for each m the random variables $m\lambda_{m^2 t}^m(Q_1), \ldots, m\lambda_{m^2 t}^m(Q_{m^2})$ are i.i.d. with the same distribution as $L_{N(t)}$, we have by Chebyshev's inequality that

$$P\left\{\left|\sum_{i=1}^{m^2} \frac{\lambda_{m^2 t}^m(Q_i)}{m} - EL_{N(t)}\right| > \epsilon\right\} \leq \frac{\mathrm{Var}(L_{N(t)})}{m^2 \epsilon^2}.$$

The assertion therefore follows from the Borel-Cantelli lemma. ∎

LEMMA 3. *There exists $C \in (0, \infty)$ such that*

$$\lim_{t\to\infty} \frac{EL_{N(t)}}{t^{1/2}} = C.$$

PROOF: Take expectations in (AI.2) and use the fact that $E(m\lambda^m_{m^2t}(Q_i)) = EL_{N(t)}$ to obtain

$$-6 + EL_{N(t)} \le \frac{EL_{N(m^2t)}}{m} \le \sqrt{5} + EL_{N(t)}.$$

It follows that for any $\epsilon > 0$ there exists t sufficiently large that

$$\left| \frac{EL_{N(m^2t)}}{mt^{1/2}} - \frac{EL_{N(t)}}{t^{1/2}} \right| < \epsilon$$

for all $m = 1, 2, \ldots$. Since L_n is nondecreasing in n, this implies that

$$\begin{aligned} E\left(\frac{L_{N(t)}}{t^{1/2}}\right) - \epsilon &\le \liminf_{s\to\infty} \frac{EL_{N(s)}}{s^{1/2}} \\ &\le \limsup_{s\to\infty} \frac{EL_{N(s)}}{s^{1/2}} \\ &\le \frac{EL_{N(t)}}{t^{1/2}} + \epsilon. \end{aligned}$$

Since $\epsilon > 0$ is arbitrary, it follows that $EL_{N(s)}/s^{1/2} \to C$ as $s \to \infty$ for some $0 \le C < \infty$. To prove that $C > 0$, note that $EL_{N(t)} \to \infty$ as $t \to \infty$, by an elementary argument. Choose t sufficiently large that $EL_{N(t)} > 4$; then (AI.4) implies that

$$\liminf_{m\to\infty} \frac{EL_{N(m^2t)}}{m} > 0. \qquad \blacksquare$$

The proof of (AI.1) may now be completed. By (AI.2), for each $t > 0$ and $m = 1, 2, \ldots,$

$$\begin{aligned} -6 + m^{-1}\sum_{i=1}^{m^2} \lambda^m_{m^2t}(Q_i) &\le \frac{L_{N(m^2t)}}{m} \\ &\le \sqrt{5} + m^{-1}\sum_{i=1}^{m^2} \lambda^m_{m^2t}(Q_i), \end{aligned}$$

so Lemma 2 implies that almost surely

$$\begin{aligned}
-6t^{-1/2} + \frac{EL_{N(t)}}{t^{1/2}} &\leq \liminf_{m\to\infty} \frac{L_{N(m^2t)}}{mt^{1/2}} \\
&\leq \limsup_{m\to\infty} \frac{L_{N(m^2t)}}{mt^{1/2}} \\
&\leq \sqrt{5}\,t^{-1/2} + \frac{EL_{N(t)}}{t^{1/2}}.
\end{aligned}$$

Now Lemma 3 implies that if t is sufficiently large then almost surely

$$\begin{aligned}
C - \epsilon &\leq \liminf_{m\to\infty} \frac{L_{N(m^2t)}}{mt^{1/2}} \\
&\leq \limsup_{m\to\infty} \frac{L_{N(m^2t)}}{mt^{1/2}} \\
&\leq C + \epsilon.
\end{aligned}$$

Since L_n is nondecreasing in n it follows that

$$\begin{aligned}
C - \epsilon &\leq \liminf_{s\to\infty} \frac{L_N(s)}{s^{1/2}} \\
&\leq \limsup_{s\to\infty} \frac{L_{N(s)}}{s^{1/2}} \\
&\leq C + \epsilon
\end{aligned}$$

almost surely. Now (AI.1) follows by letting $\epsilon \to 0$. ∎

Appendix II. Paley's Square Function Argument

This Appendix develops an argument for martingales that was introduced in Paley (1932) for Walsh functions. The only real changes made here to Paley's method are those required to provide explicit bounds on the basic constant. As one should expect, the constant is not as sharp as that given in Burkholder (1973), but the reason for reviewing Paley's argument is rather to show how the maximal function can be used to bound L_p norms of martingales. Other features of the proof are discussed at the end of the appendix.

Consider a martingale difference sequence $\{y_i : 0 \leq i \leq n\}$ with $y_0 \equiv 0$ and its associated martingale $M_k = y_1 + y_2 + \cdots + y_k$, $1 \leq k \leq n$. To keep to the essentials, we will stick to the case of even integers p. We first compute the difference sequence of pth moments:

$$\begin{aligned} E\{M_{k+1}^p - M_k^p\} &= E\{(M_k + y_{k+1})^p - M_k^p\} \\ &= E\left\{p y_{k+1} M_k^{p-1} + \binom{p}{2} y_{k+1}^2 M_k^{p-2} + \cdots + y_{k+1}^p\right\} \\ &= E\left\{\binom{p}{2} y_{k+1}^2 M_k^{p-2} + \binom{p}{3} y_{k+1}^3 M_k^{p-3} + \cdots + y_{k+1}^p\right\} \end{aligned} \tag{AII.1}$$

where only in the last inequality is the martingale property invoked. We then use Hölder's inequality on the right hand side to bring the powers of M_k up to the same level. Specifically, for $3 \le j \le p-1$ we use

$$E y_{k+1}^j M_k^{p-j} \le (E y_{k+1}^2 M_k^{p-2})^\theta (E y_{k+1}^p)^{1-\theta} \tag{AII.2}$$

where $\theta = (p-j)/(p-2)$. Since $0 < \theta \le 1$, inequality (AII.2) can be relaxed to

$$E y_{k+1}^j M_k^{p-j} \le E y_{k+1}^2 M_k^{p-2} + E y_{k+1}^p, \tag{AII.3}$$

so we can crudely bound the sum of the binomial coefficients to find

$$\left|E\{M_{k+1}^p - M_k^p\}\right| \le 2^p \left\{E y_{k+1}^2 M_k^{p-2} + E y_{k+1}^p\right\}. \tag{AII.4}$$

Finally, we sum over $0 \le k < n$ to find

$$\begin{aligned} E M_n^p &\le 2^p E\left\{\left(\sum_{k=1}^n y_k^2\right) \max_{1\le k\le n} M_k^{p-2}\right\} + 2^p E \sum_{k=0}^{n-1} y_{k+1}^p \\ &\le 2^p \left\{E\left(\sum_{k=1}^n y_k^2\right)^{p/2}\right\}^{2/p} \left(E \max_{1\le k\le n} M_k^p\right)^{(p-2)/p} + 2^p E\left(\sum_{k=1}^n y_k^2\right)^{p/2} \end{aligned} \tag{AII.5}$$

where in the first summand we used Hölder's inequality, and in the second summand we used the elementary real variable inequality for $p \ge 2$

$$a_1^p + a_2^p + \cdots + a_n^p \le (a_1^2 + a_2^2 + \cdots + a_n^2)^{p/2}.$$

Our motivation for moving from (AII.4) to (AII.5) is to use Doob's maximal inequality, or rather its consequence for $1 < p < \infty$ that

$$\left\{E\left(\max_{1\le k\le n} |M_k|^p\right)\right\}^{1/p} \le q(E M_n^p)^{1/p}, \tag{AII.6}$$

where q is the conjugate index to p (i.e. $q = p/(p-1)$). From (AII.6) we thus find

$$\|M_n\|_p^p \le q2^p \left\| \left(\sum_{k=1}^n y_k^2 \right)^{1/2} \right\|_p^2 \|M_n\|_p^{p-2} + 2^p \left\| \left(\sum_{k=1}^n y_k^2 \right)^{1/2} \right\|_p^p . \tag{AII.7}$$

Finally, we note inequality (AII.7) is of the form $x^p \le ay^2x^{p-2} + by^p$ which implies $x \le \{(2a)^{1/2} + (2b)^{1/p}\}y$, so we find our modest version of Burkholder's inequality for even integers p:

$$\|M_n\|_p \le \alpha_p \left\| \left(\sum_{k=1}^n y_k^2 \right)^{1/2} \right\|_p \tag{AII.8}$$

where $\alpha_p \le q^{1/2}2^{(p+1)/2} + 2^{(p+1)/2} \le q^{1/2}2^{(p+3)/2}$. ■

The constant α_p is larger than the $18q^{1/2}p$ we know to be sufficient, so some comment seems needed to justify our enthusiasm for this more-than-fifty year old argument. First, it uses very little about martingales; e.g. in (AII.1) we use a weak consequence of the definition, and the only other fact we need is a maximal inequality of Doob's type as given in (AII.6). Second, the differencing applied to pth powers in (AII.1) can be applied to other functions f of M_k, provided that $f(M_k + y_k) - f(M_k)$ can be bounded by a useful expression. Finally, since the argument is free of stopping times, its parts are amenable to more individual attention. In particular, the use of bounds on $\|y_i\|_p$, $1 \le p \le \infty$, can be tried out in AII.5, AII.6, or AII.7.

Added in Proof: The idea of using a spacefilling curve to sequence visits to points in the square is evidently much older than recent references seem to indicate. From the comments of R. Adler in the *Collected Works of S. Kakutani* (Kakutani 1986, V.II, p. 445), Kakutani had presented the idea as early as the spring of 1966.

Program in Statistics and Operations Research
School of Engineering and Applied Science
Princeton University
Princeton
New Jersey 08544.

The Computational Complexity of Some Classical Problems from Statistical Physics

D.J.A. Welsh

1. Introduction

The motivation of this paper is to attempt to review and classify the difficulty of a range of problems, arising in the statistical mechanics of physical systems and to which I was introduced by J.M. Hammersley in the early sixties. Their common characteristics at the time were that they all seemed hard and there was little existing mathematical machinery which was of much use in dealing with them. Twenty years later the situation has not changed dramatically; there do exist some mathematical techniques which appear to be tools in trade for this area, subadditive functions and transfer matrices for example, but they are still relatively few and despite a great deal of effort the number of exact answers which are known to the many problems posed is extremely small. Below we shall attempt to explain why this should be so by showing how the problems originally studied are special cases of a wide range of problems which can, in a well defined sense, be regarded as the most intractable enumeration problems that can sensibly be posed.

We do this by relating the problems to their position in the hierarchy of computational complexity theory. While concepts such as P (polynomial time) and NP (nondeterministic polynomial time) have rapidly become commonplace ideas in mathematics since their introduction via Cook's theorem in 1970, the counting analogue of NP, denoted by #P, introduced by Valiant (1979a) and like NP also having *complete* or hardest problems has received less attention. As we shall see, most of the natural problems arising in statistical physics can be described in this framework and as first pointed out by Valiant (1979a,b) and Jerrum (1981, 1987) this goes a long way towards explaining their apparent intractability. There are however what can almost be described as 'pockets of resistance', for example some of these problems do have 'exact' solutions for some 2-dimensional lattices, and whether or not this is a phenomenon of dimension or planarity or the very special nature of the lattice is an interesting and unanswered question to which we return at the conclusion.

2. The Statistical Physics Problems

In this section I define slightly generalised versions of the main problems from statistical physics which we shall be considering. By 'generalised' I mean that instead of formulating them as problems on one of the standard lattices the underlying structure will be a general graph. The graph theoretic notation will be standard (see Bondy and Murty 1976).

Percolation Theory

As originally propounded by Broadbent and Hammersley (1957) this is concerned with the spread of blight through a medium in which the elements of the medium independently permit or fail to permit passage. More formally this can be described as follows.

Let G be an arbitrary undirected graph and let p, $0 \leq p \leq 1$ be fixed. Suppose now that each edge of G is, independently of each other edge, removed with probability $q = 1-p$. Denote the resulting (random) subgraph of G by ω and let $P(G;p)$ denote the probability that ω is spanning, that is that in ω it is possible to move from any vertex of G to any other. We call $P(G;p)$ the *percolation probability* of G. It is clearly a measure of the reliability or vulnerability of G regarded as a communication network and has the standard S-shaped curve as p varies between 0 and 1, and is called the *all terminal reliability* by Provan and Ball (1983).

When G is a lattice we let $P(p)$ denote the probability that the component of ω which contains the origin is infinite. It is easy to see that there exists a *critical probability* p_c defined by

$$p_c = \inf\{p : P(p) > 0\}.$$

Determining p_c exactly is extraordinarily difficult; see for example the proof by Kesten (1980) that $p_c = \frac{1}{2}$ for the square lattice. As far as I am aware exact results are known only for some 2-dimensional lattices. For further details we refer to the monographs of Kesten (1982) and Grimmett (1989).

The Ising Model

This is a problem of long standing and can be defined for a general graph as follows. Let σ be an assignation of *positive* $(+1)$ and *negative* (-1) spins to the vertices of a graph G. The *interaction energy* $E(\sigma)$ is defined by

$$E(\sigma) = -J \sum_{i \sim j} \sigma_i \sigma_j - H \sum_i \sigma_i$$

where the first summation is only over i, j which are adjacent in G, H is the *external magnetic field* and J is the *coupling constant.* The *partition function* Z is then given by

$$Z(G) = \sum_{\sigma} \exp(-\beta E(\sigma))$$

where the sum is over all possible spin configurations. The fundamental problem is to determine $Z(G)$, though as far as the applications to physics are concerned it would suffice to find the thermodynamic limit of Z as G moved through an increasing sequence of subgraphs of the d-dimensional lattice. The Onsager solution (see Percus 1971) for the case $d = 2$ with zero external magnetic field is the classic result of this area. As yet there is no extension known in higher dimensions.

Self Avoiding Walks

The basic question about self avoiding walks on a lattice is to determine $f(n)$, the number of paths starting at the origin, having n edges, and visiting no point more than once. Hammersley (1957) used subadditivity to prove the existence of a constant θ such that

$$\lim_{n \to \infty} [f(n)]^{1/n} = \theta.$$

The constant θ clearly depends on the lattice, but even for the 2-dimensional square lattice its value is not known exactly; the best exact bounds give only $2.58\ldots \leq \theta \leq 2.72\ldots$. The natural generalisation of this to a general graph G is to let $W_n(G)$ be the number of paths in G of length n and which pass through each vertex at most once. When $n + 1$ equals the number of vertices of G this is the well known problem of counting the Hamiltonian paths of G.

Animals or Polyominoes

Conceptually very close to self avoiding walks, counting animals or polyominoes has been for a long time a popular if frustrating sport, see for example the article by Whittington in this volume. Although usually defined for lattices we can define an *animal* of *size* n on an arbitrary graph G to be any subset X of the vertex set $V(G)$ such that the subgraph of G induced by X is connected and $|X| = n$. We let $a_n(G)$ denote the number of such animals. When G is a lattice $\mathcal{L}$ it is easy to use subadditivity to prove that

$$\lim_{n \to \infty} [a_n(\mathcal{L})]^{1/n} = a(\mathcal{L})$$

exists, but determining this limit exactly, or merely obtaining close bounds, again seems to be extraordinarily difficult, even for the square lattice.

The Monomer-Dimer Problem

This arises in the study of physical systems involving diatomic molecules (*dimers*). For a general graph G a formal description of the problem is to ask for the number of arrangements of N_1 dimers and N_2 monomers on the edges and vertices of G such that each dimer is placed on an edge, each monomer on a vertex, and each vertex of G is either occupied by exactly one monomer or is the endvertex of exactly one dimer. Clearly for this to be possible

$$2N_1 + N_2 = |V(G)| = N.$$

The ratio N_2/N is called the *monomer density.*

When there are no monomers we have what is known as the *dimer problem*, it can be rephrased as counting the number of perfect matchings of a graph (a *matching* is a set of edges no two of which share a common vertex, it is a *perfect matching* of G if each vertex of G is the endpoint of one edge of the matching).

Similarly the *monomer dimer problem* is exactly the problem of counting the number of matchings of a given size in a graph.

Ice-Type Models

The simplest ice-type model can be described as follows. Let G be any regular 4-valent graph and let Z_{ice}, the partition function, count the number of orientations of the edges of G which satisfy the rule that the number of arrows into each vertex equals the number of arrows out. In graph theoretic terminology Z_{ice} counts the number of Eulerian orientations of G. Details of the physical motivation for this and a description of a range of ice-type models can be found in Baxter (1982). A remarkable result about the ice model is that of Lieb (1967) who showed that if $Z_{\text{ice}}(m, n)$ denotes the ice partition function on the $m \times n$ section of the square lattice then

$$\lim_{m,n\to\infty} [Z_{\text{ice}}(m, n)]^{1/mn} = (4/3)^{3/2}.$$

No extension of this to higher dimensions is known.

The q-State Potts Model

This is naturally defined for any graph G and positive integer q as follows. A *state* σ of the vertex set of G is a function which assigns to each vertex i of G a *spin* σ_i, where $\sigma_i \in \{1, 2, \ldots, q\}$. The *energy* associated with state σ is defined to be

$$E(\sigma) = -J \sum \delta(\sigma_i, \sigma_j)$$

where the summation is over all distinct $i, j \in V(G)$ which are joined by an edge and δ is the usual delta function taking values 1 and 0 depending on whether σ_i equals σ_j or not.

The *partition function* $Z(G)$ is then defined by

$$Z(G) = \sum_{\sigma} \exp\{-KE(\sigma)/J\}$$

where J, K are physical constants, see Baxter (1982). For $q = 2$ it is just the Ising model.

3. Computational Complexity

The basic notions of computational complexity are now familiar concepts in most branches of mathematics. One of the main purposes of the theory is to classify and explain the gap that seems to separate tractable computational problems from the apparently intractable. Deciding whether or not P = NP is probably the most important problem in theoretical computer science. The extension of these ideas to enumeration problems has received less attention and we will briefly review the main concepts here.

We regard a computational (enumeration) problem as a function mapping inputs to solutions, (graphs to the number of their 3-vertex colourings for example). A problem is *polynomial time computable* if there exists an algorithm which computes the function in a length of time (number of steps) bounded by a polynomial in the size of the problem instance. The class of such problems we denote by P. If A and B are two problems we say that A is *polynomial time Turing reducible* to B, written $A \propto B$, if it is possible with the aid of a subroutine for problem B to solve A in polynomial time, in other words the number of steps needed to solve A (apart from calls to the subroutine for B) is polynomially bounded.

The class #P can be described informally as the class of enumeration problems in which the structures being counted are recognisable in polynomial time. In other words there is an algorithm which runs in polynomial time and which will verify that a given structure has the form needed to be included in the count. For example counting hamiltonian paths in a graph is in #P because it is easy to check in polynomial time that a given set of edges is a hamiltonian path.

Like NP, #P has a class of 'hardest' problems called the #P-*complete* problems. They can be formally described by, problem A belonging to #P is #P-complete if for any other problem $B \in$ #P, we have $B \propto A$. The classic example of a #P-complete problem is counting truth assignments of a Boolean function. This consists of

INPUT: A Boolean formula ϕ in variables $x_1, x_2, \ldots, x_n$ and the connectives $\vee$, $\wedge$, $\neg$.

QUESTION: How many distinct assignments of truth values to the x_1, $x_2, \ldots, x_n$ make ϕ true?

The #P-complete problems tend to be the enumerative counterparts of NP-complete problems though it has to be emphasized that there is no exact formulation of this remark.

As with NP, we define a problem to be #P-*hard* if any problem in #P is polynomial time reducible to it. In other words A is #P-hard if the existence of a polynomial time algorithm for A would imply the existence of a polynomial time algorithm for any problem in #P.

It is clear from this that describing a problem as #P-hard or #P-complete is very strong evidence of its inherent intractability. There are now several thousand problems known to be #P-complete. A polynomial time algorithm for any one of them would imply #P = P and this in turn would imply NP = P. For a more precise formulation we refer to Garey and Johnson (1979).

4. The Complexity of the Physical Problems

We now turn to an examination of the status in the complexity hierarchy of the previously discussed physical problems.

Self Avoiding Walks

As defined, counting the number of self avoiding walks of n steps on a graph G of $n+1$ vertices is exactly the problem of counting the number of Hamiltonian paths in G. Recall that a path is *Hamiltonian* if it visits each vertex exactly once. This is one of the classical #P-complete problems and is known to be #P-complete even when restricted to planar graphs with maximum degree 3.

It is not surprising therefore that no exact result about self avoiding walks seems to be known except for tree like structures such as Bethe lattices.

The Dimer Problem

For a general graph this is exactly the problem of counting perfect matchings. As far as complexity is concerned it is probably the most intriguing of the problems discussed in that there is a clear cut distinction between planar and nonplanar structures. This is because of the following statements which are partial restatements of classic theorems of Kasteleyn (1967) and Valiant (1979a).

(1) Counting perfect matchings in a planar graph can be done in polynomial time.

(2) Counting perfect matchings in a general graph is a #P-complete problem.

The difference between the two statements (1) and (2) is partially explained by the following observation.

(3) For planar graphs, counting perfect matchings reduces to evaluating

the Pfaffian of a matrix and this is equivalent to evaluating a determinant. For general graphs the problem is equivalent to evaluating the permanent of a 0–1 matrix.

The *permanent* of a matrix A is just the expansion of the determinant of A with all terms having positive signs. Paradoxically (at least at the naive level) this makes it hard to compute.

The method of dealing with planar graphs is an extension of the method developed by Kasteleyn (1961) and Temperley and Fisher (1961) to show that on the 2-dimensional square lattice, if $f(N)$ denotes the number of dimer coverings of an $N \times N$ section, then

$$\lim_{n \to \infty} [f(N)]^{1/N^2} = e^{2G/\pi} = 1.791622\ldots$$

where G is Catalan's constant given by

$$G = \sum_{k=0}^{\infty} \frac{(-1)^k}{(2k+1)^2}.$$

The fact that in higher dimensions the lattices are nonplanar would suggest in view of (2), that it will be exceedingly difficult to obtain any such exact result.

The Monomer-Dimer Problem

Even for planar graphs the general version of the monomer dimer problem is #P-complete. This was first shown by Jerrum (1981, 1987), who showed that counting the total number of matchings in a graph is #P-complete. Its apparent intractability goes someway towards explaining the paucity of exact results. As far as I am aware there have not been significant improvements for the 2-dimensional square lattice over the rather weak bounds given in Bondy and Welsh (1966) and Hammersley and Menon (1970).

We next turn to the remaining problems under discussion, namely percolation, Ising, Potts and the ice problem. It turns out that they can all be regarded as specific evaluations of a well known graph polynomial. We treat this briefly first.

The Dichromate or Whitney-Tutte Polynomial

A crucial concept in what follows is the following graph polynomial introduced by Tutte (1947). It is closely related to the rank generating function introduced by Whitney (1932) and has a natural extension to vector spaces and matroids. In this context it has interpretations as the weight enumerator of a linear code (see Welsh 1976) and has recently been observed to have considerable significance in the theory of knots, see for example Kauffman (1987) or Lickorish (1988).

However in this article we will restrict attention to graphs and then it can be fairly simply defined as follows.

Let G be a graph with edge set E. For any subset A of E we define the *rank* $r(A)$ by

$$r(A) = v(A) - k(A)$$

where $v(A)$ is the number of vertices of G incident with A and $k(A)$ is the number of (connected) components of the subgraph spanned by A. Then define the *Tutte polynomial* of G to be the 2-variable polynomial

$$T(G; x, y) = \sum_{A \subseteq E} (x-1)^{r(E)-r(A)}(y-1)^{|A|-r(A)}. \tag{4.1}$$

Hence if I is an isthmus and L denotes a loop

$$T(I; x, y) = x, \ T(L; x, y) = y. \tag{4.2}$$

This, together with the recursion formulae

$$T(G; x, y) = T(G'_e; x, y) + T(G''_e; x, y) \tag{4.3}$$

whenever e is not an isthmus or a loop, effectively determines T uniquely. Here G'_e and G''_e are the graphs obtained from G by respectively deleting and contracting the edge e. When e is an isthmus or loop replacing (4.2) by

$$T(G; x, y) = \begin{cases} xT(G'_e; x, y) & e \text{ an isthmus} \\ yT(G'_e; x, y) & e \text{ a loop} \end{cases} \tag{4.4}$$

gives a complete recursion formula for calculating T for any graph G.

A more striking property of the Tutte polynomial is the following. A function f defined on the set of all graphs is an *invariant* if whenever G_1 and G_2 are isomorphic $f(G_1) = f(G_2)$. A special case of the main result of Oxley and Welsh (1979) is the following.

THEOREM 1. *Let f be a graph invariant taking values in a commutative ring R satisfying for some $a, b \in R$, the relation*

$$f(G) = af(G'_e) + bf(G''_e) \tag{4.5}$$

when e is not a loop or isthmus, and

$$f(G) = f(\{e\})f(G'_e)$$

when e is a loop or isthmus. Then f is given by

$$f(G) = a^{|E|-|V|+1}b^{|V|-1}T(G; x/b, y/a). \tag{4.6}$$

Using this theorem it is now easy to prove that all the remaining problems of statistical physics described in the last section can be reduced to evaluating the Tutte polynomial of the graph along particular curves in the x, y-plane. The proof technique is just to verify that the quantity in question, be it partition function, probability, or enumeration satisfies a recursive formula of the type (4.5) for suitable a and b.

This method gives the following interpretations of T.

Percolation

The percolation probability $P(G;p)$ is given by

$$P(G;p) = q^{|E|-|V|+1}p^{|V|-1}T(G;1,q^{-1})$$

for any connected graph G and where $q = 1 - p$.

The Ising and Potts Models

In the absence of an external magnetic field the Ising model is the special case of the Potts model defined with $q = 2$. It is straightforward to use the recursion formula (4.5) to verify that the general partition function Z is given by an evaluation of the Tutte polynomial T along the hyperbola $(x-1)(y-1) = 2$. It is perhaps easier to see this if Z is reparameterised in the following form. Let $A(\sigma)$ denote the sets of edges of G which have both endpoints the same sign under σ. Let $B(\sigma)$ be the complementary set of edges, then the generalised partition function

$$Z(G;\theta,\phi) = \theta^{|E|-|V|+1}(\theta-\phi)^{|V|-1}T\left(G;\frac{\theta+\phi}{\theta-\phi},\frac{\theta}{\phi}\right) \tag{4.7}$$

where

$$\theta = e^{\beta}, \phi = e^{-\beta}, \beta = J/kT$$

where J is the interaction strength, T is temperature and k is Boltzmann's constant.

Again using the recursion formula (4.5) it is straightforward to check that for the q-state Potts model, the partition function is given by

$$Z(G;q,v) = q^{n}T\left(G;\frac{q+v}{v},v+1\right) \tag{4.8}$$

where n is the number of vertices, and v is the parameter defined by $v+1 = \exp(-1/kT)$.

In other words the partition function of the Potts model is, up to an easily determined constant, the Tutte polynomial of G evaluated along the hyperbola $H_q \equiv (x-1)(y-1) = q$.

The relation between the above models and the Tutte polynomial seems to have been first noticed by Fortuin and Kasteleyn (1972), though their proofs are different from the method indicated above and they were using an equivalent (up to a change of variable) polynomial due to H. Whitney (1932).

The Ice Model

Lenard (see Lieb 1967) showed that determining the partition function $Z_{\text{ice}}(G)$ for any planar 4-valent graph G was equivalent to counting the 3 colourings of the faces of G in such a way that no two faces with a common edge are given the same colour. But then by a standard result linking colourings and evaluations of the Tutte polynomial we have

$$Z_{\text{ice}}(G) = T(G; 0, -2) \tag{4.9}$$

and in fact it is easy to prove directly from the recursion formula (4.3) that (4.9) holds for all (not necessarily planar) 4-valent graphs.

It follows from the above observations that whenever the evaluation of the Tutte polynomial is 'easy' then so are each of the above problems. However, for general graphs determination of the Tutte polynomial or even evaluation at a particular point has been proved to be #P-hard except in very special cases

This follows from results of Jaeger, Vertigan, and Welsh (1989), a special case of which is the following theorem.

THEOREM 2. *Evaluating the Tutte polynomial of a graph at a particular point of the complex plane is #P-hard except when either*
(a) the point lies on the hyperbola $(x-1)(y-1) = 1$, *or*
(b) the point is one of the special points $(1,1), (-1,0), (0,-1), (-1,-1), (i,-i), (-i,i), (j,j^2), (j^2,j)$ *where* $j = e^{2\pi i/3}$.
In the special cases the evaluation can be carried out in polynomial time.

As far as the physical problems are concerned the special points and special hyperbola seem to have no significance. There are combinatorial interpretations of $T(G; x, y)$ at each of the points, the most interesting being at $(1,1)$ where T counts the number of spanning trees of the graph for which there is the well known Kirchhoff determinantal formula.

5. Approximations, Monte Carlo Methods and Randomised Algorithms

One result of the evidence of intractability of most of these problems as propounded in the last section is that good approximation techniques assume even greater importance. Monte Carlo methods have long been a favoured approach to many of these; see for example Hammersley and Handscomb (1964). However, until very recently there has been very little known about the rate, or accuracy of convergence of what can loosely be described as probabilistic methods of approximation.

Very recently, however, there has been a notable breakthrough by Jerrum and Sinclair (1988) on one specific problem, and the methods used have the capability of extension to other #P-hard problems.

We now describe what we mean by a randomised approximation scheme. First, for any real numbers a, $\hat{a}$ and $r \geq 1$ we say that $\hat{a}$ *approximates* a *within ratio* r if

$$\hat{a}/r \leq a \leq \hat{a}r.$$

A *fully polynomial randomised approximation scheme*, abbreviated to fpras for a function $f : \Sigma^* \to \mathbb{N}$ is a randomised algorithm which when presented with a string $x \in \Sigma^*$ and a real number $\epsilon > 0$ runs in time which is polynomial in $|x|$ and ϵ^{-1} and with probability at least $\frac{1}{2} + \delta$ $(\delta > 0)$ its output approximates $f(x)$ within ratio $1 + \epsilon$. (By $|x|$ we mean the number of elements in or length of the string x, and as usual Σ^* is the set of strings of symbols from the finite alphabet Σ.)

It is not difficult to see that the existence of a fpras means the existence (in a precise mathematical sense) of a fast, good approximation algorithm, and what Jerrum and Sinclair have done is to show the existence of such an algorithm for determining the partition function of the monomer-dimer problem. We now sketch the ideas of their method.

Let $(X_t : 0 \leq t < \infty)$ be a finite state, ergodic, time homogeneous Markov chain M with transition matrix $P = (p_{ij})$ and having stationary distribution $\pi = (\pi_i)$.

The *relative pointwise distance* $\Delta(t)$ is defined by

$$\Delta(t) = \max_{i,j} \frac{|p_{ij}(t) - \pi_j|}{\pi_j}$$

where as usual $p_{ij}(t)$ represents the t–step transition probability. Thus $\Delta(t)$ is a measure of the rate of convergence of the Markov chain to its stationary distribution and the ideal is a situation where $\Delta(t)$ converges to zero exponentially fast as a function of time.

Now suppose that the ergodic chain M is time reversible so that it satisfies the balance condition $p_{ij}\pi_i = p_{ji}\pi_j$, $\forall i, j$. We associate with M an undirected weighted graph $G(M)$ in which the vertices are the states of M, the edges join all pairs of states with $p_{ij} > 0$ and the *weight* w_{ij} of the edge (i, j) is given by

$$w_{ij} = p_{ij}\pi_i = p_{ji}\pi_j.$$

A measure of the rate at which the Markov chain can move around its state space is the **conductance** Φ defined as follows. For any set A of vertices of $G(M)$ let ∂A denote the set of edges having exactly one endpoint in A and then let $\omega(\partial A)$ be the sum of the weights of these edges. Then

$$\Phi = \min\left(\omega(\partial A)\Big/\sum_{i\in A}\pi_i\right)$$

where the minimum is taken over all sets A of states for which

$$0 < \sum_{i\in A}\pi_i \leq \tfrac{1}{2}.$$

The main result of Sinclair and Jerrum (1988) is the following.

THEOREM 3. *If M is a time reversible ergodic chain with $p_{ii} \geq \frac{1}{2}$ for each i, then the relative pointwise distance $\Delta(t)$ satisfies*

$$\Delta(t) \leq (1-\Phi^2/2)^t/\pi_{\min}$$

where $\pi_{\min}$ is the minimum of the stationary state probabilities.

There are a few points to note about Theorem 3. First, the condition $p_{ii} \geq \frac{1}{2}$ though strange, is technical, and can be introduced into any chain by replacing P by $(I+P)/2$, this leaves the stationary distribution unchanged and reduces the conductance by a factor of $\frac{1}{2}$. This is immaterial in the sort of situations in which it is used, namely to prove:

THEOREM 4. *There exists a fully polynomial randomised approximation scheme for counting the number of weighted matchings in a graph.*

In other words the partition function of the monomer dimer problem can be approximated accurately, quickly.

The basic idea underlying the proof of Theorem 4 is to set up an appropriate Markov chain which can be proved to be rapidly mixing. For the monomer dimer (or counting matchings) problem on a graph G the Markov chain M will have as its states the matchings of G and transitions between states are carried out according to the following rules:

Let I be a particular matching or state of M and let e be any edge of G. The pair (I, e) determines a new state J by the formulae:

(a) $e \in I, J = I\backslash e$,

(b) if $e \notin I$ and $I \cup e$ is a matching then $J = I \cup e$,

(c) if $e \notin I$ but exactly one endpoint of e is covered by an edge e' of I then $J = (I\backslash e') \cup e$.

The Markov chain M is of the Metropolis type with transition probabilities determined by choosing edges of G at random and then adopting the change from I to J with probability $\frac{1}{2}$. When modified to make the self loop probabilities no smaller than $\frac{1}{2}$ as indicated, Jerrum and Sinclair show that its conductance Φ satisfies

$$\Phi \geq \frac{1}{8|E(G)|}$$

and hence using Theorem 1 the chain M is rapidly mixing.

The broad idea of the counting algorithm is given by the steps A1–3.

A1: Let the Markov chain M run for time long enough to generate a random sample of 'approximately random' members of the stationary distribution of matchings.

A2: Use the fact that for a specific edge e the partition function $Z(G)$ can be written as

$$Z(G) = Z(G^+) + Z(G^-)$$

where G^-, is the graph obtained from G by deleting $e = (u, v)$ and G^+ is the graph obtained from G by removing e, u, v and all edges incident with u, v.

A3: Let z^+, z^- be the number of members of the sample which contain the specific edge e and use these to estimate $Z(G^+)$, $Z(G^-)$. From these we can recursively estimate $Z(G)$.

The important point is that because the chain is rapidly mixing the method works in the sense that to obtain a final estimate which approximates $Z(G)$ within a ratio $1 + \epsilon$ with probability at least $\frac{3}{4}$ the sample size required is only $O(|E|^3\epsilon^{-2})$ where $E = E(G)$.

6. Conclusion

As far as I am aware all the exact results which have been proved for any of the physical problems have been for some of the 2-dimensional planar lattices. Accordingly one might suspect that it is planarity which makes things easier. However, from the viewpoint of complexity this cannot be the case, and further work extending Theorem 2 by Vertigan (1989) shows that except at a few very special points the Tutte polynomial of planar graphs is #P-hard to compute. Hence if one believes the thesis that exact results about #P-hard problems are in general almost impossible to obtain one is led to ask what additional properties of the 2-dimensional square lattices makes possible the exact results obtained for the Ising, ice, dimer and percolation problems on this particular lattice. It is doubtful if there is an easy answer to this problem. In this context it should be emphasized that calculating the asymptotic limit of a particular sequence of graph functions may be a much easier problem than the exact evaluation problem.

We close with the following:

Problem: Is there any way of extending the Jerrum-Sinclair randomised approximation approach to any of the other physical problems?

As far as I am aware the only other problems to which the method has so far been applied successfully is to estimating the volume of convex bodies (Dyer, Frieze, and Kannan 1988) and generating random graphs (Jerrum and Sinclair 1988b). Ideally we would like to be able to prove that the Metropolis type Monte Carlo methods developed in Hammersley and Handscomb (1964) for example are based on rapidly mixing Markov chains. This may be the case but proving it could be very difficult.

REFERENCES

Baxter, R.J. (1982). *Exactly Solved Models in Statistical Mechanics.* Academic Press, London.

Bondy, J.A. and Murty, U.S.R. (1976). *Graph Theory with Applications.* Elsevier, New York, and Macmillan, London.

Bondy, J.A. and Welsh, D.J.A. (1966). A note on the monomer dimer problem. *Proceedings of the Cambridge Philosophical Society* 62, 503–505.

Broadbent, S.R. and Hammersley, J.M. (1957). Percolation processes I. Crystals and mazes. *Proceedings of the Cambridge Philosophical Society* 53, 629–641.

Dyer, M., Frieze, A., and Kannan, R. (1988). A random polynomial time algorithm for approximating the volume of convex bodies. Preprint.

Fortuin, C.M. and Kasteleyn, P.W. (1972). On the random-cluster model. I. Introduction and relation to other models. *Physica* 57, 536–564.

Garey, M.R. and Johnson, D.S. (1979). *Computers and Intractability — A Guide to the Theory of NP-Completeness.* Freeman, San Francisco.

Grimmett, G.R. (1989). *Percolation.* Springer-Verlag, New York.

Hammersley, J.M. (1957). Percolation processes II. The connective constant. *Proceedings of the Cambridge Philosophical Society* 53, 642–645.

Hammersley, J.M. and Handscomb, D.C. (1964). *Monte Carlo Methods.* Methuen, London.

Hammersley, J.M. and Menon, V.V. (1970). A lower bound for the monomer-dimer problem. *Journal of the Institute of Mathematics and its Applications* 6, 341–364.

Jaeger, F., Vertigan, D.L., and Welsh, D.J.A. (1989). On the computational complexity of the Jones and Tutte polynomials. To appear.

Jerrum, M.R. (1981). *The Complexity of Evaluating Multivariate Polynomials.* Ph. D. Thesis, University of Edinburgh.

——— (1987). 2-Dimensional monomer-dimer systems are computationally intractable. *Journal of Statistical Physics* 48, 121–134.

Jerrum, M.R. and Sinclair, A.J. (1988a). Approximating the permanent. Preprint, University of Edinburgh CSR-275-88.

——— (1988b). Fast uniform generation of regular graphs. Preprint, University of Edinburgh CSR-281-88.

Kasteleyn, P.W. (1961). The statistics of dimers on a lattice. *Physica* 27, 1209–1225.

——— (1963). Dimer statistics and phase transitions. *Journal of Mathematical Physics* 4, 287–293.

——— (1967). Graph theory and crystal physics. In *Graph Theory and Theoretical Physics*, ed. F. Harary, Academic Press, London, 43–110.

Kauffman, L. (1987). *On Knots.* Princeton University Press, Princeton, New Jersey.

Kesten, H. (1980). The critical probability of bond percolation on the square lattice equals $\frac{1}{2}$. *Communications in Mathematical Physics* 74, 41–59.

——— (1982). *Percolation Theory for Mathematicians.* Birkhäuser, Boston.

Lickorish, W.B.R. (1988). Polynomials for links. *Bulletin of the London Mathematical Society* 20, 558–588.

Lieb, E.H. (1967). Residual entropy of square ice. *Physical Review* 162, 162–171.

Oxley, J.G. and Welsh, D.J.A. (1979). The Tutte polynomial and percolation. In *Graph Theory and Related Topics*, ed. J.A. Bondy and U.S.R. Murty, Academic Press, London, 329–339.

Percus, J.K. (1971). *Combinatorial Methods.* Springer-Verlag, New York.

Provan, J.S. and Ball, M.O. (1983). The complexity of counting cuts and of computing the probability that a graph is connected. *SIAM Journal of Computing* 12, 777–788.

Sinclair, A.J. and Jerrum, M.R. (1987). Approximate counting, uniform generation and rapidly mixing Markov chains. *Information and Computation*, to appear.

Temperley, H.N.V. and Fisher, M.E. (1961). Dimer problem in statistical mechanics — an exact result. *Philosophical Magazine* 6, 1061–1063.

Tutte, W.T. (1947). A ring in graph theory. *Proceedings of the Cambridge Philosophical Society* 43, 26–40.

Valiant, L.G. (1979a). The complexity of computing the permanent. *Theoretical Computing Science* 8, 189–201.

——— (1979b). The complexity of enumeration and reliability problems. *SIAM Journal of Computing* 8, 410–421.

Vertigan, D.L. (1989). The Tutte polynomial at special points. To appear.

Welsh, D.J.A. (1976). *Matroid Theory.* London Mathematical Society Monograph no. 8, Academic Press, London.

Whitney, H. (1932). A logical expansion in mathematics. *Bulletin of the American Mathematical Society* 38, 572–579.

Merton College
Oxford OX1 4JD.

Lattice Animals: Rigorous Results and Wild Guesses

S.G. Whittington and C.E. Soteros

1. Introduction

We consider the d-dimensional hypercubic lattice with vertices being the integer points in $\mathbb{R}^d$. Two points are connected by an edge if they are unit distance apart. We write $(x_1, x_2, \ldots, x_d)$ for the coordinates of a vertex v and $e = (v_1, v_2)$ for the edge joining the vertices v_1 and v_2 whose coordinates must differ by unity in exactly one coordinate.

A *bond animal* is a connected subgraph of the lattice and a *site animal* is a connected section graph of the lattice. The distinction is that for each pair of vertices v_1 and v_2 in a site animal, which differ by unity in exactly one coordinate, the edge $e = (v_1, v_2)$ must be in the site animal. That is, for site animals, edges are induced by the vertices. We shall be interested in the number of bond or site animals, with n vertices, where two animals are identical if one can be translated into the other. We write A_n for the number of site animals with n vertices and a_n for the number of bond animals with n vertices. For instance, for the square lattice (i.e. $d = 2$), $a_1 = 1$, $a_2 = 2$, $a_3 = 6$, $a_4 = 23$, $a_5 = 95$, $\ldots$ and $A_1 = 1$, $A_2 = 2$, $A_3 = 6$, $A_4 = 19$, $A_5 = 63$, $\ldots$.

In each case an interesting subset is the corresponding set of animals without cycles which we call *bond trees* and *site trees.* We write t_n and T_n for the numbers of bond trees and site trees with n vertices. Again in $d = 2$, $t_1 = 1$, $t_2 = 2$, $t_3 = 6$, $t_4 = 22$, $t_5 = 87$, $\ldots$ and $T_1 = 1$, $T_2 = 2$, $T_3 = 6$, $T_4 = 18$, $T_5 = 55$, $\ldots$.

These animals and trees have been considered as models of branched polymers with excluded volume in much the same way that self-avoiding walks have been used as models of linear polymers with excluded volume, and the techniques used to handle the animal problem are closely related to techniques in the theory of self-avoiding walks (Hammersley 1957; Kesten 1963). Lattice animals are also closely related to percolation clusters although the associated weights are different in the two problems (Broadbent and Hammersley 1957; Kesten 1982).

A good deal of the literature on site animals uses the language of polyominoes. A *cell* of the square lattice is the boundary and interior of a unit square having its vertices at lattice vertices and a *polyomino* is a connected set of cells which are joined at their edges (Golomb 1954; Klarner 1967). Because the square lattice is self-dual the number of polyominoes with n cells is precisely the number of site animals with n vertices. (Where polyominoes are regarded as distinct when one cannot be translated into another, they are sometimes called *fixed* polyominoes.)

The primary interest is in the asymptotic behaviour of a_n, A_n, t_n and T_n. There are many papers which develop methods for obtaining bounds on $A_n^{1/n}$ and we shall review some of these. Concatenation arguments easily establish the existence of the limit $\lim_{n\to\infty}(1/n)\log A_n$ once an upper bound on $A_n^{1/n}$ is available and we shall indicate the corresponding arguments to establish the existence of the limits, the growth constants,

$$\begin{aligned} \Lambda_0 &= \lim_{n\to\infty} T_n^{1/n} \\ \lambda_0 &= \lim_{n\to\infty} t_n^{1/n} \\ \Lambda &= \lim_{n\to\infty} A_n^{1/n} \\ \lambda &= \lim_{n\to\infty} a_n^{1/n}. \end{aligned} \tag{1.1}$$

In fact the only difficulty is to show that $a_n^{1/n}$ is bounded above. The inequalities $\Lambda_0 \le \Lambda$ and $\lambda_0 \le \lambda$ are immediate and we show that

$$\Lambda_0 < \Lambda < \lambda_0 < \lambda. \tag{1.2}$$

Roughly speaking, Sections 2–4 describe what is now known rigorously. Section 5 looks at the rates of approach to the limits in (1.1). There, very little is known but there are some informed guesses based on field theoretic arguments and numerical results. It seems that all four limits are approached at roughly the same rate, that this rate is characterized by a critical exponent and that the exponent is independent of the dimension of the problem for $d \ge 8$, but depends on d (but not on the particular lattice in $\mathbb{R}^d$) for $d < 8$. In Section 6 we consider animals on a lattice subset and, in particular, animals in wedge and slab geometries. Section 7 contains a collection of unsolved problems.

Like many other combinatorial problems these are closely connected to problems in physics, are easy to state and understand, but are remarkably difficult to solve.

2. Existence of Limits

We first prove that $a_n^{1/n}$ is bounded above, using a method which is an extension of an idea due to Klarner (1967).

We define the top (bottom) vertex of a set S_0 of vertices as follows. First construct the subset $S_1 \subset S_0$ such that the coordinate x_1 of every vertex in S_1 has the maximum (minimum) value over all vertices in S_0. We then recursively construct $S_k \subset S_{k-1}$ such that the coordinate x_k of every vertex in S_k has the maximum (minimum) value over all vertices in S_{k-1}. Let j be the smallest integer such that S_j contains precisely one vertex, and call this vertex t (respectively b), the top (respectively bottom) vertex of S_0.

We now construct a unique ordering of the vertices and edges of a bond animal. A vertex can have up to $2d$ edges emanating from it and we assign an order $l_1, l_2, \ldots, l_{2d}$. We specify the added constraint that the edge in the $-\hat{x}_1$ direction comes before the edges in the $\pm\hat{x}_2$ directions in this ordering. We number the bottom vertex v_1. The k edges incident on the bottom vertex are numbered $1, 2, \ldots, k$ according to their order in the list $\{l_i\}$, and the vertices connected to the bottom vertex through these edges are numbered $2, \ldots, k+1$. We now continue this numbering at vertex v_2, labelling any edges incident on v_2 and vertices connected to v_2, which have not previously been labelled, and so on through v_3, v_4, $\ldots$.

We next code the animal, proceeding through the vertices in order, using the following procedure. The vertex v_k, $k > 1$, is connected by an edge to at least one vertex v_j with $j < k$. Let j be the smallest such value and let r be the order of the edge (v_k, v_j) emanating from v_k in the ordering $\{l_i\}$ described above. Then number the ith edge (in the ordering $\{l_i\}$) emanating from v_k with the number $s_i = (i - r) \bmod (2d)$ for each $i = 1, \ldots, 2d$, $i \neq r$. Place the number zero in the $\{(2d-1)(k-1) + s_i\}$th location of a vector of length $(2d-1)n$ if the ith edge is not in the animal or if it is incident on v_m for some $m < k$. Otherwise place the number 1 in the $\{(2d-1)(k-1) + s_i\}$th location of the vector. In the special case $k = 1$, assume the edge $(v_1 - \hat{x}_2, v_1)$ is in the animal and then proceed as for v_k. In this way there is a vector of length $(2d-1)n$ associated with each animal. The total number of ones in this vector is equal to the total number of bonds in the animal. If an animal has b bonds these can be chosen in at most $\binom{(2d-1)n}{b}$ ways and the number of animals satisfies

$$a_n \leq \sum_{b=n-1}^{dn} \binom{(2d-1)n}{b}$$

$$\leq (d-1)n \binom{(2d-1)n}{\lfloor (d-1/2)n \rfloor} \tag{2.1}$$

($\lfloor x \rfloor$ denotes the greatest integer less than or equal to x) and it follows that

$$a_n^{1/n} \leq 2^{2d-1} \tag{2.2}$$

for all n.

Each animal with n vertices can be concatenated with each animal with m vertices by translating so that the coordinates $(x_1(b), x_2(b), \dots)$ of the bottom vertex of one animal and the coordinates $(x_1(t), x_2(t), \dots)$ of the top vertex of the second animal are such that

$$\begin{aligned} x_1(b) &= x_1(t) + 1 \\ x_j(b) &= x_j(t) \quad \forall j \neq 1. \end{aligned} \tag{2.3}$$

Adding an edge to join these two adjacent vertices results in an animal with $m + n$ vertices and every pair of m and n animals gives a distinct $(n + m)$-animal so that

$$a_n a_m \leq a_{n+m}. \tag{2.4}$$

From (2.2) and (2.4) it follows that

$$\sup_{n>0} n^{-1} \log a_n = \lim_{n\to\infty} n^{-1} \log a_n = \log \lambda \leq (2d - 1) \log 2, \tag{2.5}$$

where λ is called the *growth constant* of bond animals. Similar concatenation arguments establish the existence of the limits in (1.1) since T_n, t_n and A_n are all less than or equal to a_n and so $T_n^{1/n}$ etc. are all bounded above.

3. Upper and Lower Bounds on the Growth Constants

In this section we give a brief account of several methods for finding upper and lower bounds on the growth constants. Perhaps the most obvious approach for obtaining a lower bound is to use (2.5) directly since (2.5) implies that

$$a_n^{1/n} \leq \lambda \quad \forall n, \tag{3.1}$$

and with corresponding inequalities for Λ_0, Λ and λ_0. The numbers of bond and site animals and bond and site trees are known exactly for small n (see e.g. Gaunt et al. 1976, Gaunt and Ruskin 1978, Redelmeier 1981, Gaunt et al. 1982). For instance, using Redelmeier's result that $A_{24} = 5239988770268$ on the square lattice, we have $\Lambda \geq 3.388$. By noticing that the concatenation can be carried out in each of d directions (3.1) can be improved to

$$(dn)^{1/n} \leq \lambda \tag{3.2}$$

and, for site animals on the square lattice, we have $\Lambda \geq 3.487$. Similar calculations yield $\Lambda_0 \geq 3.1533$, $\lambda_0 \geq 4.1507$ and $\lambda \geq 4.3486$.

An alternative but closely related method has been discussed by Rands and Welsh (1981). This is related to an idea of Moser described in Klarner and Rivest (1973). We call an animal α *composite* if there exist two animals

α_1 and α_2 which yield α under the concatenation described in Section 2. If no such pair of animals exists we call α a *prime animal with respect to bond decomposition.* If we write p_n for the number of these prime animals with n vertices then

$$a_n = p_n + \sum_{i=1}^{n-1} p_{n-i} a_i, \quad n \geq 1. \tag{3.3}$$

If we construct generating functions

$$A(x) = \sum_{n=0}^{\infty} a_n x^n \tag{3.4}$$

and

$$P(x) = \sum_{n=1}^{\infty} p_n x^n \tag{3.5}$$

then

$$A(x) = 1 + P(x)A(x) \tag{3.6}$$

and $A(x)$ is singular when $P(x) = 1$.

If a_n is known exactly for $n \leq N$ then p_n can be determined for $n \leq N$ from (3.3). If we write $P_N(x)$ for the polynomial with degree N whose coefficients are equal to the coefficients of $P(x)$ up to x^N then Rogers (1979) shows that the unique positive zero $(1/\lambda_N)$ of $P_N(x) - 1 = 0$ is such that $\lambda_N \leq \lambda$ and converges to λ as $N \to \infty$. Using this method Rands and Welsh show that, for the square lattice, $\Lambda \geq 3.57$.

A substantial improvement results from a comparatively minor change in the concatenation operation. Instead of joining the top vertex of one animal to the bottom vertex of a second animal by adding a bond, they construct an animal by superimposing the top vertex of one and the bottom vertex of another. This gives another definition of primality (prime with respect to site decomposition) and the previous argument goes through with only minor changes. The resulting bound is improved to $\Lambda \geq 3.7355$.

The configurational data which are now known exactly allow these bounds to be improved and corresponding bounds to be determined for the other growth constants. For the square lattice, Redelmeier's results for site animals give $\Lambda \geq 3.791$. The second concatenation approach can be used for bond animals and for bond trees on the square lattice and, using the available counts (Gaunt and Ruskin 1978; Gaunt et al. 1982), we obtain the bounds $\lambda \geq 4.544$ and $\lambda_0 \geq 4.462$. This second concatenation does not work for site trees (since two site trees concatenated in this way do not necessarily yield a tree) but, using the results of Gaunt et al. (1976), the first concatenation gives $\Lambda_0 \geq 3.300$.

The first concatenation argument can be extended in the following way. With the definition of top and bottom vertex given in Section 2, an animal is composite if there exists an edge in the x_1-direction which when removed decomposes the animal into two animals whose top and bottom vertices were incident on this edge. If no such edge exists we call the animal *prime with respect to* x_1.

We now extend our definitions of top and bottom vertices to *i-top* and *i-bottom* vertices in the following way. We first construct the set of vertices such that coordinate x_i has maximum (minimum) value, and the subset of this such that x_{i+1} has maximum (minimum) value, and so on, cyclically, to x_{i-1}. This gives a unique i-top (i-bottom) vertex.

If an animal is prime with respect to x_1 we can look for a further decomposition removing an edge in the x_2-direction to give two animals whose 2-top and 2-bottom vertices were incident on this edge. If no such edge is present the animal is *prime with respect to* x_2, and so on. This implies that the p_n of (3.3) can be written in terms of the numbers $p_n(2)$ of animals prime with respect to x_2 as

$$p_n = p_n(2) + \sum_{i=1}^{n-1} p_{n-i}(2)p_i \tag{3.7}$$

and, in terms of their generating functions,

$$P(x) = \big(1 + P(x)\big)P_2(x) \tag{3.8}$$

where

$$P_2(x) = \sum_{n=1}^{\infty} p_n(2)x^n. \tag{3.9}$$

Hence

$$A(x) = \frac{\big(1 - P_2(x)\big)}{\big(1 - 2P_2(x)\big)} \tag{3.10}$$

and A has a singular point at the positive root of $P_2(x) = 1/2$.

This approach gives an improved bound for site trees on the square lattice, $\Lambda_0 \geq 3.381$, but does not improve the bounds for Λ, λ_0, or λ. We note that this approach does not generalize to the site decomposition process.

All of these methods for deriving lower bounds rely on counting animals exactly for small n. The bounds are capable of improvement by determining further terms in the series but the computational effort required is considerable.

We also mention, without much detail, several other approaches to computing lower bounds. We focus on site animals on the square lattice but the methods could be extended to other cases. If a site animal is such that the bonds in each row (column) of the lattice are contiguous we call the animal *row (column) convex*. An animal is *convex* if it is both row and column convex. Clearly these are subsets of the site animals and counting these subsets yields lower bounds on Λ, though these problems are also interesting in their own right. Row convex animals have been counted by Klarner (1965) and convex animals by Klarner and Rivest (1974) and by Delest and Viennot (1984). On a different tack, Read (1962) has used a transfer matrix method to count site animals confined between two parallel lines. Although each of these approaches has attractive features we believe that the renewal sequence method of Rands and Welsh shows most promise for calculating good lower bounds.

In his original paper on site animals Eden (1961) shows that $\Lambda \leq 27/4$ in $d = 2$. (Our argument in Section 2 that $\lambda \leq 2^{2d-1}$ is based on his approach.) Klarner and Rivest (1973) have reformulated Eden's approach in a way which allows successive improvement and we sketch their argument here. Each site animal is associated with a unique spanning tree and these trees can be regarded as a sequence of 'twigs', chosen from a fixed finite set. The number of site animals is bounded above by the number of ways of concatenating the twigs. Based on this argument, there is a particular set of twigs which gives the Eden bound $\Lambda \leq 27/4$. We note that this argument and the same set of twigs also works for bond trees and gives $\lambda_0 \leq 27/4$ in $d = 2$. Furthermore, this same set of twigs can be used to obtain the upper bound $\lambda \leq 8$ in $d = 2$ derived by us in Section 2. Klarner and Rivest describe a procedure for choosing sets of twigs which lead to successive improvements of the bound on Λ. Their best bound obtained in this way is $\Lambda \leq 4.649551$.

4. Applications of a Pattern Theorem

To motivate this section we begin by describing some work by Kesten (1963) on the number of self-avoiding walks on a lattice. Kesten defined a pattern to be any finite self-avoiding walk, i.e. any finite sequence of edges such that no vertex of the lattice is visited more than once. He proved that if there exists a self-avoiding walk on which the pattern appears three times then the pattern appears at least once on all except exponentially few sufficiently long self-avoiding walks. Kesten used this theorem to establish that, if c_n is the number of n-step self-avoiding walks, the limit $\lim_{n\to\infty} c_{n+2}/c_n$ exists. The theorem has proved useful in a variety of other areas, e.g. in studying walks confined to a subset of a lattice (Hammersley and Whittington 1985).

Recently Madras (1988) has proved a corresponding pattern theorem for lattice animals and related structures which we state as follows.

We focus on the case of bond animals though the theorem applies to certain subsets of these as well. Let L be the simple hypercubic lattice in $\mathbf{R}^d$ and let $P = (P_1, P_2)$ be a *proper pattern* if P_1 and P_2 are disjoint subsets of L such that for any n there exists an animal with $m > n$ vertices weakly embeddable in L which contains all of P_1 and none of P_2. The number $a_n(\overline{P})$ of animals with n vertices in which P does not occur is such that

$$\limsup_{n\to\infty} n^{-1} \log a_n(\overline{P}) < \log \lambda. \tag{4.1}$$

The theorem is valid if bond animals are replaced by site animals or by bond trees; i.e.

$$\limsup_{n\to\infty} n^{-1} \log A_n(\overline{P}) < \log \Lambda \tag{4.2}$$

and

$$\limsup_{n\to\infty} n^{-1} \log t_n(\overline{P}) < \log \lambda_0. \tag{4.3}$$

It is easy to prove that $\Lambda_0 \leq \Lambda \leq \lambda_0 \leq \lambda$ and Rands and Welsh (1981) conjectured that $\Lambda < \lambda_0$ while Gaunt et al. (1982) conjectured that $\Lambda_0 < \Lambda$ and $\lambda_0 < \lambda$. Each of these strict inequalities can be established by an application of Madras' pattern theorem. If we consider P_1 to be the elementary square and P_2 to be the empty set, $P = (P_1, P_2)$ occurs in both site animals and in bond animals but not in site trees or bond trees and this immediately gives

$$\Lambda_0 < \Lambda \tag{4.4}$$

and

$$\lambda_0 < \lambda. \tag{4.5}$$

Following Klarner (1967) we construct a spanning tree for each site animal. The vertices of the spanning tree are those of the animal and are numbered according to the vertex numbering scheme described in Section 2. We complete the spanning tree by adding edges as follows. We join the first and second vertices. We then consider each vertex in turn and add an edge to join this vertex to the vertex with smallest number which is adjacent to it in the lattice. These spanning trees are a subset of the bond trees and correspond 1-1 with the site animals. It is clear that the pattern in which P_1 is $\sqsupset$ and P_2 is the complement of P_1 in $\square$ can appear in a bond tree but not in the spanning tree (as defined above) of a site animal. Hence

$$\Lambda < \lambda_0. \tag{4.6}$$

Madras et al. (1988) gave an alternative proof that $\lambda_0 < \lambda$ and this argument can be strengthened to show that

$$\lambda - \lambda_0 \geq (0.00003758)\lambda_0 \geq 0.0001677. \tag{4.7}$$

5. The Subdominant Asymptotic Behaviour

The results described up to now tell us nothing about the rates of approach to the limits in (1.1). Physicists (e.g. Lubensky and Isaacson 1979) expect that

$$a_n \sim Cn^{-\theta}\lambda^n \tag{5.1}$$

which implies that the limit

$$\lim_{n\to\infty}\left(\frac{\log[a_n/\lambda^n]}{\log n}\right) = -\theta \tag{5.2}$$

exists. Proving that this limit exists would be a major advance. Similarly, it is believed that

$$t_n \sim C_0 n^{-\theta_0}\lambda_0^n \tag{5.3}$$

and there are arguments (and some numerical results) suggesting that $\theta = \theta_0$ (Lubensky and Isaacson 1979; Duarte and Ruskin 1981; Gaunt et al. 1982). The value of θ is believed to be lattice independent and to depend only on the dimension. (Notice that λ is lattice dependent.) In addition, θ is believed to be independent of d for $d \geq d_c = 8$, where d_c is called the *upper critical dimension*. For self-avoiding walks, the existence of an upper critical dimension has now been established (Slade 1987). There is an intriguing proposal (Parisi and Sourlas 1981) that θ is connected to the Yang-Lee edge singularity exponent in $d-2$ dimensions and, since this exponent is known exactly for $d = 0$ and 1, this suggests that $\theta(d=2) = 1$ and $\theta(d=3) = 3/2$. These values are certainly consistent with the available numerical evidence.

An attempt has been made to connect the results on trees with those on animals by asking for the number of animals with fixed cyclomatic index. If $a_n(c)$ is the number of bond animals with n vertices and c elementary cycles then $a_n(0) \equiv t_n$ and

$$a_n = \sum_{c\geq 0} a_n(c). \tag{5.4}$$

It is fairly easy to prove that

$$\lim_{n\to\infty} n^{-1}\log a_n(c) \equiv \log \lambda_c \tag{5.5}$$

exists and that

$$\lambda_c = \lambda_0 \quad \forall c. \tag{5.6}$$

By analogy with (5.1) one expects that

$$a_n(c) \sim C_c n^{-\theta_c}\lambda_0^n \tag{5.7}$$

and it has been shown (Soteros and Whittington 1988) that if θ_0 exists then θ_c exists (in an analogous way to (5.2)) and that

$$\theta_c = \theta_0 - c. \tag{5.8}$$

While there are heuristic arguments (Whittington, Torrie, and Gaunt 1983) that this is consistent with $\theta = \theta_0$, much remains to be done in this area.

6. Lattice Animals on Lattice Subsets

In this section we consider the number of lattice animals with n vertices, confined to lie in a subset of the square lattice. This is closely related to some work on self-avoiding walks in restricted geometries (Hammersley and Whittington 1985) and to similar problems in percolation theory (Grimmett 1983) and the Ising problem (Chayes and Chayes 1986).

We shall consider two particular cases: animals in wedges and animals in slits. We define an f*-wedge* of the square lattice to be the subset of the square lattice $\{(x, y) : x \geq 0, 0 \leq y \leq f(x)\}$ where $f(x)$ is a non-negative function of x. We now ask for the number $a_n(f)$ of animals with n vertices with one vertex at the origin and with all other vertices in the f-wedge. It is easy (following a line of argument due to Hammersley and Whittington (1985)) to show that, provided that $\lim_{x\to\infty} f(x) = \infty$,

$$\lim_{n\to\infty} n^{-1} \log a_n(f) = \log \lambda \tag{6.1}$$

independent of f. A situation of some physical interest is when

$$f(x) = \alpha x. \tag{6.2}$$

If one assumes that

$$a_n(\alpha x) \sim n^{-\theta(\alpha)} \lambda^n \tag{6.3}$$

the question is: how does θ depend on α? Of course, this is all modulo the existence of the exponent θ. There are some numerical results due to De'Bell and Lookman (1985) but nothing else. In the case of self-avoiding walks there are some definite predictions from conformal invariance arguments and these are in good agreement with the numerical results of Guttmann and Torrie (1984) and Cardy and Redner (1984). (The conformal invariance argument does not work for animals since there is no Hamiltonian formulation of the animal problem.)

The corresponding slit problem asks for the number $a_n(L)$ of animals with n vertices such that no vertex has y coordinate less than zero or greater than L and at least one vertex has y coordinate zero. Two animals are considered identical if one can be translated into the other in the x-direction. By concatenation in the x-direction it is easy to show that

$$\lim_{n\to\infty} n^{-1}\log a_n(L) = \sup_{n>0} n^{-1}\log a_n(L) = \log\lambda(L) \tag{6.4}$$

and interest focuses on the L-dependence of $\lambda(L)$.

Madras' pattern theorem readily establishes that

$$\lambda(L+1) > \lambda(L) \tag{6.5}$$

and, following Hammersley and Whittington (1985), it is easy to prove that

$$\lim_{L\to\infty} \lambda(L) = \lambda. \tag{6.6}$$

Can one say anything further about the L dependence? It is presumably the case that $\log\lambda(L)$ is a concave function of L, and a reasonable guess (supported by a scaling argument) would be that

$$\log\lambda - \log\lambda(L) \sim L^{-\phi} \tag{6.7}$$

but this seems to be difficult to prove.

7. Unsolved Problems

In this final section we list some of the unsolved problems which we have mentioned earlier.

(i) Calculate any of the growth constants for any non-trivial lattice.
(ii) Show that $a_n = \lambda^n e^{O(\log n)}$.
(iii) Prove that the $\log n$ term in (ii) has the same coefficient for trees and for animals, and for the site and bond cases.
(iv) Provide some rigorous results on the sub-dominant term for animals in a wedge of angle α.
(v) Investigate the L dependence of $\lambda(L)$ for animals in a slit geometry.

References

Broadbent, S.R. and Hammersley, J.M. (1957). Percolation processes. I. Crystals and mazes. *Proceedings of the Cambridge Philosophical Society* 53, 629–641.

Cardy, J.L. and Redner, S. (1984). Conformal invariance and self-avoiding walks in restricted geometries. *Journal of Physics A: Mathematical and General* 17, L933–L938.

Chayes, J.T. and Chayes, L. (1986). Critical points and intermediate phases on wedges in Z^d. *Journal of Physics A: Mathematical and General* 19, 3033–3048.

De'Bell, K. and Lookman, T. (1985). Lattice trees and self-avoiding walks in a triangular lattice wedge. *Physics Letters A* 112, 453–455.

Delest, M.-P. and Viennot, G. (1984). Algebraic languages and polyominoes enumeration. *Theoretical Computer Science* 34, 169–206.

Duarte, J.A.M.S. and Ruskin, H.J. (1981). The branching of real lattice trees as dilute polymers. *Journal de Physique* 42, 1585–1590.

Eden, M. (1961). A two dimensional growth process. *Proceedings of the 4th Berkeley Symposium on Mathematics, Statistics and Probability* IV, 223–239, University of California Press, Berkeley.

Gaunt, D.S. and Ruskin, H. (1978). Bond percolation processes in d-dimensions. *Journal of Physics A: Mathematical and General* 11, 1369–1380.

Gaunt, D.S., Sykes, M.F., and Ruskin, H. (1976). Percolation processes in d-dimensions. *Journal of Physics A: Mathematical and General* 9, 1899–1911.

Gaunt, D.S., Sykes, M.F., Torrie, G., and Whittington, S.G. (1982). Universality in branched polymers on d-dimensional hypercubic lattices. *Journal of Physics A: Mathematical and General* 15, 3209–3217.

Golomb, S.W. (1954). Checker boards and polyominoes. *American Mathematical Monthly* 61, 675–682.

Grimmett, G.R. (1983). Bond percolation on subsets of the square lattice. *Journal of Physics A: Mathematical and General* 16, 599–604.

Guttmann, A.J. and Torrie, G.M. (1984). Critical behaviour at an edge for the SAW and Ising model. *Journal of Physics A: Mathematical and General* 17, 3539–3552.

Hammersley, J.M. (1957). Percolation processes. II. The connective constant. *Proceedings of the Cambridge Philosophical Society* 53, 642–645.

Hammersley, J.M. and Whittington, S.G. (1985). Self-avoiding walks in wedges. *Journal of Physics A: Mathematical and General* 18, 101–111.

Kesten, H. (1963). On the number of self-avoiding walks. *Journal of Mathematical Physics* 4, 960–969.

——— (1982). *Percolation Theory for Mathematicians.* Birkhäuser, Boston.

Klarner, D.A. (1965). Some results concerning polyominoes. *Fibonacci Quarterly* 3, 9–20.

——— (1967). Cell growth problems. *Canadian Journal of Mathematics* 19, 851–863.

Klarner, D.A. and Rivest, R. (1973). A procedure for improving the upper bound for the number of n-ominoes. *Canadian Journal of Mathematics* 25, 585–602.

——— (1974). Asymptotic bounds for the number of convex n-ominoes. *Discrete Mathematics* 8, 31–40.

Lubensky, T.C. and Isaacson, J. (1979). Statistics of lattice animals and dilute branched polymers. *Physical Review A* 20, 2130–2146.

Madras, N. (1988). Personal communication.

Madras, N., Soteros, C.E., and Whittington, S.G. (1988). The statistics of lattice animals. *Journal of Physics A: Mathematical and General* 21, 4617–4635.

Parisi, G. and Sourlas, N. (1981). Critical behaviour of branched polymers and the Lee-Yang edge singularity. *Physical Review Letters* 46, 871–874.

Rands, B.M.I. and Welsh, D.J.A. (1981). Animals, trees and renewal sequences. *IMA Journal of Applied Mathematics* 27, 1–17.

Read, R.C. (1962). Contributions to the cell growth problem. *Canadian Journal of Mathematics* 4, 1–20.

Redelmeier, D.H. (1981). Counting polyominoes: yet another attack. *Discrete Mathematics* 36, 191.

Rogers, D.G. (1979). An application of renewal sequences to the dimer problem. *Lecture Notes in Mathematics* 748, 143–153, Springer-Verlag, Berlin.

Slade, G. (1987). The diffusion of self-avoiding random walks in high dimensions. *Communication in Mathematical Physics* 110, 661–683.

Soteros, C.E. and Whittington, S.G (1988). Critical exponents for lattice animals with fixed cyclomatic index. *Journal of Physics A: Mathematical and General* 21, 2187–2193.

Whittington, S.G., Torrie, G.M., and Gaunt, D.S. (1983). Branched polymers with a prescribed number of cycles. *Journal of Physics A: Mathematical and General* 16, 1695–1711.

Department of Chemistry
University of Toronto
Toronto
Ontario M5S 1A1.

Fields and Flows on Random Graphs

P. Whittle

1. Introduction

Random graphs have become objects of increasing interest over the last thirty years. Two sets of physical models which have particularly stimulated this development are those of polymerisation and of percolation.

The early polymerisation work is associated especially with the names of Flory, Stockmayer, Gordon and Good. An approach in which one sets up a reversible Markov model of association/dissociation and analyses its equilibrium properties is set out systematically in Whittle (1986). Percolation theory is firmly associated with the name of Hammersley, who initiated and so greatly developed the subject. For reviews see Hammersley and Welsh (1980) and Kesten (1982).

Finally, largely in isolation from either of the above movements, pure mathematicians developed an interest in the subject, beginning with Erdős and co-workers, and continued notably by Stepanov and Bollobás. The work of this school is systematically presented in Bollobás (1985).

However, there are now new applications developing, which will require fundamental theoretical advances. I think especially of the study of neural networks. These are viewed as random graphs, partly because their size and complexity makes the statistical approach inevitable, and partly also because (just as in communication theory), the further one penetrates into the subject, the more one realises that the statistical approach is 'right'.

The interest of a neural network is that impulses and activity of some kind are propagated around it. One is then motivated to a study which has received only sporadic attention hitherto and of which this article can only be a token: of directed dynamics on a random graph.

The view of a neural net as a random graph is explicit in the papers of Kauffman (1969), Little and Shaw (1978) and Hopfield (1982), for example. More recent papers are those by Derrida and co-workers and by the author, listed in the references, particular aspects of which we shall refer to later.

2. Random Graphs and Reversible Dynamics

Our starting point is some earlier work of the author's (see Whittle 1986 and references quoted there) which we now summarise for convenience.

The configuration $\mathcal{C}$ of a random graph on N nodes is specified by $\mathcal{C} = \{s_{ab}; a, b = 1, 2, \ldots, N\}$ where s_{ab} is the number of arcs directed from node a to node b. In a 'first-shell' Markov model this has equilibrium distribution

$$P_N(\mathcal{C}) \propto Q_N(\mathcal{C}) = \left[\prod_{a,b} \frac{h^{s_{ab}}}{s_{ab}!}\right]\left[\prod_j H_j^{N_j}\right]. \tag{2.1}$$

Here h has the form

$$h = \frac{1}{2\kappa V} \tag{2.2}$$

where κ is a constant and V is volume. Despite the fact that distance and dimension do not enter into this description, one needs an 'extension' parameter, supplied by V. The interesting results emerge in the thermodynamic limit, when N and V become infinite in constant ratio

$$\rho = N/V$$

interpretable as the 'node density'.

The quantity N_j in (2.1) is the number of nodes which have degree j (or, in polymerisation terminology, the number of units which have formed j bonds). The final factor in (2.1) then represents the component of a Gibbs distribution dependent on 'configuration energy', this energy being supposed to be dependent on 'first-shell' effects alone in this model. One can also include the effect of differing rates of arc-formation between and within components of the graph (i.e. differing rates of inter- and intramolecular association) but, for simplicity, we shall dispense with this.

Distribution (2.1) is a *consequence* of a model, but such an immediate one that we can view (2.1) as itself constituting the assumption and the model. The quantity

$$Q_N = \sum_{\mathcal{C}} Q_N(\mathcal{C})$$

is the partition function for this statistical model. We can view it as the unnormalised probability generating function (p.g.f.) of the random variables N_j, with the quantities H_j serving both as parameters of the model and as marker variables for the N_j in the p.g.f.. (To be more specific: Q_N would be the un-normalised p.g.f. with arguments z_j if H_j were replaced by $H_j z_j$ for all j.)

Define the function

$$H(\xi) = \sum_{j=0}^{\infty} \frac{H_j \xi^j}{j!}. \tag{2.3}$$

THEOREM 1. *Suppose* $\log H(\xi)$ *of less than quadratic growth at infinity. Then for model (2.1) the partition function* Q_N *has the evaluation*

$$Q_N = \sqrt{\frac{\kappa V}{2\pi}} \int_{-\infty}^{\infty} H(\xi)^N e^{-\kappa V \xi^2/2}\, d\xi. \tag{2.4}$$

The value $\bar{\xi}$ of ξ maximising the integrand of (2.4) is, in the thermodynamic limit, that maximising

$$J(\xi) = \rho \log H(\xi) - \kappa\xi^2/2. \tag{2.5}$$

This determines the statistics of node degree in that, for example

$$E(N_j) \propto \frac{H_j \bar{\xi}^j}{j!}.$$

The components of the random graph would be identified as the polymer molecules themselves in the polymerisation context. What is interesting is that the evaluation (2.4) of Q_N effectively determines the polymer statistics. The natural level of description of a polymer for model (2.1) is $r = \{r_j; j = 0, 1, 2, \dots\}$ where r_j is the number of nodes in the component of degree j. Let us term such a polymer an r-mer; it will contain

$$R = \sum_j r_j$$

nodes. Let n_r be the number of r-mers, so that necessarily

$$\sum_r R n_r = N. \tag{2.6}$$

THEOREM 2. *Suppose that* $\log\left(\sum_{N=0}^{\infty} Q_N/N!\right)$ *has the formal expansion* $\sum_r \gamma_r$ *in powers of the* H_j, *where* γ_r *is the term in* $\prod_j H_j^{r_j}$. *Then the* n_r *are distributed as independent Poisson variables with respective expectations* γ_r, *conditioned by the constraint (2.6).*

This theorem has an obvious analogue in all the variants of the model which follow.

The model demonstrates a phase transition, in that, as ρ increases through a critical value ρ_c, the assembly of polymers passes from the 'sol' to the 'gel' state (the graph changes from having many modest-sized components to having a dominant component, which includes most nodes). This transition is not revealed in the behaviour of $J(\xi)$ itself, which has a single non-negative maximising value $\bar{\xi}$ for all ρ. However, it is revealed in the representation

$$J(\xi) = \min_{\theta} \left[\theta H(\xi) - \frac{\kappa \xi^2}{2} - \rho \log \theta \right]. \tag{2.7}$$

The square bracket possesses a saddle-point (min-max in (θ, ξ)) only for $\rho \leq \rho_c$.

Suppose now that the nodes of the graph can be 'coloured' in that there is a variable α at each node which can take values $\alpha = 1, 2, \ldots, p$. Let $c(a)$ denote the value of α at node a; this can be regarded as the value at a of a field defined on the graph. The configuration $\mathcal{C}$ of the graph will now specify both the arc multiplicities $s = \{s_{ab}\}$ and the field $c = \{c(a)\}$. Under probability transition rules for $\mathcal{C}$ which are Markov and reversible but otherwise rather general one deduces the generalisation of the equilibrium distribution (2.1)

$$P_N(\mathcal{C}) \propto Q_N(\mathcal{C}) = \left(\prod_{\alpha} \sigma_{\alpha}^{M_\alpha} \right) \left(\prod_{a,b} \frac{h_{ab}^{s_{ab}}}{s_{ab}!} \right) \left(\prod_{j} \prod_{\alpha} H_{\alpha j}^{N_{\alpha j}} \right). \tag{2.8}$$

Here M_α is the number of nodes at which the field takes the value α, and h_{ab} depends on field values at a and b in that

$$h_{ab} = \frac{1}{2V \kappa_{c(a)c(b)}}.$$

The 'degree' j is now a vector of integers $j = (j_1, j_2, \ldots, j_p)$ and $N_{\alpha j}$ is the number of nodes with state value α from which j_β arcs are directed to nodes of state value β $(\beta = 1, 2, \ldots, p)$.

Model (2.8) allows field dynamics on the graph, but two points should be noted. First, these dynamics are reversible, in that (2.8) is deduced from a reversible model. Second, the model is one that allows field value and graph configuration to interact, in that each affects the transition rules of the other. This is exactly what is desired for some applications (e.g. the Ising and socio-economic models discussed in Whittle 1986). However, if one were seeking to represent a neural network then (i) dynamics on the network would not be reversible, and (ii) the interaction mentioned would be an interaction of form and function. It is natural that form (the network) should influence function (the field). However, for function to influence form represents adaptation, or learning, which is a feature one may or may not wish to incorporate.

The generalisation of the partition function evaluation (2.4) is interesting. Suppose all field values are possible, so that distribution (2.8) is subject only to the constraint

$$\sum_{\alpha} \sum_{j} N_{\alpha j} = N.$$

Define the functions of a p-vector variable $\xi = (\xi_1, \xi_2, \ldots, \xi_p)$

$$H_\alpha(\xi) = \sum_j H_{\alpha j} \prod_\beta \frac{\xi_\beta^{j_\beta}}{j_\beta!}.$$

THEOREM 3. *Suppose* $\log\left(\sum_\alpha \sigma_\alpha H_\alpha(\xi)\right)$ *of less than quadratic growth at infinity. Then for model (2.8) the partition function* Q_N *has the evaluation*

$$Q_N = \left(\prod_{\alpha,\beta} \frac{\kappa_{\alpha\beta} V}{2\pi}\right)^{1/2} \int \left[\sum_\alpha \sigma_\alpha H_\alpha(\xi)\right]^N \exp\left[-\frac{V}{2} \sum_\alpha \sum_\beta \xi_{\alpha\beta}\xi_{\beta\alpha}\right] d\eta \tag{2.9}$$

where the integral is over all real η, *and* ξ, η *are related by*

$$\xi_{\alpha\beta} = \begin{cases} \frac{1}{\sqrt{2}}(\eta_{\alpha\beta} + i\eta_{\beta\alpha}) & (\beta < \alpha) \\ \eta_{\alpha\alpha} & (\beta = \alpha) \\ \frac{1}{\sqrt{2}}(\eta_{\alpha\beta} - i\eta_{\beta\alpha}) & (\beta > \alpha). \end{cases} \tag{2.10}$$

The complex form of these integrals has considerable significance, as we shall see in part.

One may now ask whether the contribution to the integral (2.9) comes essentially from a single value $\bar{\xi}$ of ξ in the thermodynamic limit, as in the 'fieldless' case $p = 1$. Indeed, this seems to be true, at a *real* value $\bar{\xi}$ derived in the following manner. Consider the real form of transformation (2.10)

$$\xi_{\alpha\beta} = \begin{cases} \frac{1}{\sqrt{2}}(\zeta_{\alpha\beta} + \zeta_{\beta\alpha}) & (\beta < \alpha) \\ \zeta_{\alpha\alpha} & (\beta = \alpha) \\ \frac{1}{\sqrt{2}}(\zeta_{\alpha\beta} - \zeta_{\beta\alpha}) & (\beta > \alpha) \end{cases} \tag{2.11}$$

and seek for the real values of the $\zeta_{\alpha\beta}$ that maximise the integrand of (2.9) for $\alpha \geq \beta$ and minimise it for $\alpha < \beta$. Relations (2.11) determine $\bar{\xi}$ in terms of the saddle-point $\bar{\zeta}$ thus located.

It is interesting that for the case $p = 2$, when nodes can adopt exactly two states (particle and antiparticle?), one is then led to consider a function of four variables, and to seek for a value at which this is maximal with respect to three of the variables, and minimal with respect to one. That is, one has three local 'space-like' axes and one local 'time-like' axis.

Criticality shows itself as before, in that the form analogous to (2.7) may or may not possess a saddle-point of the required type. However, the integrand of (2.9) may now itself possess several saddle-points of the required type. The fact that the effective saddle-point may switch as parameters change leads to new phase transitions (see Section 3).

The spin-glass models of memory considered by various authors (see e.g. Amit et al. 1985) could also be regarded as the specification of reversible dynamics upon a random graph. A comparison deserves fuller discussion than we can afford here.

3. The 'Locally Tree-Like' Property

If we view the arcs as bonds, then models (2.1) and (2.8) permit multiple bonding, self-bonding and the formation of cycles. All these effects become less probable as one approaches the thermodynamic limit. Roughly speaking, the V^{-1} dependence in (2.2) and its vector analogue ensures that the number of bonds a node forms is roughly independent of V, for given ρ. As V increases, the probability that these bonds take place with *assigned* other nodes tends to zero.

This effect also manifests itself in that the γ_r of Theorem 2 depends upon V by a factor V^{R-L}, where

$$L = \frac{1}{2}\sum_j j r_j$$

is the number of bonds in the polymer (arcs in the component). For a polymer of given size R, polymers with cycles (i.e. $L > R-1$) are discouraged relative to the tree-form (for which $L = R-1$) as V increases. This effect can be countervailed by the combinatorial fact that the number of ways of introducing cycles increases rapidly with R.

However, one can certainly establish the following property. Consider a given node, and its *neighbourhood* of radius D (i.e. the set of nodes which are connected to the initial node by paths of length not exceeding D). Consider the subgraph $\mathcal{G}_D$ of the full graph $\mathcal{G}$ in this neighbourhood.

THEOREM 4. *$\mathcal{G}_D$ is, for given D, a tree with probability one in the thermodynamic limit.*

This we shall speak of as the 'locally tree-like' (LTL) property. The property could be stated (and will indeed be required) in stronger forms, the mildest of these being that Theorem 4 should continue to hold if D is allowed to increase to infinity at a suitable rate as the thermodynamic limit is approached.

Several authors have made an effective appeal to this property in the neural network context (see the four papers listed by Derrida and co-authors, and Hilhorst and Nijmeijer 1987). Suppose, for example, that one has directed dynamics on the graph, which do not show long-range order, in that the field values at nodes far apart on the graph are independent. Then one consequence of the LTL property and short-range order is (very roughly expressed) that field-statistics are the same whether the graph is fixed (although randomly chosen) or randomly evolving (by rules independent of the field) — the so-called *quenched* and *annealed* cases.

However, a consequence to which we shall make more explicit appeal is that, if we consider the dynamic inputs to a node via the arcs entering that node, then these will be statistically independent. This is because the subgraphs which these inputs have traversed are, with probability approaching unity, mutually disjoint out to any given radius.

As an example of this effect, suppose we consider a simplification of model (2.8) in which graph-statistics are uninfluenced by field values to the extent that the factor $\prod_j \prod_\alpha H_{j\alpha}^{N_{j\alpha}}$ is simply replaced by $\prod_j H_j^{N_j}$, as in (2.1). Let us also set

$$\kappa_{\alpha\beta}^{-1} = \psi_{\alpha\beta}$$

and define the symmetric matrix $\Psi = (\psi_{\alpha\beta})$. One can then deduce (Whittle 1989c) the following simplification of Theorem 3.

THEOREM 5. *Under the conditions stated the partition function Q_N has the evaluation*

$$Q_N \propto |\Psi|^{-1} \int \left[\sum_\alpha \sigma_\alpha H(\xi_\alpha)\right]^N \exp\left[-\tfrac{1}{2}V\xi'\Psi^{-1}\xi\right] d\xi. \tag{3.1}$$

Here $\xi = (\xi_1, \xi_2, \ldots, \xi_p)$, the function H has the definition (2.3), and the integral in (3.1) is the complex integral ensuring the u-identity

$$\int \exp\left(u'\xi - \tfrac{1}{2}\xi'\Psi^{-1}\xi\right) d\xi \propto \exp\left(\tfrac{1}{2}u'\Psi u\right). \tag{3.2}$$

We can be explicit about this integral. The symmetric matrix Ψ will have a diagonal representation

$$\Psi = U'\Lambda U$$

where U is a real orthogonal matrix and Λ a diagonal matrix, with diagonal $(\lambda_1, \lambda_2, \ldots, \lambda_p)$. Define $\eta = U\xi$. Then the integral in (3.1), (3.2) is along the whole real η_α axis if $\lambda_\alpha > 0$, and the whole imaginary η_α axis if $\lambda_\alpha < 0$ $(\alpha = 1, 2, \ldots, p)$.

We can write the integrand of (3.1) as $\exp(VJ)$ where

$$J(\xi) = \rho \log\left(\sum_\alpha \sigma_\alpha H(\xi_\alpha)\right) - \tfrac{1}{2}\xi'\Psi^{-1}\xi.$$

One can show (op. cit.) that the principal contribution to the integral comes from a *real* value $\bar{\xi}$ where $\bar{\xi}$ is an appropriate saddle-point of J. Further,

$$\rho_\alpha \propto \sigma_\alpha H(\bar{\xi}_\alpha) \tag{3.3}$$

where ρ_α is the expected density of nodes at which the field takes value α.

Consider now the symmetric two-state case

$$\psi_{11} = \psi_{22} = \psi_1$$
$$\psi_{12} = \psi_{21} = \psi_2$$
$$\sigma_1 = \sigma_2.$$

These states could represent two possible orientations of spin at the nodes of a random lattice. The quantities ψ_1 and ψ_2 then represent strengths of bonding between nodes of like or unlike spin respectively. Let us also suppose that

$$H_j = \begin{cases} 1 & j = r+1 \\ 0 & j \neq r+1. \end{cases}$$

We are then effectively considering an Ising model on a random graph whose nodes are all constrained to have degree $r+1$. Let us also define

$$\mu = (\bar{\xi}_1/\bar{\xi}_2)^r .$$

Then the saddle-point characterisation gives the equation

$$\mu = \left(\frac{\psi_1\mu + \psi_2}{\psi_1 + \psi_2\mu}\right)^r \tag{3.4}$$

for μ, and equation (3.3) gives the characterisation

$$\mu = \left(\frac{\rho_1}{\rho_2}\right)^{r/(r+1)} . \tag{3.5}$$

But equation (3.4) is exactly the equation which occurs in Spitzer's treatment (1975) of an Ising model on an r-branching tree. The fact that we recover it is a stronger manifestation of the LTL property: a graph of constant degree $r+1$ which is a tree will be an r-branching tree.

In Spitzer's case μ had the interpretation

$$\mu = \rho_1'/\rho_2' \tag{3.6}$$

where ρ_α' is proportional to the probability that the field value at the *root* of the tree is α. The difference between (3.5), (3.6) comes from the fact that ρ_α is proportional to the probability that the field value at a *randomly chosen* node is α.

As is known from Spitzer's work, equation (3.4) can have one or several real solutions, depending upon parameter values. The transition corresponds to the transition of *magnetisation*: of alignment of spins.

4. Directed Graphs

Actual neural dynamics are directed and irreversible. Consideration of directed dynamics certainly implies that one must consider a directed graph. Models (2.1), (2.8) may have seemed to be models for a directed graph, in that s_{ab} is specified as the number of arcs from a to b. However, the distribution $P_N(\mathcal{C})$ is invariant under permutation of s_{ab} and s_{ba}, and in this sense there is no real directionality.

To achieve directionality, we modify model (2.1) to

$$P_N(\mathcal{C}) \propto Q_N(\mathcal{C}) = \left(\prod_{a,b} \frac{h^{s_{ab}}}{s_{ab}!}\right)\left(\prod_{j,k} H_{jk}^{N_{jk}}\right) \tag{4.1}$$

where h has evaluation (2.2) as before, and N_{jk} is the number of nodes of degree (j,k). By this double degree we mean that j arcs leave the node and k enter it.

We now give two of the most important conclusions from Whittle (1989b). Define the function

$$H(\xi_1,\xi_2) = \sum_j \sum_k H_{jk} \frac{\xi_1^j \xi_2^k}{j!\,k!}.$$

THEOREM 6. *Suppose* $\log H(\xi_1,\xi_2)$ *of less than quadratic growth at infinity. Then the partition function* Q_N *has the evaluation*

$$Q_N = \frac{2\kappa V}{\pi} \iint_{-\infty}^{\infty} H(\eta_1+i\eta_2, \eta_1-i\eta_2)^N \exp[-2\kappa V(\eta_1^2+\eta_2^2)]\, d\eta_1\, d\eta_2. \tag{4.2}$$

The complex form of the integral is interesting and essential. It enforces the constraint

$$\sum_j \sum_k N_{jk}(j-k) = 0 \tag{4.3}$$

that the total numbers of outgoing and incoming arcs should be equal.

THEOREM 7. *In the thermodynamic limit, the dominant contribution to integral (4.2) comes from the value* $\bar{\xi} = (\bar{\xi}_1, \bar{\xi}_2)$, *this being the real value that simultaneously maximises*

$$J(\xi) = \rho \log H(\xi_1,\xi_2) - 2\kappa V \xi_1 \xi_2$$

with respect to $\xi_1\xi_2$ *and minimises it with respect to* ξ_1/ξ_2.

So we find, for example, that

$$E(N_{jk}) \propto H_{jk} \frac{\bar{\xi}_1^j \bar{\xi}_2^k}{j!\,k!} \tag{4.4}$$

to within terms of smaller order in V. Expression (4.4) then defines the distribution p_{jk} of the random variable (j, k), where this is the degree of a randomly chosen node. Relation (4.3) will have the implication

$$E(j - k) = 0.$$

5. Directed Dynamics

The study of directed dynamics is not yet advanced. Consider, for example, the case of a purely linear graph, for which $a = 1, 2, 3, \ldots$ represent the consecutive nodes. This might constitute a discrete model of a nerve fibre. As ever, $c(a)$ is the value of field at node a.

A simple specification of directed dynamics would be to say that transitions in $c(a)$ were conditioned by the values of $c(a)$ and $c(a-1)$. Suppose now one wishes to evaluate the equilibrium distribution of $c(a)$ conditional on $c(1)$ for large a. This evaluation is not simple, because the relevant Markov process is infinite-dimensional, not p-dimensional. Might one expect $c(a)$ to become independent of $c(1)$ as a becomes infinite? In other words, is there long-range order or not? If one is speaking of the nerve fibre model then one would hope for long-range order, for faithful propagation along the fibre of impulses injected at $a = 1$ would imply dependence at all distances.

One model for which one can conjecture conclusions is a random Jackson network. Let us consider the simplest form of such a network, for which $c(a)$ represents the number of 'quanta' at node a, and the transition $(c(a), c(b)) \to (c(a) - 1, c(b) + 1)$, in which a quantum passes from node a to node b has intensity $c(a)s_{ab}$. In a large, fixed, closed network the stream of quanta from node a to node b will then be a Poisson stream of rate $f_{ab} = w_a s_{ab}$. Here the parameters w_a are subject to the balance conditions

$$\sum_b (f_{ab} - f_{ba}) = 0 \qquad (a = 1, 2, \ldots, N).$$

Suppose we consider a random such network with directed network statistics specified by (4.1).

CONJECTURE. *In the thermodynamic limit the quantum stream along a randomly chosen arc is Poisson with random rate f, where the characteristic function $\phi(\zeta) = E(e^{\zeta f})$ of f satisfies*

$$\phi(\zeta) = \frac{\sum_j \sum_k j p_{jk} \phi(\zeta/j)^k}{\sum_j \sum_k j p_{jk}}. \tag{5.1}$$

Here p_{jk} is the distribution of node degree (j, k) determined in Section 3, and both summations exclude $j = 0$.

'PROOF': Suppose the arc chosen emanates from a node of degree (j, k). By the LTL property the inputs to the node will be independent, each Poisson with a rate having characteristic function $\phi(\zeta)$. Since the sum of the k independent inputs is then divided into j streams of equal rate, this rate on the output arcs will have characteristic function $\phi(\zeta/j)^k$. Averaging over j, k we deduce the identity (5.1). In averaging we use a distribution proportional to jp_{jk}, because the fact that we have chosen an arc randomly will weight the distribution p_{jk} by the factor j. In particular, the value $j = 0$ is excluded — a point worth making, since inverse powers of j will occur if we consider coefficients of powers of ζ in expression (5.1).

Relation (5.1) gives determining equations for the moments of f. One finds that $f = 0$ unless $P(j = 0) = 0$; nodes should have zero probability of being 'absorbing' if there is to be a continuing flow. If this condition is satisfied then there is an undetermined parameter in the f-distribution: the number of quanta contained in the graph component in which the chosen arc lies.

Our 'proof' is of course not a proof, because it appeals to a stronger form of the LTL property than has been established.

REFERENCES

Amit, D.J., Gutfreund, H., and Sompolinsky, H. (1985). Spin-glass models of neural networks. *Physical Review A* 32, 1007–1018.

——— (1985). Storing infinite numbers of patterns in a spin-glass model of neural networks. *Physical Review Letters* 55, 1530–1533.

Bollobás, B. (1985). *Random Graphs.* Academic Press, London.

Derrida, B., Gardner, E., and Zippelius, A. (1987). An exactly solvable asymmetric neural network model. *Europhysics Letters* 4, 167–173.

Derrida, B. and Nadal, J.P. (1987). Learning and forgetting on asymmetric, diluted neural networks. *Journal of Statistical Physics* 23, 993–1011.

Derrida, B. and Pomeau, Y. (1986). Random networks of automata; a simple annealed approximation. *Europhysics Letters* 1, 45–49.

Derrida, B. and Weisbuch, G. (1986). Evolution of overlaps between configurations in random Boolean networks. *Journal de Physique (Paris)* 47, 1297–1303.

Hammersley, J.M. and Welsh, D.J.A. (1980). Percolation theory and its ramifications. *Contemporary Physics* 21, 593–605.

Hilhorst, H.J. and Nijmeijer, M. (1987). On the approach of the stationary state in Kauffman's random Boolean network. *Journal de Physique (Paris)* 48, 185–191.

Hopfield, J.J. (1982). Neural networks and physical systems with emergent collective computational abilities. *Proceedings of the National Academy of Science (USA)* 79, 2554–2558.

Kauffman, S.A. (1969). Random genetic nets. *Journal of Theoretical Biology* 22, 437–467.

Kesten, H. (1982). *Percolation Theory for Mathematicians.* Birkhäuser, Boston.

Little, W.A. and Shaw, G.L. (1978). Analytic study of the memory storage capacity of a neural network. *Mathematical Biosciences* 39, 281–290.

Spitzer, F. (1979). Markov random fields on an infinite tree. *Annals of Probability* 3, 387–398.

Whittle, P. (1986). *Systems in Stochastic Equilibrium.* Wiley, Chichester.

——— (1989a). The antiphon: a device for reliable memory from unreliable elements. *Proceedings of the Royal Society A*, to appear.

——— (1989b). The statistics of random directed graphs. *Journal of Statistical Physics*, to appear.

——— (1989c). Random fields on random graphs. In preparation.

Statistical Laboratory
University of Cambridge
16 Mill Lane
Cambridge CB2 1SB.

Bond Percolation Critical Probability Bounds for the Kagomé Lattice by a Substitution Method

John C. Wierman[1]

Abstract

A new substitution method improves bounds for critical probabilities of the bond percolation problem on the Kagomé lattice, $\mathcal{K}$. The method theoretically produces a sequence of upper and lower bounds, in which the second pair of bounds establish

$$.5182 \leq p_c(\mathcal{K}) \leq .5335.$$

1. Introduction

Percolation processes were introduced by Broadbent and Hammersley in 1957 as models for the flow of a fluid through a random medium. A *bond percolation model* is comprised of an infinite lattice graph G, with each bond independently designated as open with probability p, $0 < p < 1$, and closed with probability $q = 1 - p$. The *open cluster* containing a specific vertex $v \in G$, denoted C_v, is the set of all vertices that can be reached from v through a path of open bonds. Let P_p denote the probability measure corresponding to parameter value p. The *critical probability* of the graph G, denoted by $p_c(G)$, is defined by $p_c(G) = \inf\{p : P_p[|C_v| = \infty] > 0\}$, which is independent of the vertex v if G is connected.

Since the seminal papers on mathematical percolation theory (Broadbent and Hammersley 1957; Hammersley 1957), there has been considerable interest in determining exact values of the critical probability for specific lattices. Sykes and Essam (1964) gave a heuristic determination of exact critical probability values for the square, triangular, and hexagonal lattice bond percolation models, and the value $\frac{1}{2}$ for the triangular lattice site

[1]Research supported in part by the National Science Foundation through grants DMS-8403646 and DMS-8801209.

percolation model. Verifying these values was a focus of research for nearly two decades. Even now, their method has not been completely justified.

The first rigorous determination of the critical probability of a periodic graph was due to Kesten (1980), who proved that the critical probability of the square lattice bond model is $\frac{1}{2}$. Wierman (1981) verified the values conjectured by Sykes and Essam for the triangular and hexagonal lattice bond model critical probabilities. The critical probabilities of two lattices for which no values were previously conjectured were found by Wierman (1984). The key aspects of these proofs were the use of planar graph duality, and (for graphs that are not self-dual) use of the star-triangle transformation. However, only these few cases currently have rigorous solutions, and there is no general method for rigorously determining critical probability values. There are only a few techniques for generating bounds on the critical probability of other graphs, and these provide unsatisfying results. In no case do they provide bounds that completely determine the leading digit of the critical probability value. For example, prior to this work, the Kagomé lattice bond model was known to satisfy $.4045 \leq p_c(\mathcal{K}) \leq .6180$. (See Wierman 1988.) The purpose of this paper is to introduce a rigorous method for determining much more accurate bounds for two-dimensional bond percolation models.

A key aspect of the proofs of Wierman (1981, 1984) is use of the star-triangle transformation, which was a crucial tool in Sykes and Essam's derivation for the triangular and hexagonal lattice bond models. Ottavi (1979) also used the star-triangle transformation to compute upper and lower bounds for the Kagomé lattice, by a non-rigorous argument. In each case, two lattices are related by a substitution of portions of one lattice into the other, while applying an appropriate transformation to the parameters in the percolation model. In the cases considered by Wierman (1981, 1984), transformations provide the equivalence of two models, while in Ottavi's case an exact transformation does not exist. Motivated by this previous work, we propose a modified 'substitution method' which can be rigorously verified. The method has the advantage of providing both upper and lower bounds. In fact, by considering larger portions of the lattices as the basic units of substitution, sequences of upper and lower bounds are obtained. While the computations become increasingly unwieldy, the first two bounds have been computed for the Kagomé lattice, and suggest that the bounds converge rapidly to the true critical probability value, although there is no proof of convergence.

To illustrate the method, and familiarize the reader with the basic ideas, we begin §2 by calculating the first of the sequence of upper and lower bounds for the Kagomé lattice: $.5182 \leq p_c(\mathcal{K}) \leq .5413$. The best previous upper bound, derived by the contraction principle of Wierman (1988), was $p_c(\mathcal{K}) \leq .6180$, while the best lower bound was by containment

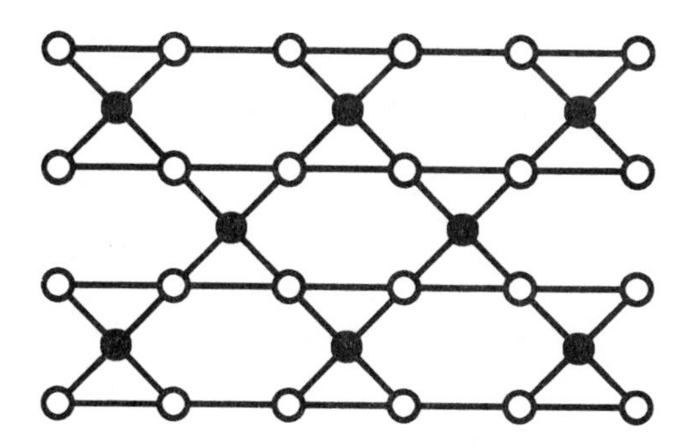

FIG. 1. The Kagomé lattice.

in the bowtie lattice, giving $p_c(\mathcal{K}) \geq .4045$. Later in §2, the upper bound is improved further, to $p_c(\mathcal{K}) \leq .5335$, by computing the second bound in the sequence.

The bounds obtained contradict an early Monte Carlo estimate of $.449 \pm .032$ by Dean (1963), and a renormalization group method estimate of .4697 by Murase and Yuge (1979). A Monte Carlo estimate of .526 by Neal (1972) is consistent with these bounds. It is rare to have sufficiently accurate rigorous bounds to rule out such estimates.

The substitution method introduced and applied to the Kagomé lattice in §2 may also be applied to other lattices. Preliminary work on the pentagon lattice bond percolation model indicates substantial improvement over the best previous bounds, obtained by Wierman (1988) by the contraction principle. Research in progress is investigating the possibility of extending the method to site percolation models, and suggests that the best current bounds for the square lattice site model may be improved.

Some necessary definitions and background material are included in the description of the substitution method presented in §2. The proof itself is given in §3. It shows the equivalence of two partial orders on the set of probability measures on a partially ordered set.

2. Kagomé Lattice Computations

We begin by illustrating the computation of bounds for the Kagomé lattice bond percolation critical probability, deferring the justification for the computations until §3. The Kagomé lattice is shown in Figure 1. It arises as the dual graph of the dice lattice, and also as the covering graph or line graph of the hexagonal lattice.

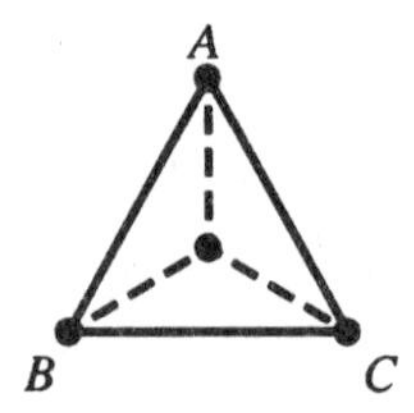

FIG. 2. Superposition of a triangle and a three-star.

Comparison with the Hexagonal Lattice

Note that the Kagomé lattice may be partitioned into disjoint triangles. By substituting a three-star for each triangle in $\mathcal{K}$, we obtain the graph $\mathcal{H}^\star$, which may be recognized as a subdivision of the hexagonal lattice $\mathcal{H}$ with one vertex subdividing each edge of $\mathcal{H}$. Since the critical probability of bond percolation on $\mathcal{H}$ is exactly $1-\sin\frac{\pi}{18}$ (see Wierman 1981), the bond percolation critical probability of $\mathcal{H}^\star$ is $q_0 = \{1 - 2\sin\frac{\pi}{18}\}^{1/2} \approx .807901$. In the remainder of this section, we first construct two probability measures on partitions of the vertices on the boundary of the triangle and the three-star, derived from the bond percolation models on the Kagomé and subdivided hexagonal lattices. We introduce a concept of stochastic ordering which allows us to compare these probability measures, and determine a parameter value p_L so that the probability measure corresponding to percolation on $\mathcal{K}$ is stochastically smaller than that associated with $\mathcal{H}^\star$ with parameter .807901, and a parameter value p_U such that the reverse holds. In §3, we show that p_L and p_U are in fact lower and upper bounds for $p_c(\mathcal{K})$.

Boundary Partitions

Consider a three-star and a triangle superimposed as in Figure 2, denoting the vertices on the boundary by A, B, and C. Any configuration (a designation of bonds as open and closed) on the triangle partitions the boundary vertices $\{A, B, C\}$ into clusters of vertices which are connected by open bonds, and similarly for configurations on the three-star. Each such *boundary partition* may be denoted by a sequence of vertices and vertical bars, where vertices are in distinct open clusters if and only if they are separated by a vertical bar.

The percolation model on the Kagomé lattice with parameter p induces a probability measure on the set of boundary partitions, in which the probability of a particular boundary partition is the sum of the probabilities of all configurations on the triangle which produce that boundary partition. Simple calculations show that this probability measure, which we denote by P_p^K, is given by

$$P_p^K[ABC] = 3p^2(1-p) + p^3,$$
$$P_p^K[AB \mid C] = P_p^K[AC \mid B] = P_p^K[A \mid BC] = p(1-p)^2,$$
$$P_p^K[A \mid B \mid C] = (1-p)^3.$$

Similarly, a different probability measure on the boundary partitions is determined by the percolation model on the subdivision of the hexagonal lattice with parameter q. We denote this probability measure by P_q^H, and compute that

$$P_q^H[ABC] = q^3,$$
$$P_q^H[AB \mid C] = P_q^H[AC \mid B] = P_q^H[A \mid BC] = q^2(1-q),$$
$$P_q^H[A \mid B \mid C] = 3q(1-q)^2 + (1-q)^3.$$

We will compare these two probability measures to derive the critical probability bounds for the Kagomé lattice.

The Partition Lattice

For two boundary partitions π and σ, we say that σ *dominates* π, denoted $\pi \leq \sigma$, if any two elements u and v that are in a cluster in π are also in a cluster in σ. Equivalently, $\pi \leq \sigma$ if and only if every cluster of π is wholly contained in a cluster of σ, or, conversely, every cluster of σ fully decomposes into clusters of π. If this is the case, π is called a *refinement* of σ. The set of boundary partitions on a given graph, when ordered by refinement, is a partially ordered set which is in fact a lattice, called the *partition lattice.*

In our example above, we see that ABC dominates each of $AB \mid C$, $AC \mid B$, and $A \mid BC$ in the refinement ordering, which in turn each dominate $A \mid B \mid C$.

Stochastic Ordering of Probability Measures on Partially Ordered Sets

Let $(S, \leq)$ be a finite partially ordered set (also called a *poset*). A function $P : S \to [0,1]$ is a *probability measure on* S if $P(s) \geq 0$ for all $s \in S$ and $\sum_{s \in S} P(s) = 1$. P may be defined on subsets of S by $P[A] = \sum_{s \in A} P(s)$. (Note that we use the same notation for the probability measure and its frequency function.)

A *filter* in a partially ordered set S is a subset $F \subset S$ such that if $g \geq f$ and $f \in F$, then $g \in F$. For $A \subset S$, the set $\{x \in S : x \geq a$ for some $a \in A\}$ is a filter, denoted $F(A)$, called the *filter generated by* A.

For two probability measures P and Q defined on a partially ordered set S, we say that P is *stochastically smaller* than Q, denoted $P \leq_S Q$, if for each filter F, $P[F] \leq Q[F]$. If the partially ordered set is a subset of the real line with the usual ordering, this definition agrees with the usual concept of stochastic ordering. Note that two probability measures may be incomparable with respect to the stochastic partial ordering $\leq_S$. Any filter is generated by its minimal elements. Thus, it is not necessary to check the inequality for filters generated by all subsets, but only for the filters generated by sets called anti-chains (in which all elements are incomparable).

Computation of Bounds

In §3, it is shown that if $P_p^K \leq_S P_{q_0}^H$, then p is a lower bound for the critical probability of the bond percolation model on the Kagomé lattice, and that if $P_{q_0}^H \leq_S P_p^K$, then p is an upper bound for the critical probability of the bond percolation model on the Kagomé lattice. Therefore, to compute a lower bound for $p_c(\mathcal{K})$, we solve for the largest p satisfying the following four inequalities: from the filter consisting of only ABC,

$$3p^2(1-p) + p^3 \leq q_0^3;$$

from the filters generated by one, two, or all three of $AB \mid C$ and $AC \mid B$ and $A \mid BC$,

$$\begin{aligned} p(1-p)^2 + 3p^2(1-p) + p^3 &\leq q_0^2(1-q_0) + q_0^3 \\ 2p(1-p)^2 + 3p^2(1-p) + p^3 &\leq 2q_0^2(1-q_0) + q_0^3 \\ 3p(1-p)^3 + 3p^2(1-p) + p^3 &\leq 3q_0^2(1-q_0) + q_0^3. \end{aligned}$$

(The filter generated by $A \mid B \mid C$ consists of the entire partially ordered set, which has probability one in both measures.) Numerical solution of the inequalities produces the upper bound

$$p_c(\mathcal{K}) \leq .5413.$$

Reversing all four inequalities and solving, we obtain the lower bound

$$p_c(\mathcal{K}) \geq .5182.$$

Sequences of Bounds

By partitioning the Kagomé lattice into larger regions, and carrying out the same process of substitution of corresponding regions of the $\mathcal{H}^*$ lattice, a sequence of bounds may be obtained.

To illustrate, a second pair of bounds for $p_c(\mathcal{K})$ may be obtained by considering a region consisting of two adjacent triangles, shown in Figure 3.

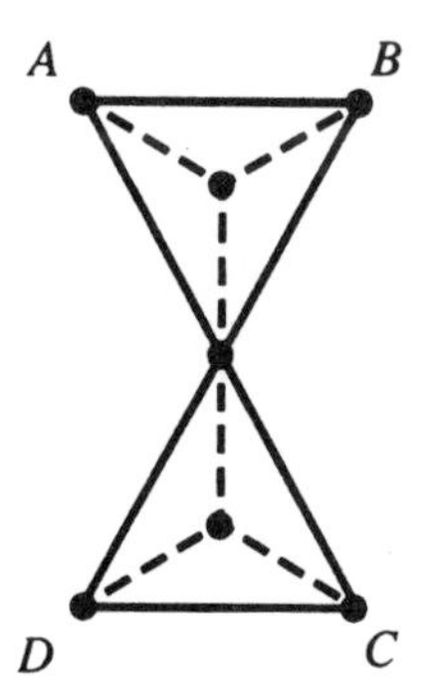

FIG. 3. Superposition of a pair of triangles and a pair of three-stars, used in determining the second set of critical probability bounds.

For the percolation model on $\mathcal{K}$ with parameter p, the probability measure P_p^K is given by:

$$P_p^K[ABCD] = [3p^2(1-p)+p^3]^2,$$

$$P_p^K[ABC \mid D] = P_p^K[ABD \mid C] = P_p^K[ACD \mid B] = \\ = P_p^K[BCD \mid A] = [3p^2(1-p)+p^3]p(1-p)^2,$$

$$P_p^K[AB \mid CD] = 2[3p^2(1-p)+p^3]p(1-p)^2 + p^2(1-p)^4,$$

$$P_p^K[AC \mid B \mid D] = P_p^K[AD \mid B \mid C] = P_p^K[BD \mid A \mid C] = \\ = P_p^K[BC \mid A \mid D] = p^2(1-p)^4,$$

$$P_p^K[AB \mid C \mid D] = P_p^K[CD \mid A \mid B] = \\ = p(1-p)[4p(1-p)^3 + (1-p)^4 + p^2(1-p)^2] + (1-p)^4p^2.$$

For the percolation model on $\mathcal{H}^*$ with parameter q_0, we determine $P_{q_0}^H$ to be:

$$P_{q_0}^H[ABCD] = q_0^6,$$

$$P_{q_0}^H[ABC \mid D] = P_{q_0}^H[ABD \mid C] = P_{q_0}^H[ACD \mid B] = \\ = P_{q_0}^H[BCD \mid A] = q_0^5(1-q_0),$$

$$P^H_{q_0}[AB \mid CD] = q_0^4(1 - q_0^2),$$

$$P^H_{q_0}[AC \mid B \mid D] = P^H_{q_0}[AD \mid B \mid C] = P^H_{q_0}[BD \mid A \mid C] = \\ = P^H_{q_0}[BC \mid A \mid D] = q_0^4(1 - q_0)^2,$$

$$P^H_{q_0}[AB \mid C \mid D] = P^H_{q_0}[CD \mid A \mid B] = \\ = q_0^2[(1 - q_0)^2 + 2q_0(1 - q_0)(1 - q_0^2)].$$

Numerical solution of the inequalities generated by all filters provide the bounds

$$.5182 \leq p_c(\mathcal{K}) \leq .5335.$$

The upper bound is an improvement over the first calculation, but the lower bound is identical, since the inequality generated by the filter consisting of $ABCD$ is equivalent to that in the first step, and is the active constraint. The smallest region that will produce an improvement in the lower bound is a ring of six triangles in $\mathcal{K}$, for which the computations have not been completed.

If one considers a sequence of larger regions, each a union of copies of its predecessor, one will obtain monotone sequences of upper and lower bounds. It is not known if these sequences converge to the true critical probability value, although the computational results suggest rapid convergence.

3. Justification

The Flow Ordering

Let Q be a probability measure on a partially ordered set $(S, \leq)$. If $s \in S$ with $Q[s] > 0$, and $t \in S$ with $t < s$, we may construct a new probability measure P by moving probability mass x, $0 < x < Q[s]$, from s to t. Formally, define P by letting $P[s] = Q[s] - x$, $P[t] = Q[t] + x$, and $P[u] = Q[u]$ for $u \neq s, t$. Since the construction moves probability downward in the poset (from one element to another that it dominates), we say that P is constructed from Q by a *downward flow.*

The concept of downward flow leads to a partial ordering on the set of probability measures on the poset $(S, \leq)$. We call a finite sequence of downward flows a *flow sequence.* A flow sequence on a subset F of S moves probability only between elements of F, leaving the probability measure unchanged elsewhere. Define the *flow ordering*, $\leq_F$, by letting $P \leq_F Q$ if and only if P may be constructed from Q using a flow sequence.

The downward flow operation may be interpreted in terms of random variables. The probability measure Q is the distribution of an S-valued random variable X. A downward flow moving probability mass x in Q from s to t, where $s > t$, produces a probability measure which is the distribution of the S-valued random variable which takes the value X if $X \neq s$ or if $B_{x,s,t} = 0$, and the value t if $X = s$ and $B_{x,s,t} = 1$, where $B_{x,s,t}$ is a Bernoulli random variable, independent of X, with $P[B_{x,s,t} = 1] = x/Q(s)$. We will denote this random variable by $X * B_{x,s,t}$. Thus, if $P \leq_F Q$ and X has distribution Q, there exist Bernoulli random variables B_{x_i,s_i,t_i}, $i = 1, 2, \ldots, n$, (corresponding to a flow sequence) such that $X * B_{x_1,s_1,t_1} * \cdots * B_{x_n,s_n,t_n}$ has distribution P.

Inequalities Between Percolation Probabilities

Consider probability measures P_p^K and P_q^H on a partition lattice derived from percolation models on the Kagomé and $\mathcal{H}^*$ lattices. Fix the parameter values p and q so that $P_p^K \leq_F P_q^H$. We will construct related bond percolation models on $\mathcal{K}$ and $\mathcal{H}^*$ by the following procedure.

Superimpose $\mathcal{K}$ on $\mathcal{H}^*$ so that each triangle in $\mathcal{K}$ exactly contains one three-star of $\mathcal{H}^*$. Construct the percolation model on $\mathcal{H}^*$ as usual, by declaring each edge to be open with probability q, independently of all other edges. This creates a random boundary partition X_σ with probability distribution P_q^H on the set of boundary vertices of each star-triangle pair σ in $\mathcal{H}^*$.

Since $P_p^K \leq_F P_q^H$, there is a flow sequence which produces P_p^K from P_q^H. Thus, for each star-triangle pair σ there exist independent Bernoulli random variables (independent of those for other star-triangle pairs) B_{σ,x_i,s_i,t_i} such that $Y_\sigma = X_\sigma * B_{\sigma,x_1,s_1,t_1} * \cdots * B_{\sigma,x_n,s_n,t_n}$ has distribution P_p^K exactly.

Define vertices u and w of $\mathcal{K}$ to be in the same open cluster if and only if there exists a sequence of vertices $u = v_0, v_1, v_2, \ldots, v_{k-1}, v_k = w$ for some k, such that for each $i = 1, 2, \ldots, k$, v_{i-1} and v_i are on a common star-triangle pair σ and are in the same boundary partition set in Y_σ. With this definition, for each connected subgraph C in $\mathcal{K}$, the probability that C is an open cluster is identical in the model just described and in the bond percolation model on the Kagomé lattice with parameter p.

Let $C(\mathcal{H}^*)$ denote the open cluster in $\mathcal{H}^*$ containing a fixed vertex $v \in \mathcal{K}$, and let $C(\mathcal{K})$ denote the open cluster in $\mathcal{K}$ containing v. By construction, the boundary partition of each triangle in $\mathcal{K}$ is a refinement of the boundary partition of the corresponding three-star in $\mathcal{H}^*$. Therefore, $u \in C(\mathcal{K})$ implies $u \in C(\mathcal{H}^*)$, so $C(\mathcal{K}) \subset C(\mathcal{H}^*)$. If $q < p_c(\mathcal{H}^*)$, then $P_q^H[|C(\mathcal{H}^*)| = \infty] = 0$, so $P_p^K[|C(\mathcal{K})| = \infty] = 0$ also. Thus, $q < p_c(\mathcal{H}^*)$ implies $p < p_c(\mathcal{K})$.

It follows that to compute a lower bound for $p_c(\mathcal{K})$ it is sufficient to find the largest value of p such that P_p^K is smaller than $P_{q_0}^H$ in the flow ordering. Similar reasoning applies to the determination of upper bounds. In order to conveniently compute such values of p, in the following we show that the flow ordering is equivalent to the stochastic ordering.

Combinations of Flow Sequences

Let f and g be flow sequences on a poset $(S, \leq)$. For $\lambda \in [0,1]$, define λf to be the sequence of flows which moves probability mass λa from s to t whenever f moves a from s to t. Define $f+g$ to be the downward flow sequence which moves $a+b$ from s to t whenever f moves a and g moves b from s to t.

Suppose Q_f and Q_g are the probability measures obtained from Q by applying the downward flow sequences f and g respectively. Then the measure Q_λ, corresponding to $f_\lambda = \lambda f + (1-\lambda)g$, satisfies $Q_\lambda[s] = \lambda Q_f[s] + (1-\lambda)Q_g[s]$ for all $s \in S$, so is a linear function of λ at each element of S.

Equivalence of Stochastic Ordering and Flow Ordering

It is easy to see that if P is obtained from Q by a flow sequence, then P is stochastically smaller than Q. (It suffices to check this for a single downward flow.) We next show that the converse is true, so, in fact, the partial orders $\leq_S$ and $\leq_F$ on the set of probability measures are equivalent.

Let P and Q be probability measures on a partition lattice S, satisfying $P \leq_S Q$. The *excess* of Q relative to P at a set V of elements of S is defined by $e_{Q/P}(V) = \max\{Q[V] - P[V], 0\}$. Similarly, the deficit of Q relative to P on V is defined by $d_{Q/P}(V) = \max\{P[V] - Q[V], 0\}$.

Suppose that there does not exist a flow sequence which produces P from Q. Consider the set $\mathcal{F}$ of filters F for which no flow sequence on F can produce a probability Q' from Q which satisfies $Q'[s] \geq P[s]$ for all $s \in F$. $\mathcal{F}$ is non-empty, since the entire poset S is in $\mathcal{F}$ by hypothesis.

Let A be a minimal filter in $\mathcal{F}$ (when $\mathcal{F}$ is ordered by set inclusion). Consider the probability measure Q' obtained from Q by a flow sequence on A which minimizes the sum of deficits (of Q' relative to P) of elements in A. If there exist elements of A which are not minimal and have positive excesses, allow the excess to flow downward so that only minimal elements of A have positive excesses. This cannot increase the deficit sum, since it removes probability only from elements with positive excesses. It also cannot decrease the deficit sum, since the original flow sequence minimizes the deficit sum. Denote the resulting flow sequence by f_0 and the corresponding probability measure by Q_0.

If Q_0 has no positive excess relative to P at any element of A, then since by hypothesis there is an element with a positive deficit, the filter A satisfies $P[A] > Q_0[A]$. However, the flow sequence f_0 kept all Q-probability in A while constructing Q_0, so $Q_0[A] = Q[A]$, which implies that $P[A] > Q[A]$. This exhibits a filter for which the stochastic ordering condition fails, so $P \leq_S Q$ does not hold.

If Q_0 has a positive excess relative to P at an element of A, let E denote the set of elements with positive excesses and D denote the set of elements with positive deficits.

Note that the set $A \setminus E$ is a filter, since all elements of E are minimal in A. By the minimality of A, there exists a flow sequence f_1 on $A \setminus E$ which produces a probability measure Q_1 satisfying $Q_1[s] \geq P[s]$ for all $s \in A \setminus E$.

Consider the flow sequences $f_\lambda = \lambda f_0 + (1 - \lambda) f_1$, where $0 < \lambda < 1$, denoting the resulting probability measures by Q_λ.

If $s \in A \setminus (E \cup D)$, we have $P[s] = Q_0[s] \leq Q_1[s]$ and thus by linearity $Q_\lambda[s] \geq P[s] = Q_0[s]$ for all λ.

If $s \in E$, then by definition $Q_0[s] > P[s]$. Although $Q_1[s]$ may be less than $P[s]$, by linearity there exists $\epsilon_s > 0$ such that $Q_\lambda[s] \geq P[s]$ for all $\lambda < \epsilon_s$. Since E is a finite set, there exists ϵ_E such that for all $s \in E$ and for all $\lambda \leq \epsilon_E$ we have $Q_\lambda[s] \geq P[s]$.

If $s \in D$, then $Q_0[s] < P[s]$ and $Q_1[s] \geq P[s]$, so $Q_\lambda[s]$ is a strictly increasing function of λ, and thus the deficit of Q_λ relative to P at s is strictly decreasing as a function of λ.

Thus, for $\lambda \leq \epsilon_E$, Q_λ has no positive deficits relative to P except at elements of D, at which the deficits are strictly decreasing functions of λ. Hence, each f_λ, $0 < \lambda < \epsilon_E$, is a flow sequence on A which produces a smaller sum of deficits at elements of A than f_0. Since this is a contradiction, there actually is no element s with $e_{Q_0/P}(s) > 0$.

Therefore, if there is no flow sequence that can obtain P from Q, there exists a filter A such that $P[A] > Q[A]$, so $P \leq_S Q$ does not hold. Hence, the partial orderings $\leq_F$ and $\leq_S$ on the set of probability measures on a poset S are equivalent.

REFERENCES

Broadbent, S.R. and Hammersley, J.M. (1957). Percolation processes I: crystals and mazes. *Proceedings of the Cambridge Philosophical Society* 53, 629–641.

Dean, P. (1963). A new Monte Carlo method for percolation problems on a lattice. *Proceedings of the Cambridge Philosophical Society* 59, 397–410.

Hammersley, J.M. (1957). Percolation processes: lower bounds for the critical probability. *Annals of Mathematical Statistics* 28, 790–795.

Kesten, H. (1980). The critical probability of bond percolation on the square lattice equals $\frac{1}{2}$. *Communications in Mathematical Physics* 74, 51–59.

Murase, C. and Yuge, Y. (1979). A renormalization group approach for two-dimensional percolating systems: hexagonal bond lattice and Kagomé lattice. *Journal of Physics A: Mathematical and General* 12, 1541–1546.

Neal, D.G. (1972). Estimates of critical percolation probabilities for a set of two-dimensional lattices. *Proceedings of the Cambridge Philosophical Society* 71, 97–106.

Ottavi, H. (1979). Majorant et minorant du seuil de percolation de lien du Kagomé. *Journal de Physique* 40, 233–237.

Sykes, M.F. and Essam, J.W. (1964). Exact critical percolation probabilities for site and bond problems in two dimensions. *Journal of Mathematical Physics* 5, 1117–1127.

Wierman, J.C. (1981). Bond percolation on the honeycomb and triangular lattices. *Advances in Applied Probability* 13, 298–313.

——— (1984). A bond percolation critical probability determination based on the star-triangle transformation. *Journal of Physics A: Mathematical and General* 17, 1525–1530.

——— (1988). Bond percolation critical probability bounds derived by edge contraction. *Journal of Physics A: Mathematical and General* 21, 1487–1492.

Department of Mathematical Sciences
The Johns Hopkins University
Baltimore
Maryland 21218.

Brownian Motion and the Riemann Zeta-Function

David Williams

My intention when I went to Oxford as an undergraduate was to be a physicist, but to do some mathematics first. In the first undergraduate year of mathematics, John Hammersley gave a course which included the quaternion proof of the Four-Squares Theorem, the 'elementary' proof of the Prime Number Theorem, and an introduction to Operational Research. It was strong meat for first-year undergraduates, and I'm sure that there was little which I understood fully. But it was marvellous material, conveyed with style and infectious excitement; and, more than anything, it persuaded me to stay with mathematics.

My intention when I was invited to contribute to this volume was to submit something on one or other of two concrete problems of interest to John Hammersley. But, unfortunately, while John had been able to convey enthusiasm to us students, it was of course impossible for him to grant us some of his creativity with hard problems; and, left therefore to my own devices, I have failed to make progress with either problem. Meanwhile ...

It seems that number-theorists have recently become interested in path-integral representations of the Riemann ζ-function. Such representations have for a long time been familiar to aficionados of Brownian excursion theory — I am sure that Kai Lai Chung and many others have known them as long as I have. C.M. Newman (1975) had explained that if it could be shown that a certain probability density function is 'ferromagnetic', then the Riemann Hypothesis would follow. The fact that this density function arises fairly naturally in the study of Brownian motion (of which more, I hope, in a later paper with Tim Mortimer) therefore has a certain entertainment value, though perhaps nothing more. Here — with thanks, apologies for not being brighter, and very best wishes, to John Hammersley — is a talk I gave recently to some of the number-theorists at Cambridge.

The theory of Brownian motion contains many remarkable identities. Many now have a complete explanation, though even in certain of these cases, there was a time when they were regarded as 'coincidences'. Amongst the identities for which a proper explanation remains to be found are some which are closely related to Riemann's ξ-function.

Contents

1. What is Brownian Motion?

For $t > 0$ and $x, y \in \mathbb{R}$, define

$$p_t(x, y) = (2\pi t)^{-1/2} \exp\{-(y - x)^2/2t\}. \tag{1.1}$$

Thus $p_t(x, \cdot)$ is the density of the *normal distribution* of mean x and variance t. The fact that p solves the *heat equation*:

$$D_t p = \tfrac{1}{2} D_{xx} p = \tfrac{1}{2} D_{yy} p \tag{1.2}$$

is best regarded as expressing the formula

$$P_t = \exp(t\tfrac{1}{2}\Delta), \text{ where } P_t f(x) := \int_{\mathbb{R}} p_t(x, y) f(y) dy, \tag{1.3}$$

which is made precise by Hille-Yosida theory. Let $\mathcal{C}$ be the smallest σ-algebra on *'path-space'* $C[0, \infty)$ such that, for each $t > 0$, the evaluation map $w \mapsto w(t)$ on $C[0, \infty)$ is $\mathcal{C}$ measurable.

WIENER'S THEOREM. *For $x \in \mathbb{R}$, there exists a unique measure W^x on $(C[0,\infty), \mathcal{C})$ such that for $n \in \mathbb{N}$, for $0 < t_1 < t_2 < \ldots < t_n$ and for $A_1, A_2, \ldots, A_n \in \mathcal{B}(\mathbb{R})$,*

$$W^x(\{w \in C[0,\infty) : w_{t_i} \in A_i \ (1 \le i \le n)\})$$
$$= \int_{x_1 \in A_1} \cdots \int_{x_n \in A_n} \prod_{i=1}^{n} \{p_{t_i - t_{i-1}}(x_{i-1}, x_i) dx_i\} \qquad (1.4)$$

where $t_0 = 0, x_0 = x$.

The probability measure W^x is called *Wiener measure corresponding to starting position x.*

Suppose that we have a set-up $(\Omega, \mathcal{F}, \mathbb{P}^x, B)$ where Ω is a set, $\mathcal{F}$ is a σ-algebra on Ω, each $\mathbb{P}^x$ is a probability measure on $(\Omega, \mathcal{F})$, and

$$B : \Omega \to C[0,\infty), \qquad B^{-1} : \mathcal{C} \to \mathcal{F},$$
$$\omega \mapsto (t \mapsto B_t(\omega)).$$

Then B is called a *Brownian motion* if $\mathbb{P}^x \circ B^{-1} = W^x \ (x \in \mathbb{R})$ on $\mathcal{C}$.

Canonical Brownian motion is the set-up:

$$(\Omega, \mathcal{F}, \mathbb{P}^x, B) = (C[0,\infty), \mathcal{C}, W^x, \mathrm{id}).$$

Properties (1.3) and (1.4) say: *Brownian motion is Markovian with transition density function p and with generator $\frac{1}{2}\Delta$.*

Expectation. If $Z : \Omega \to \mathbb{R}$ is $\mathcal{F}$-measurable, we define

$$\mathbb{E}^x Z = \int_\Omega Z(w) \mathbb{P}^x(dw).$$

Example. For Borel function f on $\mathbb{R}$, $\mathbb{E}^x f(B_t) = \int f(y) \mathbb{P}^x(B_t \in dy) = P_t f(x)$.

2. Cauchy's Proof of the Functional Equation

Let Γ be the circle $\Gamma = \mathbb{R}/\mathbb{Z}\sqrt{2\pi}$, and let π (no confusion possible!) be the projection $\pi : \mathbb{R} \to \Gamma$. If B is $BM(\mathbb{R})$ (a Brownian motion on $\mathbb{R}$), then $B^\Gamma := \pi \circ B$ is a $BM(\Gamma)$, Markovian with transition density function

$$p_t^\Gamma(x, y) = \sum_{\{z : \pi z = y\}} p_t(x, z) \qquad (2.1)$$

and generator $\frac{1}{2}\Delta^\Gamma$. Now, $\frac{1}{2}\Delta^{\mathcal{G}}$ has

normalized eigenfunctions: $(2\pi)^{-1/4}e^{in\sqrt{2\pi}\theta}$ $(n \in \mathbb{Z})$,
corresponding eigenvalues: $-n^2\pi$.

Hence $P_t^\Gamma := \exp(t\frac{1}{2}\Delta^\Gamma)$ has eigenvalues $e^{-n^2\pi t}$, so that

$$\text{Trace}\left(P_t^\Gamma\right) = \theta(t) := \sum_{n\in\mathbb{Z}} e^{-n^2\pi t}.$$

But, using (2.1) and the obvious fact that $\text{Trace}(P_t^\Gamma) = \sqrt{2\pi}p_t^\Gamma(0,0)$, we see that

$$\text{Trace}(P_t^\Gamma) = \sqrt{2\pi}\sum_{n\in\mathbb{Z}}(2\pi t)^{-1/2}\exp(-n^2\cdot 2\pi/2t) = t^{-1/2}\theta(t^{-1}).$$

So, we have

$$\theta(t) = t^{-1/2}\theta(t^{-1}). \qquad \text{(Jacobi)}$$

As everyone knows, $\zeta(z) = \sum_{n\in\mathbb{N}} n^{-z}$ extended analytically from $\{\mathcal{R}z > 1\}$ to $\mathbb{C}\backslash\{1\}$. It was already known to Riemann that Jacobi's functional equation for θ implies the functional equation

$$\xi(z) = \xi(1-z)$$

for ξ (or ζ), where ξ is the entire function:

$$\xi(z) = \tfrac{1}{2}z(z-1)\pi^{-z/2}\Gamma(\tfrac{1}{2}z)\zeta(z).$$

The Riemann Hypothesis says: *if* $\xi(z) = 0$, *then* $\mathcal{R}z = \frac{1}{2}$.

3. Brownian Bridges and Bessel Bridges

(a) The 1-dimensional case. Intuitively, Brownian bridge with values in $\mathbb{R}$, $BB(\mathbb{R})$, is $BM(\mathbb{R})$ with time-parameter set [0,1] conditioned to be at 0 at times 0 and 1. Rigorously, there is unique measure $W^{0,0}$ on $C[0,1]$ with obvious σ-algebra $\mathcal{C}[0,1]$ $(= \mathcal{B}(C[0,1])!)$ such that for every $h \in C_b(C[0,1])$,

$$\int_{C[0,1]} h(w)W^{0,0}(dw) = \lim_{\epsilon\downarrow 0}\frac{\int_{C[0,\infty)\cap\{w:|w(1)|<\epsilon\}} h\left(w|_{[0,1]}\right)W^0(dw)}{W^0\{|w(1)|<\epsilon\}}.$$

A set-up $(\Omega, \mathcal{F}, \mathbb{P}, BB)$, where $BB : \Omega \to C[0,1]$ etc, is called a $BB(\mathbb{R})$ if

$$\mathbb{P}\circ BB^{-1} = W^{0,0} \text{ on } \mathcal{C}[0,1].$$

Example. Suppose that $(\Omega, \mathcal{F}, \mathbb{P}^x, B)$ is a $BM(\mathbb{R})$, and that we set

$$BB_t(\omega) := \begin{cases} tB\left(\frac{1-t}{t},\omega\right) & \text{if } 0<t<1, \\ 0 & \text{if } t\in\{0,1\}. \end{cases}$$

Then $(\Omega, \mathcal{F}, \mathbb{P}^0, BB)$ is a $BB(\mathbb{R})$.

(3.1) THEOREM. *Let BB be a $BB(\mathbb{R})$, and set*

$$R(\omega) := \sqrt{\frac{2}{\pi}}\left(\sup_{t\le 1} BB_t(\omega) - \inf_{t\le 1} BB_t(\omega)\right).$$

Then, for all z in $\mathbb{C}$,

$$\xi(z) = \tfrac{1}{2}\mathsf{E}(R^z) = \tfrac{1}{2}\int_\Omega R(\omega)^z \mathsf{P}(d\omega).$$

This, or some equivalent, has been known for some time. It appears in a fine paper by Biane and Yor (1987). I would like to say something about how the result relates to *'interchanging space and time'*, and also to the following *Fourier expansion of* $BB(\mathbb{R})$.

(b) Let $(\Omega, \mathcal{F}, \mathsf{P})$ carry independent random variables $G_1, G_2, \ldots$ each normally distributed with mean 0 and variance 1. Define

$$BB_t(\omega) := \sum_{n\ge 1} \frac{G_n(\omega)}{n\pi}\sqrt{2}\sin(n\pi t), \quad 0 \le t \le 1.$$

Then $(\Omega, \mathcal{F}, \mathsf{P}, BB)$ is a $BB(\mathbb{R})$. Note that Parseval says:

$$\int_0^1 BB_t(\omega)^2 dt = \sum \frac{G_n(\omega)^2}{n^2\pi^2}. \tag{3.2}$$

(c) Brownian motion in $\mathbb{R}^n$, $BM(\mathbb{R}^n)$. We build $BM(\mathbb{R}^n)$ by making the component processes *independent* $BM(\mathbb{R})$ processes. Since

$$C([0,\infty); \mathbb{R}^n) = \prod_{i=1}^n C([0,\infty); \mathbb{R}) \text{ canonically,}$$

we can define, for $\mathbf{x} \in \mathbb{R}^n$,

$$\left(\mathbf{W}^{\mathbf{x}} \text{ on } (C[0,\infty); \mathbb{R}^n)\right) = \prod_{i=1}^n \left(W^{x_i} \text{ on } C([0,\infty); \mathbb{R})\right). \tag{3.3}$$

A $BM(\mathbb{R}^n)$ is a set-up $(\Omega, \mathcal{F}, \mathsf{P}^{\mathbf{x}}: \mathbf{x} \in \mathbb{R}^n, \mathbf{B})$, $\mathbf{B}: \Omega \to C([0,\infty); \mathbb{R}^n)$ such that $\mathsf{P}^{\mathbf{x}} \circ \mathbf{B}^{-1} = \mathbf{W}^{\mathbf{x}}$. A $BM(\mathbb{R}^n)$ is Markovian with generator

$$\tfrac{1}{2}\Delta = \sum \tfrac{1}{2}\frac{\partial^2}{\partial x_i^2} = \tfrac{1}{2}\frac{\partial^2}{\partial r^2} + \frac{n-1}{2r}\frac{\partial}{\partial r} + \frac{1}{r^2}\cdot\tfrac{1}{2}\Delta^{S^{n-1}}. \tag{3.4}$$

The first formula for Δ only reiterates the product measure structure (3.3).

(d) Bessel process $BES(n)$ on $[0,\infty)$. Invariance of the family $\mathbf{W}^{\cdot}$ under $O(n)$ implies that the radial part $r = |\mathbf{B}|$ of a $BM(\mathbb{R}^n)$ $\mathbf{B}$ is Markovian with generator $\frac{1}{2}\frac{d^2}{dr^2} + \frac{n-1}{2r}\frac{d}{dr}$; and we say that r is $BES(n)$, Bessel process on $[0,\infty)$ associated with dimension n. The second formula for $\frac{1}{2}\Delta$ at (3.4) means:

$$dr = d\beta + \frac{n-1}{2r}dt, \qquad \beta \text{ a } BM(\mathbb{R}),$$

$$\frac{B_t}{r_t} = BM^{S^{n-1}}\left(\int_0^t r_s^{-2}ds\right).$$

Making the first of these precise: if $(\Omega, \mathcal{F}, \mathbb{P}^{\mathbf{x}}, \mathbf{B})$ is a $BM(\mathbb{R}^n)$, then, for $\mathbf{x} \neq 0$, $\mathbb{P}^{\mathbf{x}} \circ \beta^{-1} = W^{|x|}$, where

$$\beta_t(\omega) := |\mathbf{B}_t(\omega)| - \int_0^t \frac{n-1}{2|\mathbf{B}_s(\omega)|}ds.$$

(e) Brownian bridge in $\mathbb{R}^n$, $BB(\mathbb{R}^n)$. If $(\Omega, \mathcal{F}, \mathbb{P}^x, \mathbf{B})$ is a $BM(\mathbb{R}^n)$, and $\mathbf{BB}_t(\omega) := t\mathbf{B}\left(\frac{1-t}{t}\right)$, then $\mathbb{P}^0 \circ \mathbf{BB}^{-1} = \mathbf{W}^{0,0} := \prod_{i=1}^n W^{0,0}$. So, we say that $(\Omega, \mathcal{F}, \mathbb{P}^0, \mathbf{B}B)$ is a $BB(\mathbb{R}^n)$.

(f) $BES(n)BR$ with values in $[0,\infty)$. If $(\Omega, \mathcal{F}, \mathbb{P}, r)$ is a $BES(n)$ starting at 0, then $(\Omega, \mathcal{F}, \mathbb{P}, tr\left(\frac{1-t}{t}\right))$ is a $BES(n)BR$.

Pythagoras says: if $(\Omega^i, \mathcal{F}^i, \mathbb{P}^i, BB^i)$ is a $BB(\mathbb{R})$ $(1 \le i \le n)$ and we set $(\Omega, \mathcal{F}, \mathbb{P}) = \prod_{i=1}^n (\Omega^i, \mathcal{F}^i, \mathbb{P}^i)$, then $(\Omega, \mathcal{F}, \mathbb{P}, \hat{r})$ is a $BES(n)BR$, where

$$\hat{r}_t(\omega) := \left(\sum_{i=1}^n BB_t^i(\omega)^2\right)^{1/2}.$$

4. The Excursion Picture of Reflecting Brownian Motion

Consider a reservoir which can hold any volume of water from $-\infty$ up to 0. Suppose input to reservoir is a $BM(\mathbb{R})$ process B starting at 0. This is represented by upper curve. Then

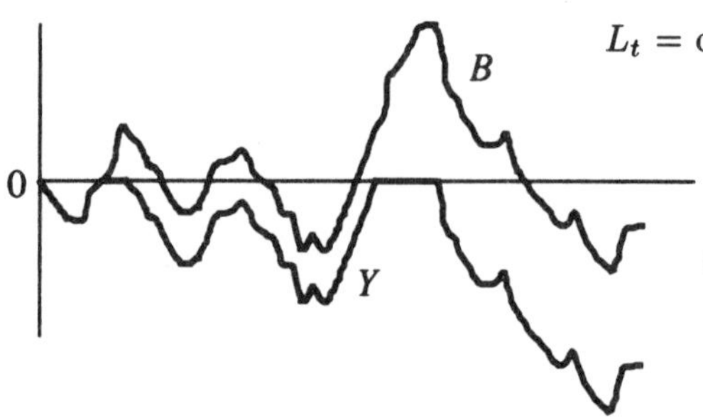

L_t = overflow by time $t = \sup_{s \le t} B_s$.

Actual volume at time t $= Y_t = B_t - L_t$.

$(X_t) = (-Y_t)$ is reflecting BM (RBM) on $[0,\infty)$, with the same law as $(|B_t|)$.

New picture

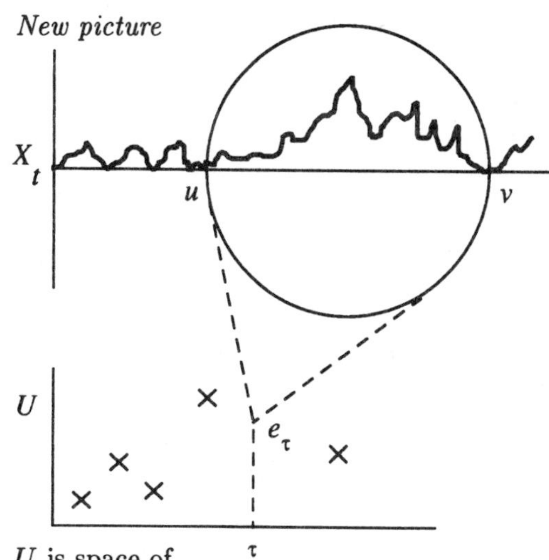

X: RBM, L local time at 0 for X (as above).

Define

$$\gamma_\tau := \inf \{t: L_t > \tau\}$$

For circled excursion of X away from 0,
$L_u = L_v = \tau$ (say),
$\gamma(\tau-)=u$, $\gamma(\tau)=v$.

Path in circle:

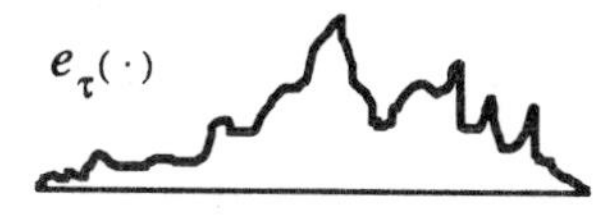

U is space of excursion paths

is excursion at local time τ:
$e_\tau(t) = X(t + \gamma_{\tau-})$, $0 \leq t \leq \gamma_\tau - \gamma_{\tau}$.

ITÔ'S THEOREM. *The points $(\tau, e_\tau(t))$ in $[0,\infty) \times U$ are the points of a Poisson point process. (N.B. We have point (τ, e_τ) if and only if $\gamma_\tau > \gamma_{\tau-}$.) Numbers falling in disjoint regions of $[0,\infty)\times U$ are* **independent** *variables. There exists a sigma-finite measure n on U such that for (measurable) $\Gamma \subseteq [0,\infty) \times U$, the number N_Γ of points in Γ has Poisson distribution parameter $\lambda(\Gamma)$, that is,*

$$\mathbb{P}(N_\Gamma = k) = e^{\lambda(\Gamma)}\lambda(\Gamma)^k/k! \text{ where } \lambda = \text{Lebesgue} \times n.$$

The measure n on the space of excursion paths is called the **Itô excursion law**. *Given n, we can build X from its excursions.*

5. The Itô Excursion Law: Bessel Descriptions

Picture of an excursion

M: maximum (at time V_1).
$V = V_1 + V_2$: lifetime.

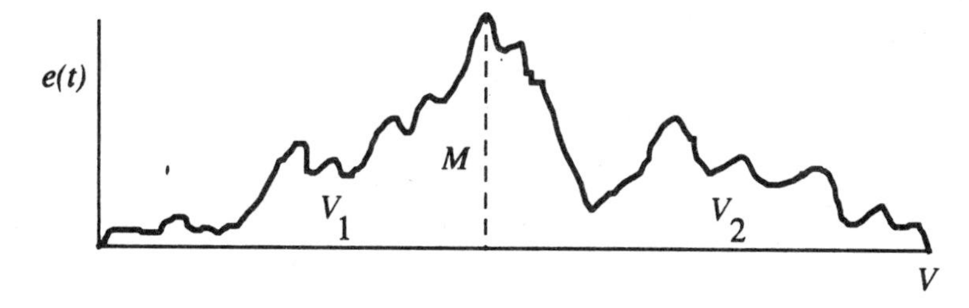

Description I.

I(a): $n(V \in dv) = \frac{1}{\sqrt{2\pi v^3}}dv$ (Lévy);

I(b): *under* n, *conditional on* V, $\{V^{-1/2}e(tV) : 0 \leq t \leq 1\}$ *is a* $BES(3)BR$ (Williams (1970) — after Lévy, Itô and McKean);

Description II.

II(a): $n(M \in dm) = m^{-2}dm$ (Lévy);

II(b): *under* n, *conditional on* M, $\{e_t : t \leq V_1\}$ *and* $\{e_{V-t}: t \leq V_2\}$ *are independent* $BES(3)$ *processes started at 0 and run until they hit* M (Williams (1970)).

Now consider the pictures (which relate to I(b) with $V = 1$ and II(b) with $M = 1$ — both 'scaled' versions):

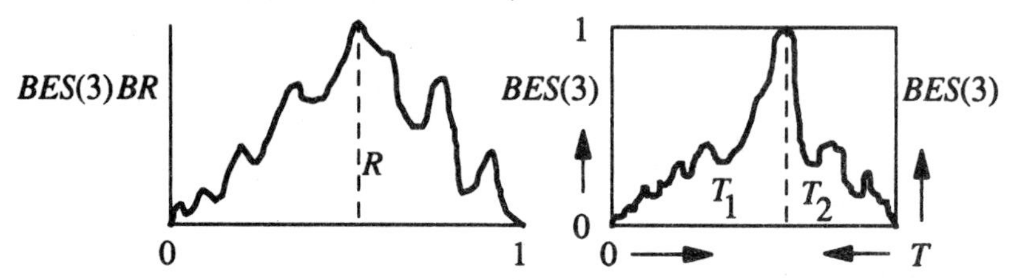

Biane and Yor (1987) explain (see Appendix for a more direct proof) the initially-surprising fact that agreement of Descriptions I and II implies:

$$\tfrac{1}{2}\mathbb{E}\left\{\left(\sqrt{\frac{2}{\pi}}R\right)^{z}\right\} = \tfrac{1}{2}\mathbb{E}\left\{\left(\frac{\pi T}{2}\right)^{\frac{1}{2}-\frac{1}{2}z}\right\}. \tag{5.1}$$

It has been known for a long time (and will now be proved) that the right-hand side of (5.1) equals $\xi(z)$. The fact that the left-hand side of (5.1) equals $\xi(z)$ is (because of a result of Vervaat (1979)) equivalent to Theorem 3.1.

A calculation. Consider the following picture:

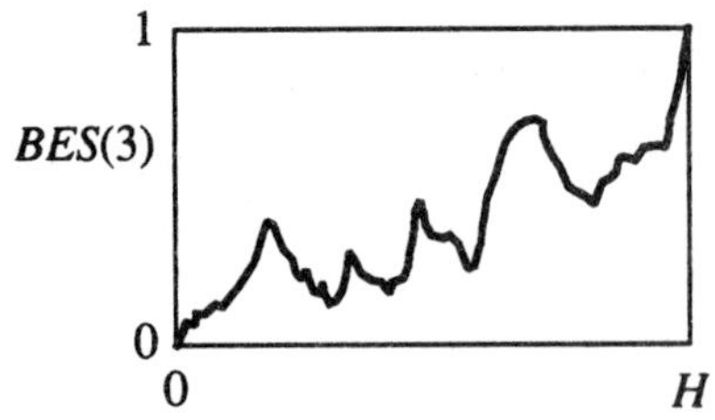

H is the hitting time of position 1 for $BES(3)$

Recall that $BES(3)$ has generator radial $\left(\frac{1}{2}\Delta^{\mathbb{R}^3}\right) = \frac{1}{2}\frac{d^2}{dx^2} + \frac{1}{x}\frac{d}{dx}$.

Standard theory says that $u(x) = \mathbb{E}^x e^{-\lambda H}$ satisfies

$$\tfrac{1}{2}u'' + x^{-1}u' = \lambda u \text{ on } (0,1), \ u(1) = 1,$$
$$u \text{ bounded near } 0.$$

Hence $u(x) = \frac{\sinh \gamma x}{\gamma x} \cdot \frac{\gamma}{\sinh \gamma}$, $\gamma = (2\lambda)^{1/2}$.

It follows that if $T = T_1 + T_2$, where T_1 and T_2 have the same distribution as H under $\mathbb{P}^0$, then

$$\mathbb{E}e^{-\lambda T} = \left(\frac{\gamma}{\sinh \gamma}\right)^2.$$

For $\mathcal{R}z > 0$, we have

$$\mathbb{E} \int_{\lambda=0}^{\infty} e^{-\lambda T} \lambda^{z-1} d\lambda = \mathbb{E} \int_{u=0}^{\infty} e^{-u} \left(\frac{u}{T}\right)^{z-1} \frac{du}{T} = \Gamma(z)\mathbb{E}(T^{-z}),$$

so that

$$\Gamma(z)\mathbb{E}(T^{-z}) = \int_0^{\infty} \left(\frac{\gamma}{\sinh \gamma}\right)^2 \lambda^{z-1} d\lambda = \int_0^{\infty} \frac{8\lambda}{(e^{\gamma} - e^{-\gamma})^2} \lambda^{z-1} d\lambda$$

$$= 8 \int_0^{\infty} e^{-2\gamma} \sum_{n=0}^{\infty} (n+1) e^{-2n\gamma} \lambda^z d\lambda$$

$$= 8 \sum_{n=1}^{\infty} \int_0^{\infty} n e^{-2n\sqrt{2\lambda}} \lambda^z d\lambda = 8 \sum_{n=1}^{\infty} \int_0^{\infty} n e^{-u} \left(\frac{u^2}{8n^2}\right)^z \cdot \frac{u}{4n^2} du$$

$$= 2 \sum_{n=1}^{\infty} n^{-(1+2z)} \Gamma(2z+2) 8^{-z} = 2\Gamma(2z+2)\zeta(1+2z)8^{-z}.$$

Using duplication formula for Γ, we find that, initially for $\mathcal{R}z > 1$,

$$\tfrac{1}{2}\mathbb{E}\left\{\left(\frac{\pi T}{2}\right)^{\frac{1}{2}-\frac{1}{2}z}\right\} = \xi(z) \quad \left(= \tfrac{1}{2}\mathbb{E}\left\{\left(\frac{\pi T}{2}\right)^{\frac{1}{2}z}\right\} \text{ if we assume functional equation} \right).$$

6. Integrated Local Time

Hints that interchanging time and space might be relevant have already been given. One of the standard ways of achieving such an interchange is via the celebrated Ray-Knight theorem.

Again consider the picture

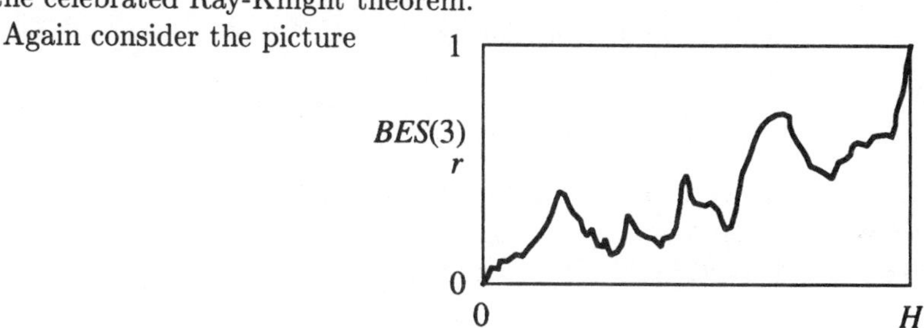

but now insist that the $BES(3)r$ starts at 0.

THEOREM (RAY, KNIGHT). *For $\Lambda \in \mathcal{B}[0,1]$,*

$$\text{measure}\{t < H: r(t) \in \Lambda\} = \int_\Lambda \hat{r}_2(x)^2 dx,$$

where $\hat{r}_2$ is a $BES(2)BR$.

Hence

$$H = \int_0^1 \hat{r}_2(x)^2 dx = \sum_n \frac{G_{1,n}^2 + G_{2,n}^2}{n^2\pi^2}, \quad G\text{'s independent } N(0,1),$$

using Pythagoras to tell us that a $BES(2)BR^2$ is the sum of the squares of two independent $BB(\mathbb{R})$ processes and also using the Parseval result (3.2).

Now the sum of the squares of two independent $N(0,1)$ variables is exponential with mean 2, so

$$\mathbb{E}\exp\left(-\lambda\frac{G_{1,n}^2 + G_{2,n}^2}{n^2\pi^2}\right) = \frac{1}{1 + \lambda \cdot \frac{2}{n^2\pi^2}}.$$

Hence

$$\mathbb{E}e^{-\lambda H} = \prod_{n=1}^{\infty} \frac{1}{1 + \frac{\gamma^2}{n^2\pi^2}} = \frac{\gamma}{\sinh\gamma}, \quad \gamma = (2\lambda)^{1/2},$$

giving another explanation for the $\gamma/\sinh\gamma$ term and hence of ξ.

7. Ferromagnetism and the Lee-Yang Theorem

I end with a result from another branch of probability theory which it would be fascinating to combine with results of earlier sections.

By an *isolated ferromagnetic spin-$\frac{1}{2}$ system on N sites* is meant the following set up:

ρ is measure $\{\frac{1}{2}, \frac{1}{2}\}^N$ on $\{-1,1\}^N$,
$\beta \geq 0$ (β is inverse temperature), $J_{ij} \geq 0 (i < j)$ (interaction),
for $\mathbf{x} \in \{-1,1\}^N$, $H(\mathbf{x}) = -\sum\sum_{i<j} J_{ij}x_ix_j$ (Hamiltonian),
ν is a probability measure on $\{-1,1\}^N$ (Gibbs measure) with

$$\frac{d\nu}{d\rho} = \exp(-\beta H)/Z,$$

where $Z = \int \exp(-\beta H) d\rho$ (partition function).

Define spins X_i $(1 \leq i \leq n)$ via $X_i(\mathbf{x}) = x_i$. Call a variable Y *special mean-zero ferromagnetic* if for some non-negative numbers λ_i $(1 \leq i \leq N)$,

$$Y = \sum_i \lambda_i X_i.$$

Now call a random variable Y *mean-zero ferromagnetic* if there exists a sequence $Y^{(n)}$ of special mean-zero ferromagnetic random variables such that

(i) $Y^{(n)} \Rightarrow Y$, that is, $\mathbb{E}h(Y^{(n)}) \to \mathbb{E}h(Y)\ \forall h \in C_b(\mathbb{R})$,

(ii) $\mathbb{E}((Y^{(n)})^2 \to \mathbb{E}(Y^2)$.

Examples.

(i) If $Y \sim N(0,\sigma^2)$, then Y is mean-zero ferromagnetic, and $\mathbb{E}e^{zY} = e^{\sigma^2 z^2/2}$.

(ii) If $Y \sim U[-\gamma,\gamma]$, then Y is mean-zero ferromagnetic, and $\mathbb{E}e^{zY} = \frac{\sinh \gamma z}{\gamma z} = \prod_{n=1}^{\infty}\left(1+\frac{\gamma^2 z^2}{n^2\pi^2}\right)$.

THEOREM (LEE-YANG-NEWMAN). *If Y is mean-zero ferromagnetic, then*

$$\mathbb{E}e^{zY} = e^{bz^2}\prod_j\left(1+\frac{z^2}{\alpha_j^2}\right), \quad \textit{all } \alpha_j \textit{ real.}$$

Appendix. Proof of (5.1)

Notation

True Itô excursion: Lifetime V, maximum M.

'Scaled' excursion of duration $1 = BES(3)BR$: R maximum height.

Excursion of height 1; T duration, so that $T = T_1 + T_2$, etc, as before.

Descriptions I and II show that *under the Itô excursion law n*,

$$(M,V) \sim (M, M^2T) \sim (RV^{1/2}, V).$$

Biane and Yor remind us (and some of *us* needed reminding!) that, because n has infinite total mass, we can**NOT** conclude that $T^{-1/2} \sim R$.

Correct analysis using $n(M \in dx) = x^{-2}dx$, $n(V \in dv) = (2\pi v^3)^{-1/2}dv$: for a test function g

$$\begin{aligned}
n(g(M,V)) &= \int_x\int_y \frac{1}{x^2}g(x,y)\mathbb{P}(x^2T \in dy)\,dx \\
&= \int_x\int_y g(x,y)f_T\left(\frac{y}{x^2}\right)\frac{1}{x^2}\cdot\frac{1}{x^2}\,dxdy \\
&= \int_x\int_y \frac{1}{\sqrt{2\pi y^3}}g(x,y)\mathbb{P}(y^{1/2}R \in dx)\,dy \\
&= \int_x\int_y g(x,y)f_R\left(\frac{x}{y^{1/2}}\right)\cdot\frac{1}{y^{1/2}}\cdot\frac{1}{\sqrt{2\pi y^3}}\,dxdy,
\end{aligned}$$

whence, on taking $r = xy^{-1/2}$,

$$f_R(r) = \sqrt{2\pi}\frac{1}{r^4} f_T\left(\frac{1}{r^2}\right).$$

(This is trivially equivalent to (5.1).)

Acknowledgement

Thanks to Helen Rutherford and (especially) Sarah Shea-Simonds for their TEXpertise.

REFERENCES

All relevant background (except that for §7) is in Rogers and Williams (1987).

Biane, P. and Yor, M. (1987). Valeurs principales associées aux temps locaux Browniens. *Bulletin de Science Mathématique* 111, 23–101.

Newman, C.M. (1975). Gaussian correlation inequalities for ferromagnets. *Zeitschrift für Wahrscheinlichkeitstheorie und Verwandte Gebiete* 33, 75–93.

Rogers, L.C.G. (1989). A guided tour through excursions. *Bulletin of the London Mathematical Society*, to appear.

Rogers, L.C.G. and Williams, D. (1987). *Diffusions, Markov Processes, and Martingales, Volume 2: Itô Calculus.* John Wiley and Sons, Chichester.

Vervaat, W. (1979). A relation between Brownian bridge and Brownian excursion. *Annals of Probability* 7, 143–149.

Williams, D. (1970). Decomposing the Brownian path. *Bulletin of the American Mathematical Society* 76, 871–873.

Yang, C.N. and Lee, T.D. (1952). Statistical theory of equations of state and phase transitions, I. Theory of Condensation. *Physical Review* 87, 404–409.

Statistical Laboratory
16 Mill Lane
Cambridge CB2 1SB.

Index